Environmental
Toxicology
and
Chemistry

TOPICS IN ENVIRONMENTAL CHEMISTRY
A SERIES OF ADVANCED TEXTBOOKS AND MONOGRAPHS

Environmental Toxicology and Chemistry

DONALD G. CROSBY
Department of Environmental Toxicology
University of California, Davis

New York   Oxford
OXFORD UNIVERSITY PRESS
1998

OXFORD UNIVERSITY PRESS

Oxford   New York
Athens   Auckland   Bangkok   Bogota   Bombay   Buenos Aires
Calcutta   Cape Town   Dar es Salaam   Delhi   Florence   Hong Kong
Istanbul   Karachi   Kuala Lumpur   Madras   Madrid   Melbourne
Mexico City   Nairobi   Paris   Singapore   Taipei   Tokyo   Toronto   Warsaw

and associated companies in
Berlin   Ibadan

Library of Congress Cataloging-in-Publication Data
Crosby, Donald G.
Environmental toxicology and chemistry / Donald G. Crosby.
p. cm.—(Topics in environmental chemistry)
Includes bibliographical references and index.
ISBN 0-19-511713-1 (cloth)
1. Environmental toxicology.   2. Environmental chemistry.
I. Title.   II. Series.
RA1226.C76 1997
615.9′02—dc21   97-22438
                                             CIP

5 7 9 8 6

Printed in the United States of America
on acid-free paper

About the Cover. Topographic projection of acetylcholine receptors obtained by cryoelectron microscopy of crystals grown from postsynaptic membranes of the electric ray, *Torpedo marmorata* [see N. Unwin, C. Toyoshima, and E. Kubalek, *J. Cell Biol.* **107,** 1123–38 (1988)]. Each rosette is about 90 Å across and consists of a fluid-filled pore surrounded by five rod-like glycoprotein macromolecules, two of which reversibly bind the acetylcholine. A bulky molecule of highly toxic nicotine alkaloid can bind to the receptor in place of acetylcholine, blocking the pore to the diffusion of the Na^+ and K^+ required for nervous system function. Thus, a familiar neurotoxic response to tobacco is seen to be a matter of chemistry (see Section 9.1).

To my students,
here and abroad, this book is
dedicated
with
affection and respect.

Contents

Special Topics

Preface

Environmental toxicology and environmental chemistry are inseparable, two sides of the same coin. I have tried to pursuade students of this fact for over 30 years, usually with success. When I started teaching environmental toxicology at the University of California in 1964, the only toxicology text available was the medically oriented *Textbook of Toxicology* by DuBois and Geiling, and there was no single text on environmental chemistry. Over the years, my colleagues and I grumbled about the lack of a book suited to our style of coursework, produced large volumes of lecture handouts, and directed students to reference works such as Casarett and Doull's *Toxicology: The Basic Science of Poisons* and Schwarzenbach's *Environmental Organic Chemistry*. A bit late, I am offering a text that seeks to combine what often are seen as two separate subjects.

The subjects are based on principles often poorly understood, or misunderstood, even by some of those who must apply them. Almost daily, news media describe some problem caused by toxic chemicals, but they seldom provide the necessary basic background. The people who are, or will be, responsible for uncovering, clarifying, and solving such problems deserve to share a common base of fundamentals and their applications. That is what this book is intended to provide.

In addition to the third- and fourth-year undergraduates and first-year graduate students for whom this book is primarily intended, it will also be appropriate for the growing range of professionals who require a refresher in environmental toxicology and/or environmental chemistry. I find that many toxicologists, chemists, ecologists, engineers, attorneys, journalists, educators, and government officials never had an opportunity to view toxicology and chemistry in a connected way. This treatment makes no attempt to be exhaustive, and so the references most often cite books and articles that review and expand the subject. It assumes that the reader has an acquaintance with general biology and chemistry but requires no previous knowledge of toxicology or environmental science.

The book is divided into three roughly equal parts: After an introductory chapter comes a review of principles that govern the occurrence of chemicals in the environment, then some principles of toxicology applied to those chemicals, and finally application of the principles to specific examples of environmental chemicals. Each chapter is expanded by a special topic of particular current interest, and most are further supplemented by appendices of pertinent data. Some important terms are indicated in bold type and collected in a glossary at the end, but the subject's scope and nontraditional presentation may require readers to use the index to locate all the places their favorite subject is to be found. The writing is intended only as a framework upon which the instructor or interested individual can impose additions and interpretations; no problem sets or questions are included.

No book comes into existence in isolation. I gratefully acknowledge the help and thoughtful suggestions of Mike Denison, Deanna Dowdy, Lynn Jaeger, Scott Mabury, Glenn Miller, Marion Miller, Randy Magdalena, Ken Ngim, Mike Stimman, and Ron Tjeerdema. Figure 9.1 is reproduced from Adrian Albert's *Selective Toxicity* (p. 29) by permission of Chapman and Hall, London. I am especially indebted to Jan Chambers of Mississippi State University, Richard Lee of the Skidaway Institute of Oceanography, and Scott Mabury of the University of Toronto for reviewing separately the entire manuscript and providing important suggestions. However, I alone bear responsibility for the interpretation, coverage, and accuracy of the topics presented. Special thanks go to editor Bob Rogers for his guidance, and to my wife, Nancy, for continual encouragement and moral support.

D.G.C.

Davis, California
February, 1997

Environmental
Toxicology
and
Chemistry

Environmental Toxicology and Chemistry

1

1.1 | POISONS

Toxic means poisonous. People today are surrounded by, and often dependent on, a wide array of poisonous chemicals that require understanding and control if one is to live in safety. The purpose of this chapter is to introduce a few of those problem chemicals, outline their environmental toxicology and chemistry, and offer a preview of the rest of the book.

Popularly known as "toxics," toxic substances make news almost daily, often leaving the impression that they are something novel and manmade. Our ancestors, too, were confronted daily and often fatally by poisonous microorganisms, plants, animals, and minerals; poisoning is nothing new. Primitive people used poisons in hunting and warfare, and the term *toxic* comes from the ancient Greek word *toxikon* referring to a bow (poisoned arrows). Toxic chemicals also helped to catch fish, protect against diseases, and fight off insects. The poisons were all natural, of course, but there is no toxicological distinction between natural and synthetic chemicals.

People have always been fascinated by poisons. The Ebers papyrus of 1550 B.C. described common poisons such as hemlock *(Conium maculatum)*, the poison used over 1000 years later to execute the Greek philosopher Socrates and still killing the unwary today. Poisons continued to be employed over the centuries for social and political purposes, and the names Borgia and de'Medici from medieval Europe are still associated with poisoning. In seventeenth-century France, Catherine Deshay's victims were said to have included over 2000 unwanted babies, and an Italian woman named Toffana made her living and reputation by selling arsenic-laced *Aqua Toffana* to wives who longed to become widows. Interesting and entertaining accounts of historical poisonings are provided by Thompson (1931) and Casarett (1975).

Nature employs toxic chemicals as well. Plants, microorganisms, and wild animals use poisons for defense or to expand their living space. Microorganisms produce

TABLE 1.1
Causes of Human Death in the United States[a]

Cause	1970	1975	1980	1985	1990
Motor vehicles	54,633	45,853	53,272	45,901	46,814
Falls	16,926	14,896	13,294	12,001	12,313
Fire	6,718	6,071	5,822	4,938	4,175
Guns	2,406	2,380	1,955	1,649	1,416
Poisons, accidental	5,299	6,271	4,331	5,170	5,803
Poisons, suicide	6,584	6,426	5,453	5,704	5,424
Total poisons[b]	**11,883**	**12,697**	**9,784**	**10,874**	**11,227**

[a]USDC, 1993.
[b]Excludes homicide, adverse therapy, alcoholism, and venomous animals.

antibiotics, many plants release allelochemicals that prevent other plants from growing near them, and invertebrate animals often resort to toxic defense substances, as anyone who has been bitten by a fire ant or stung by a jellyfish can attest. The human tendency to get in the way of these natural battles sometimes has disastrous results, as in 994 A.D. when over 40,000 Europeans died from eating food contaminated with the toxic ergot fungus *Claviceps purpurea*. James Cook's eighteenth-century voyages in the South Pacific were stalked by death and illness from fish tainted with Ciguatera toxin, and members of Abe Lincoln's family, including his mother, died by drinking milk from cows that had eaten white snakeroot *(Eupatorium urticaefolium)*. See Chapter 12 for more about toxic plants.

Poisons still present problems. In a typical year, thousands of people die from accidents, suicide, or homicide due to poisoning (Table 1.1), and many others are killed by tobacco and alcohol (Section 1.4.2), which, too, are poisons. Although the exact numbers vary among sources, human death by poison appears to be on the upswing, the government's Statistical Abstract listing a U.S. total of 9784 in 1980 and 11,227 a decade later (USDC, 1993). However, as only a small percentage of poisoning incidents are fatal, the total of toxic exposures is actually much greater. Reports to U.S. poison control centers exceeded 715,000 in 1992 (Litovitz et al., 1993), and worldwide, pesticides alone cause over 3 million poisonings annually (WHO, 1990). Accidental poisoning is widespread among both wild and domestic animals (Sections 1.4.3 and 1.4.4), and toxic chemicals are used intentionally and widely against undesirable weeds, rodents, and insects. The growing interest in environmental toxicology is well deserved.

1.2 ENVIRONMENTAL TOXICOLOGY

Toxicology is the science of poisons. It deals with the adverse effects of chemical agents on biological systems. The American Heritage Dictionary defines poison as "a substance that causes illness, injury, or death . . . by chemical means." In their 1959 textbook, DuBois and Geiling further define toxicology as dealing with the occurrence, physical and chemical properties, effects, and detection of poisons. Note that these definitions do not restrict toxicology to studies in animals. Scientific toxicology started during the historic period of 1810–1850, when great advances in chemistry, physics, and biology also were being made. The name appears to have been coined by M. J. B. Orfila (1787–1853), the Spanish "father of toxicology."

Subdivision of any field of knowledge is tempting. Zakrzewski (1991) describes only three branches of toxicology—clinical, forensic (legal), and environmental—but Ballantyne et al. (1995) suggest eight more, including veterinary, occupational, and regulatory. Although each has advanced rapidly in recent years, *environmental toxicology,* so named in the early 1960s, currently receives the major share of public attention.

From these definitions, it follows that environmental toxicology is the branch of science concerned with the nature, properties, effects, and detection of toxic substances in the environment and in *any* environmentally exposed species, including humans. The key word is *environment,* according to the dictionary "the complex of physical, chemical, and biotic factors that act upon an organism and ultimately determine its form and survival" which certainly includes poisons. The subject does not have rigid boundaries, and it merges with occupational toxicology in the work environment, clinical toxicology in the home environment, and so on.

However, the principal realm of environmental toxicology remains out of doors. The emphasis is on (1) the occurrence, availability, and form of toxic chemicals, (2) exposure to such chemicals and the accompanying dangers, and (3) comparative effects and mechanisms of action in the broad range of exposed species, again including but not focused on humans. The toxic chemicals of concern often are manmade *xenobiotics,* that is, substances foreign to living organisms, but natural poisons also are included, as are toxic levels of normal body constituents such as vitamins and hormones. While sometimes equated with "wildlife toxicology" or "pollution toxicology," environmental toxicology obviously represents much more.

Where traditional toxicology usually deals with relatively large, measured doses of known chemicals such as drugs, environmental toxicology usually must consider variable low levels of dispersed and often unidentified substances. This is where environmental chemistry comes in.

1.3 | ENVIRONMENTAL CHEMISTRY

Toxicity is based on chemistry. Although often treated as separate subjects by college courses, professional organizations, and even many practitioners, toxicology and chemistry actually are inseparable. Poisons are chemicals (Chapter 2), exposure is governed by chemical forces (Chapter 10), and both the action of poisons (Chapter 9) and an organism's ability to protect itself (Chapter 6) are largely a matter of chemistry. Such practical problems as predicting what happens to toxic chemicals (Chapter 16) and remediating toxic wastes (Special Topic 15) require a knowledge and use of chemical principles.

Environmental chemistry is concerned with the sources, identity, levels, reactions, transport, and fate of chemical species in water, soil, and air environments (Manahan, 1994). It applies the principles of inorganic, organic, analytical, and physical chemistry to environmental chemicals and processes. Much of its current emphasis is placed on manmade pollutants, but as with toxicology, natural chemicals originating from animals, plants, or minerals must be included. As the environmental chemistry focus frequently involves the harmful effects of a substance, environmental toxicology and chemistry form a continuum.

From toxicology's viewpoint, perhaps the most important function of environmental chemistry is to provide the exposure information necessary for evaluation of toxicity and risk. This includes identification of toxic chemicals, reactions that reduce their availability or convert them into more toxic or less toxic forms, and quantitative measurement of both their environmental and available concentrations. As soon as any chemical enters air, soil, water, or a living organism, it starts to move and change (Chapters 3 to 6). Take smog, for example. Petroleum hydrocarbons volatilize from the earth's surface into the atmosphere, become dispersed, and are photooxidized by sunlight into products very different from, and toxicologically more interesting than, the original alkanes and alkenes. Structure, reactivity, availability, and exposure form the basis of intoxication.

1.4 | TOXICITY

1.4.1. Intoxication

Intoxication is the scientific term for poisoning. It results from the interaction of a chemical "poison" with some biochemical entity or process that sustains life (Chapter 9). Like other chemical processes, intoxication is mass driven; that is, it is dependent on the degree of *exposure* or *dose,* and this *dose–response relationship* is perhaps the most significant feature of toxicity. The degree of subsequent harm is also controlled by the organism's ability to absorb, degrade, and eliminate the *toxicants,* as well as its particular biochemistry and physiology (Chapter 7). The balance of exposure to and removal of the chemical from the organism determines the toxic outcome. As the saying goes, **"The dose makes the poison."**

These characteristics also set the stage for *selective toxicity,* where one species can be affected by a poison while another seemingly is not. Selectivity is a necessary feature of most medicines and pesticides, and many weed killers, for example, are almost nontoxic to mammals because they kill plants by such nonanimal processes as photosynthesis. Nonetheless, intoxication is common to all living organisms, from people to bacteria, as illustrated in the next sections.

TABLE 1.2
Deaths Due to Poisoning in the United States, 1991[a]

	Accidental	Intentional[b]	Intent Unknown	Suicide	Total
Drugs and Medicines	5,232	163	1,049	3,095	9,539
Pain killers[c]	1,278	8	285	560	
Psychotropics[d]	269	5	147	988	
All others	3,685	150	617	1,547	
Solids and Liquids	1,203	6	45	173	1,427
Alcohols[e]	331				
Cleaners[f]	80				
Acids, alkalies	10		1	17	
Metallics	16		1	3	
All others	766		43	153	
Gases	736	15	102	2,230	3,083
Utility gas	86		5	29	
Carbon monoxide[g]	528		89	2,179	
Other gases	122		8	22	
Pesticides	10	0	3	46	59
Plants, Animals, Food	96	0	0	0	96
Other Poisons	17	1	0	0	18
Total Poisoning	7,294	185	1,199	5,544	14,222

[a] USPHS, 1996.
[b] Homicide, adverse therapy.
[c] Analgesics, antipyretics, narcotics, sedatives, etc., but not illegal drugs (1189 deaths).
[d] Tranquilizers, other psychotropics.
[e] Includes alcoholic beverages (15 deaths), but not alcoholism (17,102 deaths).
[f] Includes disinfectants, paints, and varnishes.
[g] Includes motor vehicle exhaust.

1.4.2. Human Intoxication

Most people are concerned primarily with poisoning in humans, a frequent occurrence. Legal drugs and medicines cause 70% of both accidental and intentional poisoning deaths in humans (Table 1.2), and the poisoning homicides in a single year in the United States represent twice the lifetime record of Toffana and her customers. Over a third of the fatalities are suicides, 65% of them in males, with tranquilizers and automobile exhaust the agents of choice. The 17,000 deaths due to alcoholism in 1991 were not included in Table 1.2, nor were the almost 1200 deaths caused by illegal drugs. Emphysema (lung damage) and cancer of the lungs and mouth, due largely to smoking, provided yet another 170,000 fatalities.

Of the 700,000 nonfatal accidental poisonings reported each year in this country, over 60% occur in children under the age of six (Litovitz et al., 1993), mostly in their own homes at times when parents are busy. The age of greatest danger is from one to two, and boys are poisoned more frequently than are girls (Ottoboni,1991).

Medicines, especially pain relievers such as aspirin, are responsible in about half the cases, while household cleaners, cosmetics, and toxic plants account for about 10% each. Contrary to public perception, pesticides are responsible for less than 5%. Data from poison control centers in other parts of the world confirm this picture, although pesticides are involved more frequently. In 1992, a total of 1,864,188 human exposures were reported to U.S. poison control centers (Litovitz et al., 1993).

Fewer than 1% of fatal human poisonings take place in the outdoor environment, but nonfatal intoxication there is commonplace. Poison oak dermatitis continues to be a major cause of lost-time accidents in California, at least 480 pesticide-related illnesses were reported among agricultural workers in California in 1990, and there are some 10,000 victims of *Staphylococcus* food poisoning in the United States each year. Water pollution, especially by nitrate and and other inorganic toxicants, affects large numbers of people, and the nonlethal action of atmospheric ozone and photochemical oxicants on the eyes and respiration of smog victims is well known.

1.4.3. Domestic Animals

Unlike humans, domestic animals such as dogs, cats, cows, and horses are poisoned primarily by pesticides. Of almost 42,000 animal poisonings reported to U.S. poison control centers in 1990, a third were due to insecticides, herbicides, fungicides, or rodenticides. Veterinary drugs added another 25%, and toxic plants provided 12% (Hornfeldt and Murphy, 1992). Of the 425 cases (1%) that proved fatal, insecticides were responsible for 21%, rodenticides 14%, and plants and ethylene glycol (antifreeze) 10% each. Deaths from eating plastics, chocolate, and poisonous toads also were listed. Dogs and cattle were the most frequent victims, almost all poisoned by accident.

While mass fatalities due to manmade chemicals have occurred over the years, plant poisonings in the West became so important at one time that the U.S. Department of Agriculture (USDA) established a poisonous plants laboratory near Logan, Utah. In fact, the USDA owes its start to a serious outbreak of ergot poisoning among Kansas cattle in 1884. Other culprits include locoweed *(Astragalus),* tansy ragwort *(Senecio),* and lupine *(Lupinus).* St. Johnswort *(Hypericum perforatum)* was estimated to cover 2.3 million acres of California rangeland in 1951 and to have caused greater financial losses to cattlemen than any other factor (Kingsbury, 1964).

1.4.4. Wildlife

The term wildlife, as used here, includes any undomesticated animals such as deer, birds, fish, insects, and earthworms. Poisoning among the large terrestrial mammals usually thought of as wildlife is hard to document, although it must occur. Rats, mice, and ground squirrels usually are poisoned intentionally, although foxes, coyotes, and other predators also become victims. Birds previously were hit hard by chlorinated hydrocarbons such as DDT in their food—pelicans via fish, robins via earthworms, and falcons via rodents. Some deaths among songbirds eventually were traced to other causes such as natural cyanide in the seeds they ate, but organophosphate insecticides still affect birds through both their diet and their surroundings, migratory waterfowl have been killed by mistaking granular pesticides for seeds, and many baby birds have died as a result of poisoning by natural selenium such as that at Kesterson Reservoir in California.

Insects seem to be the worldwide target of intentional poisoning, with freshwater fish and other aquatic animals probably the most frequent unintended victims. Massive fish kills have become so common that they now excite little attention, and many natural waters today cannot support aquatic life. Cyanide and hypochlorite still are used to capture fish in some parts of the world, and natural rotenone is employed by government agencies to destroy "trash" fish. People tend to focus on economically and esthetically obvious wildlife, but little or nothing is known about the effects of chemicals on the myriad tiny or uncuddly species that inhabit the earth with us.

1.4.5. Plants and Microorganisms

Green plants, fungi, and bacteria are all bound by the same principles of toxicology that apply to animals. They undergo widespread and often intentional poisoning, frequently by selective herbicides, fungicides, and bactericides, but their deaths usually are reported gleefully as the killing of nasty weeds and germs. The lethal toxic effects of smog on crops and forest trees is apparent near cities such as Los Angeles, and lichens are sensitive indicators of air pollution. Also, plants often employ natural chemical defenses against each other, a phenomenon called *allelopathy.* For example, many broad-leaved crop plants cannot grow near black walnut (*Juglans niger*) trees due to the toxic 5-hydroxynaphthoquinone secreted into the soil. Bacteria and fungi likewise form antibiotics, natural antimicrobial chemicals such as penicillin that have been put to good use by humans for combatting the same microbial competitors.

1.4.6. Ecosystems

An ecosystem represents a community of organisms interacting with each other and with their environment. It is composed of diverse producers represented mostly by green plants, consumers such as herbivores and carnivores, bacterial and fungal decomposers, and the abiotic (nonliving) environment. It is resilient and resistant to change, seeks to restore and maintain equilibrium when perturbed, and is defined by the flow of energy and materials through it rather than by geographical boundaries.

Classical toxicology has focused almost exclusively on adverse effects in individuals or small populations of a very limited number of readily replaceable species. However, the ability of toxic chemicals to affect the structure or function of an entire ecosystem, such as species diversity or energy flow, has received scant attention until recent years. An example, fortunately not yet realized, is the possible chemical inhibition of photosynthesis in the marine phytoplankton which produce much of the oxygen we breathe. The branch of toxicology that deals with ecosystems and their components is called *ecotoxicology* and is so important that it will be the subject of Special Topic 1 at the end of this chapter.

1.5 | HAZARD AND RISK

One often finds the terms *toxic* and *hazardous* used synonymously. Actually, *hazard* is defined as the *potential* for harm, and its magnitude is equated with the severity of the expected consequences (Chapter 10). ***Risk*** is a function of the magnitude of a hazard and the *probability* of its occurrence. A community that expects an oil spill once a year faces the same hazard but at 10 times the risk as one that has a spill only once every 10 years. While exposure and toxicity should both be amenable to objective measurement (Chapter 8), as in such well-defined situations as life-cycle toxicity tests on fish, the relationship becomes much more complex when applied to practical situations. The assessment of risk is important for decision-making at both a personal and government level, but it proves to be remarkably difficult.

Risk is a human concept. It involves an understanding and consideration of causes, identity of victims, degree of control, and personal values. With toxic risk, much of the present public concern is based on the results of animal tests, and test species become a controversial issue. Should they be those closest to humans? Easiest to standardize? The most sensitive? Economically most important? Those of greatest current popular interest? Those with greatest ecological relevance? Who decides and on what basis? People's perception of personal and environmental risk drives much of the decision process. Many Americans harbor a real fear of toxic chemicals that often works against the acceptance of accurate risk assessments, especially when risks are low but public agitation is high. For the basics of risk analysis, see Cohrssen and Covello (1989).

Exposure, hazard, and risk are modified by the environmental dissipation of chemicals. As each chemical vaporizes, dissolves in water, becomes bound to soil, or reacts to give other forms, its concentration and availability change. At the same time, important routes of exposure also may change, for example, from skin to lung as a liquid toxicant volatilizes into the atmosphere. Chapters 3 to 6 discuss the movement and breakdown of toxic chemicals, whose absorption and bioconcentration (Chapter 3) become the key to both intoxication and risk.

The presence of multiple risks makes evaluation of a specific toxic threat difficult. Risk often is measured as odds, for example, the number of chances in a million of an event occurring. Using the chance of being struck by lightening in a given year (about one in a million) as a yardstick, Table 1.3 illustrates that familiar hazards such as medical x-rays and smoking actually may outweigh those of pollution. The ultimate challenge for modern toxicology is providing an accurate and scientific assessment of the human and ecological risk presented by poisons.

TABLE 1.3
Cancer Risks in Everyday Living

Activity	Frequency	Cancer Risk[a]	Cause
1 Transcontinental flight	Per year	1	Cosmic radiation
Medical x-ray	Each	20	x-radiation
Drink 1 pint of milk daily	Per year	2	Radioactive strontium
Smoke 1 cigarette	Each	1	PAH
Share a smoker's room	Per year	20	PAH
Drink 1 diet cola daily	Per year	10	Artificial sweetener
Drink 1 ppb of TCE in water	Per year	<0.1	Pollution

[a] Probability per million over a lifetime.

1.6 | MAJOR TOXIC HAZARDS

Early in the sixteenth century, the German alchemist–scientist Theophrastus, also called Paracelsus, published a statement destined to become the cornerstone of toxicology: "What is it that is not a poison? All things are poison, and nothing is without poison. Only the dose makes a thing not a poison." Although every substance must be considered toxic at some level, some obviously are more toxic than others at a given dose.

Actually, hazard—the *likelihood* of harm—should concern us more than toxicity as such, and hazard involves exposure. For example, although tranquilizers are not especially toxic, exposure to them is widespread and so they present a distinct hazard (Table 1.2). *Epidemiology* helps to identify who or what actually has been poisoned by some particular chemical (Special Topic 8), and Table 1.4 provides a list of what experience shows are some of the most important *hazardous* chemicals, especially with respect to humans. In making this selection, intrinsic toxicity must be modified by exposure, which varies according to age, occupation, and geographic location. It is both interesting and instructive to put together a personal list of chemicals to which one is regularly exposed.

For example, *polycyclic aromatic hydrocarbons,* the first class of chemical carcinogens (cancer producers) to be recognized, are found in smoked and broiled foods, tar, used motor oil, and asphalt (Section 13.3.2). *Natural toxicants* from plants include urushiol, the rash-producing biotoxin of poison oak *(Toxicodendron diversilobum)* and poison ivy *(T. radicans),* and highly toxic nicotine found in both wild and cultivated tobacco (Chapter 12). Paralytic shellfish poison (PSP) and toadstool poisons take their toll of food gatherers and pets every year.

Reactive gases rate their own section of the mortality tables (Table 1.2). Carbon monoxide alone is responsible for about 3000 human deaths and many more illnesses in our country every year. Its effect on respiration can be cumulative; getting warm by a leaky gas heater followed by sitting in a closed car in rush-hour traffic while smoking cigarettes would almost guarantee a trip to the hospital (Section 15.8). Com-

TABLE 1.4
Some Major Toxic Hazards

Class	Examples
Polycyclic aromatic hydrocarbons	Benzo[a]pyrene
Biotoxins	Urushiols, nicotine, shellfish poisons, aflatoxins, toadstool toxins
Reactive gases	CO, HCN, Cl_2, SO_2/SO_3, H_2S, O_3
Metal and metalloid compounds	As, Pb, Hg, Cd, Ni, Cr, Cu, Sn
Cholinesterase inhibitors	Organophosphate insecticides, carbamate insecticides, solanine
Halogenated hydrocarbons	Carbon tetrachloride, chlordane, chloroethylenes, DDT, dichloromethane, dieldrin, HCB, methyl bromide, TCDD
Solvents	Benzene, methanol, MIBK
Corrosives	NaOH, KOH, H_2SO_4, bleach (NaOCl), asbestos
Miscellaneous	Aspirin, pentachlorophenol, phenol, formaldehyde, ethylene glycol

pounds of many *metals* and *metalloids* are classic poisons: Toffana employed arsenic oxide in her trade, lead acetate ("sugar of lead") used as a sweetener in wine helped bring down the Roman Empire, and eating mercury-contaminated fish resulted in hundreds of deaths and injuries in Minamata, Japan, in the late 1960s (Chapter 11).

Pesticides that affect nerve impulse transmission *(cholinesterase inhibitors)*, especially the organophosphorus insecticides, poison at least 100 people each year in California alone (Section 9.4.1), and over 1000 people reported illness from eating California watermelons illegally treated with the carbamate insecticide aldicarb in 1985. However, animals and people are poisoned likewise by the natural cholinesterase inhibitor solanine in sprouting potatoes (Section 12.2).

People are surrounded by other common hazardous chemicals, including *halogenated hydrocarbons, solvents,* and *corrosive household cleaners* (sodium and potassium hydroxide), as well as by ethylene glycol antifreeze, bleach (sodium hypochlorite), battery acid (sulfuric acid), and a host of others (Chapter 2). The next few chapters will discuss these chemicals, their distribution, and their environmental breakdown as a prelude to discussion of the principles and examples of their toxic action.

1.7 PERSPECTIVE

The opening of this chapter expressed the need for people to recognize and understand toxic chemicals and toxicity. The fact remains that everyone is exposed every day to toxic chemicals at home, at work, at school, and in the outdoor environment. Other animals, plants and microorganisms encounter many of these same substances as environmental pollutants or as natural agents of defense or aggression.

The fields of environmental toxicology and chemistry are still young, active, and growing. The fate of chemicals in the environment, toxic effects in nonmammals, ecotoxicology, and molecular mechanisms of toxicity are popular areas of today's research. Important applications include protection of Earth's ozone layer, development of less persistent chemical products, and disposal of toxic waste. Participation cannot be left only to research scientists; the understanding and prevention of intoxication, in ourselves and our coinhabitants of earth, is everyone's business.

1.8 | REFERENCES

Ballantyne, B., T. Marrs, and P. Turner. 1995. *General and Applied Toxicology,* Macmillan, New York, NY.

Cairnes, J., Jr. 1989. Will the real ecotoxicoogist please stand up? *Environ. Toxicol. Chem.* **8:** 843–44.

Casarett, L. J. 1975. Origin and scope of toxicology, in *Toxicology: The Basic Science of Poisons* (L. J. Casarett and J. Doull, eds.), Macmillan, New York, NY, pp. 1–10.

Cohrssen, J. J., and V. T. Covello, 1989. *Risk Analysis: A Guide to Principles and Methods for Analyzing Health and Environmental Risks,* Council on Environmental Quality, Office of the President, Washington, DC.

DuBois, K. P., and E. M. K. Geiling. 1959. *Textbook of Toxicology,* Oxford University Press, New York, NY.

Hornfeldt, C. S., and M. J. Murphy. 1992. Poisonings in animals: A 1990 report of the American Association of Poison Control Centers. *J. Am. Vet. Med. Assoc.* **200:** 1077–80.

Kingsbury, J. M. 1964. *Poisonous Plants of the United States and Canada,* Prentice-Hall, Englewood Cliffs, NJ.

Litovitz, T. L., K. C. Holm, C. Clancy, B. F. Schmitz, L. R. Clark, and G. M. Oderda. 1993. 1992 Annual report of the American Association of Poison Control Centers toxic exposure surveillance system. *Am. J. Emerg. Medicine* **11:** 494–555.

Manahan, S. E. 1994. *Environmental Chemistry,* 6th Ed., CRC Press, Boca Raton, FL.

Moriarity, F. 1988. *Ecotoxicology: The Study of Pollutants in Ecosystems,* 2nd Ed., Academic Press, New York, NY.

National Research Council. 1981. *Testing for the Effects of Chemicals on Ecosystems,* National Academy Press, Washington, DC.

Ottoboni, M. A. 1991. *The Dose Makes the Poison,* Van Nostrand Reinhold, New York, NY.

Römbke, J., and J. F. Moltmann. 1996. *Applied Ecotoxicology,* Lewis Publishers, Boca Raton, FL.

Thompson, C. J. S. 1931. *Poisons and Poisoners. With Some Historical Accounts of Some Famous Mysteries in Ancient and Modern Times,* H. Shaylor, London, U.K.

Truhaut, R. 1977. Ecotoxicology: Objectives, principles, and perspectives. *Ecotoxicol. Environ. Safety* **1:** 151–73.

USDC. 1993. *Statistical Abstracts of the United States,* 107th Ed., U.S. Department of Commerce, Bureau of the Census, Washington, DC.

USPHS. 1996. *Vital Statistics of the United States 1991,* U.S. Public Health Service, National Center for Health Statistics, Hyattsville, MD. Vol. 2: Mortality Part A.

WHO. 1990. *Public Health Impact of Pesticides Used in Agriculture,* World Health Organization, Geneva, Switzerland.

Zakrzewski, S. F. 1991. *Principles of Environmental Toxicology,* American Chemical Society, Washington, DC.

Special Topic 1: Ecotoxicology

Ecotoxicology is concerned primarily with the release of toxic pollutants into the environment, their distribution and fate in the biosphere and especially in food chains, and qualitative and quantitative measurement of toxic responses in ecosystems and ecosystem components. It deals with microorganisms, plants, and animals (including humans) of all kinds in relation to their abiotic (nonliving) environment in an *integrated way* (Truhaut, 1977). Ecotoxicology is environmental toxicology from an ecological perspective.

Traditional toxicology focusses primarily on the effects of toxicants in individual organisms and small populations, almost always mammals, because they are easiest and cheapest to deal with and are seen as related to humans. Conditions and doses are well defined, as with drugs, for example. By contrast, ecotoxicology concentrates on the effects of dispersed and mixed pollutants and their transformation products, as well as the toxicological relationships among a wide range of organisms. Ecotoxicology also includes the nonliving environment, while traditional toxicology seldom does. Ecotoxicity tests with individual species are not just extensions of traditional assays to wildlife but relate to larger objectives (Section 8.4), especially the structure and functions of communities and ecosystems (National Research Council, 1981).

Ecosystems have structure, representing not only living individuals but also their populations, communities, and nonliving habitats. Important characteristics include diversity and stability (Section 1.3), the more diverse in species and habitats, the more stable. Stability denotes resistance to perturbation under stress and, if perturbed, the resiliency to return to some equilibrium state. This does not mean that ecosystems are static but that each follows a defined *trajectory.* They evolve, often over vast periods of time, and undergo a succession of stages where productivity at first is high, food chains linear, biomass small, and diversity and stability low. As they mature, net production declines, food webs predominate, biomass and individual organisms tend to be large, and both diversity and stability are high.

Although they may contain familiar habitats such as forests, lakes, or deserts, ecosystems actually represent a flow of energy and materials (Fig. 1.1) that arise in plant photosynthesis by *primary producers,* move through *macroconsumers* (herbivores and carnivores) and on to microbial *microconsumers* (scavengers). Materials finally return to the plants in inorganic form, and nutrients such as C, O, N, P, and trace elements are conserved through this biogeochemical cycling. If intoxication fundamentally alters any of these basic ecosystem structures and functions, serious environmental consequences eventually follow.

For example, a weedy vacant lot constitutes an ecosystem of primary producers (grass and weeds), herbivorous insects, carnivores such as insect-eating birds, and scavengers to decompose the dead remains. If broadleaf weeds are killed with a selective herbicide such as 2,4-D, much of the primary production is destroyed, many insects starve, and the birds move to someplace else. However, once the herbicide has dissipated, weed seeds sprout, insect eggs hatch, the birds come back, and the ecosystem returns to equilibrium. Individuals may suffer catastrophe, but basic ecosystem structure and function remain unaffected in the long term.

However, if *all* the microbes, plants, and invertebrates in the lot were to be eradicated by a soil sterilant such as metam-sodium, the necessary primary production, community structure, and food chains would be destroyed and the ecosystem cease to exist. It will eventually be replaced, perhaps by a new one of lawn and flowers more to *human* liking. One of the purposes

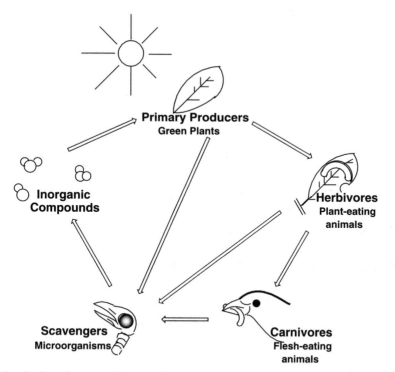

Figure 1.1. Cycling of energy and materials in an ecosystem.

of ecotoxicology is to seek methods, such as ecological risk assessment (Section 10.4), by which future ecosystem destruction or unwanted change can be forestalled or at least predicted. The science is still young: To the prominent ecologist, John Cairnes, Jr., "the term 'ecotoxicologist' is an aspiration rather than a reality" (Cairnes, 1989, p. 843). Good ecotoxicology textbooks are available, such as those by Moriarity (1988) and Römbke and Moltmann (1996), but the following chapters also provide a background required for further study of this important subject.

C H A P T E R

Environmental Chemicals

2

2.1 | CHEMICALS IN THE ENVIRONMENT

Toxic chemicals in the environment are why environmental toxicology exists. All organisms, including humans, are surrounded by natural and synthetic chemicals that can become hazardous upon sufficient exposure. Some of the most dangerous examples already have been cited, but this chapter places them—and others like them—into the context of the air, water, soil, and organisms where they are found.

Many natural chemicals are both highly toxic and readily contacted by humans and other creatures. Polycyclic aromatic hydrocarbons (PAH), urushiol, nicotine, and paralytic shellfish poison do a lot of harm. However, manmade pollutants have received far more attention. Almost obscure at the beginning of World War I, synthetic chemicals now dominate our lives and underlie almost all the material items so familiar to us (Table 2.1). Escape of these chemicals during manufacture, handling, use, or disposal results in "environmental pollution."

Chemical pollution usually is blamed on manufacturing industries, often with justification. The U.S. Environmental Protection Agency (EPA) requires manufacturers to provide data for its annual Toxic Release Inventory (TRI), illustrated in Appendix 2.1. The TRI attempts to identify and locate the release of chemical wastes into water, air, and soil, and perhaps its most striking feature is that the top 22 offenders in 1989—over 80% of all wastes reported—were not those ordinarily expected by the public. Another surprise is that over 40% of the total was released into the atmosphere, and another 20% released underground, while only about 3% went into surface water (Figure 2.1). The 16% labeled "off-site" left the factory but became "lost." States with the largest 1989 toxic releases were Texas (14% of the U.S. total), Louisiana (12%), and Ohio (6%). This hazardous waste is the subject of Special Topic 15.

The air, water, and soil referred to in Appendix 2.1 may be pictured as physically separate environmental "compartments," together with the aggregate of living organisms known as the "biota." The importance of this concept will become apparent as the environmental distribution, transport, and transformations of chemicals are dis-

TABLE 2.1
U.S. Synthetic Chemical Applications, 1990[a]

Application	Production Volume (10^6 kg)	
	1989	1992
Cyclic and acyclic chemicals[b]	48,871	54,925
Plastics and resins[c]	26,995	31,419
Cyclic intermediates	24,756	26,864
Miscellaneous products	13,503	17,286
Surfactants	3,085	3,174
Elastomers[d]	2,091	2,614
Plasticizers	976	915
Pesticides	572	492
Rubber processing chemicals	176	157
Dyes	174	148
Medicinals	130	149
Flavors and fragrances	64	101
Organic pigments	50	57
Total synthetic organics	121,443	138,301

[a] USITC (1994).
[b] Includes many solvents.
[c] Includes synthetic fibers.
[d] Primarily synthetic rubbers.

cussed in subsequent chapters. Compartments also represent physical states. the atmosphere gaseous, water liquid, soil solid, and biota thought of as a lipid "phase."

2.2 | DETECTION AND MEASUREMENT

Accounting records such as the TRI provide important information about chemicals released into each environmental compartment and where the releases occur. However, the records do not reveal environmental concentrations, the fate of each chemical, and the exposure levels of people and other organisms. Such information comes only from chemical analysis. Environmental analyses have been conducted for many decades to determine the levels of arsenic in food, detect the pollution of water and air, and measure the levels of persistent contaminants in human milk. If erroneous, such data would be worthess and even dangerous.

One tends to believe any analytical data that appear in books or newspapers. However, analytical measurements are inherently erroneous, although the degree of error can be estimated and minimized. Chemical analysis normally requires sampling, purification, detection, measurement, and data interpretation, and errors may be introduced at any step. In a typical example, analysis of the same water sample for the insecticide malathion by three reputable laboratories produced answers of 0.13, 0.62, and 1.33 μg/L, respectively. Why the discrepancy? Isn't there only one "true" number? Probably so, but that number is remarkably difficult to obtain and recognize.

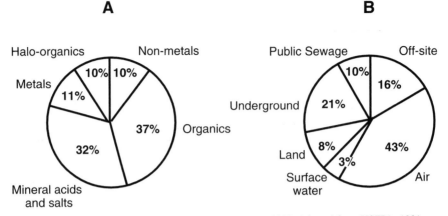

A

Halo-organics   Non-metals

Metals

10% 10%

11%

37%   Organics

32%

Mineral acids
and salts

B

Public Sewage   Off-site

Underground

10%

16%

21%

Land   8%   43%

3%

Surface   Air
water

Figure 2.1. The type (A) and placement (B) of TRI releases, 1989. Adapted from USEPA, 1991.

A sample that accurately represents its particular substrate is essential. Samples are best collected at several adjacent locations and then combined and thoroughly mixed. Replicate samples should be collected at as close to the same time as possible; consecutive samples of river water collected from a boat, for example, can produce very different analytical results as the river's composition changes from minute to minute.

Deciding what to analyze for is not a trivial question. For example, in order to analyze a sample of bird tissue from Kesterson Reservoir for selenium, one must understand that Se may exist there as selenate, selenite, elemental Se, and/or covalently bonded organoselenium compounds, each exhibiting its own physical and chemical characteristics. Most environmental substrates (the *matrix*) are themselves complex mixtures of chemicals from which the particular substance to be measured (the *analyte*) must be separated. Thus, an ability to distinguish between matrix and analyte, that is, the *specificity* of the analytical procedure, becomes essential.

The separation of matrix and analyte often begins with cleanup steps to remove background material such as protein, carbohydrate, and fat. The analyte must then be purified by a physical process such as volatilization, selective solubility (extraction), or selective adsorption (chromatography) or be converted to another chemical that possesses more desirable properties *(derivatization).* For example, bird tissue could be oxidized to convert all forms of selenium into selenate, which then is reduced to hydrogen selenide and isolated by volatilization. Common separation methods include gas chromatography (GLC), liquid chromatography (LC or HPLC), and thin-layer chromatography (TLC).

Once purified, the analyte must be *detected,* either visually by color or precipitation, or electronically based on some physical or chemical property (Table 2.2). An electronic signal from the detector is converted to a graph (Fig. 2.2) or to a digital or analog reading on a meter. The signal or reading must be proportional to the amount or concentration of analyte and allow *measurement* in relation to a known amount of some standard substance. On occasion, some other chemical with similar properties will be detected instead of, or in addition to, the analyte, creating an *interference.*

TABLE 2.2
Some Modern Analytical Detectors

Type	Separation[a]	Specificity	Sensitivity (pg)
Alkali flame ionization	GLC	N, P	1
Electron capture	GLC	e-rich	0.1
Flame photometric	GLC	P, S, Sn	1
Hall conductivity	GLC	Cl	0.1
Ultraviolet absorption	LC	UV-absorbing	50
Fluorescence	LC	Fluorescent	5
Color reagent	TLC	Varies	>1000

[a] GLC = gas chromatography, LC = liquid chromatography, TLC = thin-layer chromatography.

For example, in some gas chromatographic analyses, commonly encountered phthalate ester plasticizers give exactly the same signal as chlorinated hydrocarbon analytes, and some early analytical values published for DDT were later shown to represent only plasticizer and so to be completely erroneous.

Sometimes an analyte may be present in a sample but at concentrations below the ability of the detector to see it. However, technological advances have provided enough variety in detectors so that a more sensitive or selective alternative often may be available. Many analytes are now measured down into the picogram (10^{-12} g) range, but interferences and mechanical errors increase with sensitivity such that the most sensitive analytical method is not always the best one. Units commonly used in trace analysis, listed in Table 2.3, include the milligram (1 mg = 10^{-3} g), microgram (1 μg = 10^{-6} g), and nanogram (1 ng = 10^{-9} g).

Environmental analytical methods continue to evolve, and they are increasingly biochemical. An example is ***immunoassay,*** in which the analyte generates specific antibodies when injected into a rabbit, and these can then be measured colori-

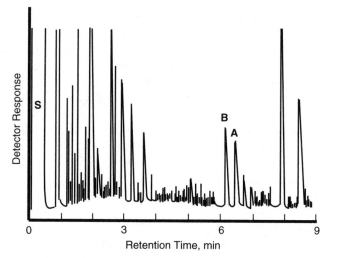

Figure 2.2. Gas chromatogram of the insecticide fipronil (A) and a degradation product (B) from field soil. Glass capillary column, N–P detector, S = solvent. Courtesy of K. Ngim.

TABLE 2.3
Environmental Concentration Equivalents

	Water	Air[a]	Soil and Biota
1 Part per million (ppm)	1mg/L	40.9MW μg/m^3	1 mg/kg
1 Part per billion (ppb)	1 μg/L[b]	40.9MW ng/m^3	1 μg/kg
1 Part per trillion (ppt)	1 ng/L[c]	40.9MW pg/m^3	1 ng/kg

[a] MW = molecular weight.
[b] 1 microgram (μg) = 0.001 mg.
[c] 1 nanogram (ng) = 0.001 μg.

metrically with speed, sensitivity, and accuracy (Van Emon and Lopez-Avila, 1992). Another method employs enzymatic reactions to generate a ***secondary analyte,*** the species actually measured. Enzyme-catalyzed hydrolysis of acetylcholine forms acetic acid as one product; inhibition of the enzyme by an organophosphate causes a quantifiable increase in pH in a cholinesterase-containing electrode.

The accurate interpretation of analytical data is not always simple. If possible, there should be enough replicate samples and analyses so that data can be treated statistically; the identity of the substance providing the signal must be established to avoid interferences; and a known amount of standard should be put through the procedure to determine what proportion survives to the end *(recovery).* Normally, only recoveries of at least 80% are acceptable. The reader of an article or report often does not have access to this essential information and so should always be a bit skeptical of published analytical results.

2.3 | THE ATMOSPHERE

2.3.1. The Natural Atmosphere

The atmosphere is the principal recipient and transporter of pollutants, mostly in the troposphere. The troposphere extends from the earth's surface to an altitude of about 10 km (33,000 ft) and contains all the dust, wind, and clouds. The tropospheric volume of 5.1×10^9 km^3, or 5.1×10^{18} m^3 is so large that a seemingly negligible concentration of a component such as carbon tetrachloride (1 μg/m^3) would total 5 billion kilograms if evenly distributed. As every substance is presumed to volatilize to some extent (Section 3.4), each molecule on earth should, in theory, pass through the atmosphere if given enough time.

Even pristine air is a complex mixture of nitrogen (78.09%), oxygen (20.94%), rare gases (0.93%), carbon dioxide (0.03%) and the toxicologically more interesting trace constituents listed in Table 2.4 (0.01%). The amount of water vapor is large but variable. The most abundant trace gas is methane, which, although toxicologically unimportant itself, is degraded by atmospheric oxidation to the toxic and carcinogenic formaldehyde. Most natural methane is produced by anaerobic microbial fermentation in wetlands (its old name is "marsh gas") and in the rumen of cattle (Table 2.5); anthropogenic sources provide less than 10% of the total. The tropospheric concentra-

TABLE 2.4
Trace Constituents of the Normal Troposphere[a]

Constituent	Formula	Conc. (ppb)[b]	Conc. ($\mu g/m^3$)
Carbon compounds			
Methane	CH_4	1400	900
Carbon monoxide	CO	60–200	70–230
Terpenes	$(C_5H_8)_n$	1–10 (C)[c]	3–30[d]
Formaldehyde	CH_2O	< 10	< 12
Halogen compounds			
Methyl chloride	CH_3Cl	0.5	1
Carbon tetrachloride	CCl_4	0.1–0.25	0.6–1.6
Freon 12	CF_2Cl_2	0.2	1
Freon 11	$CFCl_3$	0.1	0.7
Oxygen compounds			
Ozone	O_3	10–30	20–60
Nitrogen compounds			
Nitrous oxide	N_2O	330	600
Ammonia	NH_3	6–20	4–14
Nitric acid	HNO_3	3	7.5
Nitrogen oxides	NO/NO_2	1	~1.6
Sulfur compounds			
Sulfur dioxide	SO_2	1–4	3–11
Hydrogen sulfide	H_2S	< 0.2	< 0.3

[a] Adapted from Holland (1978).
[b] Concentration in volume per volume.
[c] Based on carbon content.
[d] Based on isoprene.

tion of methane has approximately doubled over the past 100 years (Findlayson-Pitts and Pitts, 1986), presumably due to increased numbers of cattle and rice fields. Natural terpene hydrocarbons and their derivatives, including isoprene (2-methyl-butadiene), are responsible for the odors of flowers and forests and are emitted by deciduous and coniferous trees in quantities sufficient to cause the blue haze, or "natural smog," seen in mountainous regions throughout the world (Duce et al., 1983).

TABLE 2.5
Sources of Atmospheric Methane[a]

Source	Emissions (Tg/yr)[b]
Wetlands	150
Cattle	120
Paddy fields	95
Direct anthropogenic	40
Burning of biomass	25
Oceans	13
Tundra	12
Lakes	10
Other sources	88
Total	553

[a] Adapted from Khalil and Rasmussen (1983).
[b] 1 teragram (1 Tg) = 10^{12} g = 1 billion kg.

Both natural and synthetic organohalogen compounds are normal constituents of the atmosphere. In addition to those shown in Table 2.4, oceans and volcanos produce chloro-, bromo-, fluoro-, and iodoalkanes and alkenes including CCl_4, CH_3Br, C_2H_5I, and $F_2C=CF_2$ (Gribble, 1994). Except for the relatively inert N_2O, the principal nitrogen compound in the atmosphere is ammonia derived from fuel combustion, fertilizer, and microbial activity. Ammonia and volatile amines are especially noticeable near stockyards and, being water soluble, will appear in a dish of water set outside even many miles away. Ammonia reacts with atmospheric nitric and sulfuric acids to produce solid and liquid inorganic particles responsible for much of the Northern Hemisphere's airborne haze.

Sulfur dioxide and hydrogen sulfide also are normal air constituents, but H_2S is so rapidly oxidized to sulfur oxides that it normally remains undectable. Most natural H_2S results from the microbial reduction of sulfate, although volcanos also are significant contributors. Varying low levels of carbon disulfide and carbon oxysulfide (COS) are present in the atmosphere, and the natural organosulfur compounds include methyl mercaptan (CH_3SH), dimethyl sulfide (CH_3SCH_3), dimethyl disulfide (CH_3S-SCH_3), and other alkyl sulfides up to C_{10} (Adams et al., 1981). The ocean contributes about 80 Tg/yr (80 billion kg/yr) of $(CH_3)_2S$, three times as much as does the land (Andreae and Raemdonck, 1983). Total natural emissions of sulfur and its compounds into the atmosphere are estimated at ~150 Tg/yr (S content), and anthropogenic pollution adds another 104 Tg annually (Cullis and Hirschler, 1980). One teragram (Tg) is one billion kg.

2.3.2. Polluted Atmospheres

Air pollution is nothing new. An English treatise of the 1600s referred to "this horrid Smoake which obscures our Church and makes our Palaces look old, which fouls our Cloth and corrupts our Waters, so as the very Rain. . . ." Pollution episodes in London killed thousands of people as recently as 1952 and were characterized by high levels of SO_2 and particulate matter (smoke) combined with fog ("smog"). Historically, industrial smoke with particle diameters of 10–100 μm provided the most obvious visual contribution to smog, but the very fine (<2.5 μm diameter) respirable particles currently cause the greatest health concerns. They can contain cadmium compounds from burning rubber, lead from automobile exhaust, and PAH from combustion of all kinds (Findlayson-Pitts and Pitts, 1986).

The coarse particles of smoke, dust, sea spray, and volcanic emissions are primarily inorganic and are removed from the troposphere by gravity. The smallest and most numerous particles (Aitken nuclei), with diameters <0.08 μm, are transient and soon coalesce. The particles of greatest toxicological interest have diameters between ~0.1 and 1 μm, called the "accumulation range," and although they represent only ~5% of the total number of airborne particles, they contain about 50% of the mass and often are largely organic in composition. For example, up to 3% of the fine particles from Los Angeles outdoor air can be traced to cigarette smoke and contain at least 117 identified organic compounds, including hydrocarbons, organic acids, phenols, nicotine and derivatives, and PAH congeners (Rogge et al., 1994). However, cooking meat was the source of 20% of the fine particles, while another 21% came from vehicle exhaust.

volumes changed slightly

The inventory of industrial chemicals released into the environment (TRI, Section 2.1) shows that over 40% are introduced directly into the atmosphere, and much of the remainder eventually appears there by volatilization. Although the emissions of even major chemical products such as methanol (0.09 Tg/yr) or toluene (0.12 Tg/yr) may seem small compared with those of some natural constituents (Table 2.5), the locally high atmospheric concentrations and toxicity of the synthetics often are enough to generate concern. Pesticides, such as the herbicide molinate, may be readily detected by odor even miles from the site of application, and anyone who has been in the vicinity of a paper mill recognizes the foul odor of dimethyl sulfide. Why are we not inundated by such huge volumes of chemicals?

The answer is that almost all organic chemicals react rapidly with atmospheric oxidants such as ozone and hydroxyl radicals to form small fragments, especially under the influence of ultraviolet (UV) radiation. When the initial concentrations of organic air pollutants such as petroleum alkenes become high enough, their atmospheric reactions result in *photochemical smog* (Section 5.3.3), a mix of chemicals like peroxyacetyl nitrate (PAN) and acrolein that can be quite toxic.

2.3.3. Indoor Air

Have you ever noticed a peculiar odor when you enter a new building? Buildings now are so airtight that fumes from construction materials, cleaners, and copiers accumulate to levels far above those in the outside environment. Formaldehyde from wallboard, styrene and phthalate esters from plastics, and nitrous acid from cooking are examples. "New car" odor is due to vinyl chloride and plasticizers, and the odor in your clothes closet may come from chlorinated cleaning solvents. Tobacco smoke presents a clear indoor hazard, as its toxic and carcinogenic constituents including PAH, nitrosamines, cadmium, nickel, nicotine, and carbon monoxide are not free to dissipate. The persistent tobacco odor that clings so tenaciously to clothing, furniture, and carpets is due largely to nicotine oxidation products such as myosmine, whose chronic toxicity potential remains unknown (Section 12.2).

One of the most serious indoor pollution problems is radon, ^{222}Rn, a gaseous, radioactive element that seeps into houses and other "energy-efficient" structures from building materials and from soil (Section 11.7). Another major problem is the carbon monoxide produced by gas stoves and furnaces, fireplaces, and automobile exhaust which, trapped indoors during cold weather, leads to many human deaths and injuries as listed in Table 1.2 (Turiel, 1985).

2.4 | WATER

2.4.1. Natural Waters

Like the atmosphere, natural waters are chemical mixtures, although the principal component always is hydrogen oxide, H_2O. For its molecular weight and size, this substance is unusual in that it has a relatively high boiling point (100°C), melting point (0°C), viscosity, and solvent capacity (Section 3.2). Because of its pronounced dissolving power, any sample of natural water contains detectable organic and inorganic solutes such as natural acids, carbohydrates, carbonic acid, and carbonates, salts, and oxygen. Natural inorganic components sometimes reach concentrations high enough to make them hazardous. Wells in parts of California have been closed because of high arsenic levels, dissolved nitrate has caused livestock poisoning, and the toxicity of natural selenium to wildlife at the Kesterson Reservoir in central California eventually led to abandonment of the entire facility.

2.4.2. Polluted Waters

Water also receives and dissolves numerous manmade chemicals. Among their sources are industrial waste (Appendix 2.1), domestic sewage, deposition from the atmosphere, and runoff from roads, parking lots, and farms. In 1979, the EPA released its selection of the 129 most important chemical pollutants found in U.S. industrial waste waters (Appendix 2.2), including the frequency with which they had been detected in over 2600 samples (Keith, 1979). While not all-inclusive, the list appears to cover the most prominent offenders and has never been superseded. A comparison of Appendices 2.1 and 2.2 reveals that only six of the priority pollutants appear among the top 22 chemicals on the TRI. The priority pollutant list was based on what was actually identified by water analysis, while the TRI represents an accounting on paper of highly soluble inorganics and volatile solvents that actually disperse too readily to be detectable in effluents.

Most priority pollutants were found in fewer than 8% of the wastes tested, although some inorganics, aromatic hydrocarbons, chloromethanes and -ethanes, and phthalate esters occurred more frequently. A 1974 survey of drinking water from five major cities across the United States showed surprisingly high levels of chloroform, bromodichloromethane, chlorodibromomethanes, and nitrotrichloromethane derived not from industrial pollution but from "purification" of the water by chlorination. The chlorination process indeed kills pathogenic microorganisms, but it also halogenates natural dissolved organic matter to generate halomethanes and other chlorinated and brominated compounds.

A 1986 survey of organic contaminants in large public well-water supplies in California (Appendix 2.3) likewise showed prominent halomethanes, suggesting that the samples had been collected *after* chlorination. However, there is ample evidence that chemical pollutants occasionally leach through the soil and enter groundwater. The same California study detected the nematicide DBCP (1,2-dibromo-3-chloropropane) in about 5% of the wells and the herbicides atrazine and simazine in about 1%. The National Pesticide Survey subsequently found these and a number of other chemicals in wells across the United States (USEPA, 1990).

Rain and fog wash impurities out of the atmosphere (Section 4.4.3). Analysis of rainwater from over Lake Michigan revealed ng/L levels of PCB, PAH, chlorinated hydrocarbon insecticides, and phthalate esters (Eisenreich et al., 1981). Airborne chemicals also precipitate as dry fallout, so it should not be surprising that seawater and marine sediments often contain pollutants, although usually at levels too low to be measured directly in the ocean. The California Mussel Watch Program routinely collects shellfish from about 200 sites along the Pacific coast and detects a wide variety of pesticides and other substances that have bioaccumulated (Table 14.2). Oil is probably the most frequent marine pollutant, some of it from natural seeps but most from tanker "spills" (Special Topic 6). The water-soluble fraction of petroleum contains phenols, N and S compounds, and some PAH (Table 13.3), which can be very toxic to marine animals (NRC,1985). Paints used on boat hulls often contain organotin compounds such as tributyltin oxide that are lethal to many forms of marine life, especially molluscs (Section 11.4.4).

Marine algae generate huge amounts of halogenated hydrocarbons (Gschwend et al., 1985). The principal ones are CH_2Br_2, $CHBr_2Cl$, and $CHBr_3$, but CH_2I_2, C_2H_5I, and various others have been reported. The 10^7 kg of organobromine compounds released annually to the atmosphere by algae is similar to that released by human activities, and both are dwarfed by the 10^9 kg/yr (1 Tg/yr) of algal methyl iodide.

2.5 | SOILS

2.5.1. Soil Composition

Gravity assures that soils are the repository for many kinds of chemicals. For centuries, soil has been used for the storage and disposal of chemical wastes, pesticides and fertilizers are applied intentionally to soil, and chemicals inadvertently reach soil from the atmosphere and from such human activities as construction, mining, and transportation. Soils are composed of sand, silt (rock particles <0.05 mm in diameter), clay, water, and air in varying proportions, as well as an organic fraction. Sand and silt are primarily silicon dioxide, often containing carbonate minerals, while clays are hydrated aluminum silicates. Other minerals may include oxides of iron, manganese, and titanium, Fe_2O_3 often giving the soil a red color.

In addition to debris such as sticks and leaf parts, the organic fraction contains amines, amino acids, sugars, and phenols as well as macromolecular lignins, polysaccharides, and humic substances (Table 2.6). The humic acids (soluble only in base)

TABLE 2.6
Properties of Typical U.S. Soils[a]

Soil Type	Location	pH	% Water[b]	% Sand/silt	% Clay	% OM[c]
Sand	CA	7.2	13.6	98	2	0.7
Sandy loam	CA	7.7	13.1	88	12	0.8
Loam	PA	6.9	13.1	76	24	1.8
Silty clay loam	TX	7.8	14.6	64	36	2.5
Sandy clay loam	TX	8.3	13.2	68	32	2.6
Clay	CA	8.1	13.1	46	54	1.2

[a] Yih et al. (1970).
[b] Water included in other percentages.
[c] OM = organic matter.

and fulvic acids (soluble in both acid and base) contain many oxygenated compounds and their acidic character comes from the multiple carboxylic and phenolic groups that give soil an overall negative charge. Chemical characteristics of soils are discussed further in Chapter 5.

2.5.2. Toxic Waste

According to the TRI, about a third of industrial chemical pollutants are released directly into or onto soil. However, most serious and persistent soil pollution problems relate to chemicals other than those listed in Appendix 2.2. The infamous chlorinated wastes of Love Canal, the metals from widespread mine and smelter wastes, and the hundreds of sites of petroleum leaks and spills are examples. The search for secure storage of radioactive waste has been well publicized.

Probably the most widespread soil pollution is from oil or gasoline leaked from underground storage tanks and pipelines. Here, aromatic hydrocarbons, especially the carcinogen benzene, are the principal concern. Contamination from chlorinated solvents used for dry cleaning, degreasing, and paint stripping of metal parts also is widespread and involves tetra- and trichloroethylenes, dichloromethane (methylene chloride), and 1,1,1-trichloroethane (methylchloroform). The soil at McClellan Air Force Base in California, for example, contains these solvents in addition to electroplating chemicals, low-level radioactive waste, jet fuel, oil, and polychlorinated biphenyls (PCB) and serves as a continual reservoir of pollutants for the groundwater beneath it. The same chlorinated solvents, leaked from underground storage tanks in the Silicon Valley near San Jose, California, caused extensive and widely publicized contamination of local groundwater, and they were found to have formed an extensive layer on the bottom of the Zürichsee in Switzerland.

Toxic metals are often are found in soils. Land even some distance away from copper mines often is devoid of vegetation due to the toxicity of mine tailings and drainage, and some plants accumulate sufficient selenium from soil to make them toxic to range animals (Section 12.5). An epidemic of illness and death in the 1970s among horses grazing on hillsides near Benicia, California, was traced to lead carried on smoke from a smelter located many miles upwind.

2.5.3. Pesticides

Pesticide chemicals have been employed by humans for at least 3000 years to control unwanted organisms. Insecticides kill insects, herbicides control plants ("weeds"), fungicides inhibit fungi, bactericides kill bacteria, and so on. Even before 1000 B.C., the Greek writer Homer described the use of sulfur in pest control. The ancient Romans employed the powdered roots of hellebore (*Veratrum alba*) against rats and insects, and the Chinese were applying arsenic minerals to control garden insects before 900 A.D. Carbon disulfide (1869) and dinitrocresol (1892) were among the earliest synthetic pesticides, and hydrogen cyanide was used as a grain fumigant during the late 1800s. Heavy applications of such chemicals were required, for example, 700 lbs of toxic and explosive carbon disulfide were required on each acre of soil for nematode control. World War II led to the discovery of new classes of synthetic pesticides, a chemical industry capable of producing them, and, especially, the realization that relatively small amounts of the new chemicals could control insects, weeds, and plant diseases that had been only marginally affected by the natural pesticides used previously. Pesticide chemicals continue to evolve, and recent ones, such as londax and glyphosate (Section 9.5), are effective at only a few grams per hectare and are almost nontoxic to mammals.

New pesticide structures generally are found by biological testing of large numbers of synthetic compounds often prepared for other purposes, by modifying new structural types discovered by competitors or university scientists, or by altering the structure of an active natural pesticide such as pyrethrins. At present, there are roughly 1000 pesticide chemicals in use worldwide—far too many to describe here— but older ones are withdrawn as they are seen to be environmentally undesirable or economically unprofitable, and new ones with improved characteristics are introduced. More than a decade of time and over $20 million are required to bring a new pesticide to market, so only a few major corporations still develop them.

California remains first among the states in the volume of pesticides used, and the 90.7 million kilograms (almost 200 million pounds) applied in California agriculture in 1994 represent hundreds of active ingredients (Appendix 2.4) formulated into thousands of commercial products. A total of 285.4 million kg (627 million pounds) of pesticide active ingredients was sold in California in 1994, much of it in the form of cleaning agents and bactericides (Cal EPA, 1996).

Pesticides are almost never applied as pure chemical. In addition to active ingredients, the *formulation* contains a carrier as well as *adjuvants* that alter physical properties, and it is further diluted into a large proportion of water or oil before application. The solid formulation one buys at the store usually is a dust (designated D) or wettable powder (W or WP), and the carrier often is talcum powder or sand. Liquid formulations usually are solutions of the pesticide in xylene or oil that form an emulsifiable concentrate (EC), or they are simply aqueous solutions (S). Adjuvants include adhesives ("stickers") such as polyvinyl acetate, surfactant spreaders, and emulsifiers. "Sevin 50W®" would be a wettable powder formulation that contains 50% active ingredient, the insecticide Sevin® (carbaryl), plus 50% carrier and adjuvants.

Pesticide chemicals provide a broad range of toxic effects and mechanisms (Appendix 9.1). However, fewer than 100 pesticides receive extensive use, and most are discussed somewhere in this book. In addition to the reference works cited for specific chemical groups such as the organophosphates and pyrethrins, the physical, environmental, and toxic properties of hundreds of pesticides are summarized by Tomlin (1994), and other valuable literature resources cover pesticide mechanisms of action (Corbett et al., 1984) and pesticide chemistry (Matolcsy et al., 1988).

Some pesticides are applied directly to the soil, most others reach soil via atmospheric deposition or during applications to foliage, and four of California's top 12— methyl bromide, metam sodium (Vapam®), 1,3-dichloropropene (Telone®), and chloropicrin—are injected directly into the soil as fumigants to control plant-parasitic nematodes. Many pesticides are "systemic"; that is, they are absorbed from the soil into plant roots and then translocated upward into the leaves, and examples common in home gardening are the herbicides 2,4-D (Weedone®) and trifluralin (Treflan®) and insecticides acephate (Orthene®) and disulfoton (Disyston®). It is important to recognize that most pesticides go by more than one name: Each has a registered **trade name,** such as Sevin®; a common or **generic name,** carbaryl in this case; and a **chemical name** (1-naphthyl *N*-methylcarbamate).

Although neither systemic nor applied directly to soil, DDT and its degradation products are still detectable in soils almost anywhere in the world. They are said to be **persistent.** Many other types of pesticides, including urea and triazine herbicides as well as others originally thought to have limited longevity (Kearney et al., 1969), also persist over long periods, depending on temperature, moisture, and soil type. Decades ago, highly toxic inorganic pesticides containing copper, lead, or arsenic saw extensive use, but they proved to be essentially indestructible and today contaminate large areas once in agricultural production but now often in housing developments.

2.6 │ BIOTA

2.6.1. Toxic Organisms

Poisonous snakes and spiders are what most people think of when the term *biohazard* is used. Actually, there are relatively few **poisonous** animals, that is, animals whose flesh is toxic. Toads are probably the most common example. However, **venomous** animals, which inject their toxins via fangs or stingers, include rattlesnakes, black widow spiders, bees, and red ants. While the toxicants of ant venoms can be simple chemicals such as formic acid and alkaloids, those of snakes and spiders are complex mixtures of proteins (Section 12.5).

Although pathogenic bacteria and fungi represent by far the most widespread biotic hazards, common plants also can be very toxic and often are responsible for poisoning in young children. Houseplants such as dumbcane (*Dieffenbachia* species), oleander *(Nerium oleander),* and lily-of-the-valley *(Convallaria majalis)* in the garden, and field plants such as death camus (*Zigadenus* sp.) and larkspur (*Delphinium* sp.) present a distinct hazard to people and other mammals. Some plants are nontoxic to humans but affect other animals: *Tephrosia* paralyzes fish, marigold *(Tagetes)* repels nematodes, and pyrethrum *(Chrysanthemum)* flowers kill insects.

Other wild plants produce natural poisons that can be transferred to man and other animals. An historical example, the "milk sickness" produced by drinking milk from cows that had consumed white snakeroot *(Eupatorium rugosum),* was mentioned in Section 1.1, and anagyrine from lupine species causes deformities in calves whose dams eat the plants early in gestation and in human babies whose pregnant mothers drink the cow's milk (Section 7.4.6). People often are fatally poisoned by eating shellfish containing saxitoxin absorbed from marine dinoflagellates or by eating fish that fed on marine algae containing ciguatoxin.

However, the greatest exposure to toxic chemicals comes from eating ordinary food. Food is just a mixture of chemicals, not all of which are necessarily good for us. Most major food components—carbohydrates, fats, proteins, and water—are nutritious; it is the *secondary substances,* including flavors, vitamins, and other trace chemicals, that cause trouble. The secondary substances also include natural isothiocyanates from garlic, the cholinesterase-inhibiting alkaloid solanine from potatoes, polycyclic aromatic hydrocarbons from smoked or charbroiled meats, and phototoxic coumarins from citrus oils and celery. Microorganisms also produce a wide variety of the toxic chemicals found in food, from the polypeptides that cause "food poisoning" to aflatoxins that induce cancer (Chapter 12).

Most environmental chemicals can be bioconcentrated to some degree into living plants and animals (Section 3.3), the greater the fat solubility the higher the concentration, and breakdown products of the chemicals often may be present as a result of biodegradation (Chapter 6). An extreme example is DDT, which can be bioconcentrated more than 10,000-fold and degraded to persistent residues of DDD, DDE, and other metabolites. Despite over two decades of restricted DDT use, residues are still detectable in the fat of almost every animal in the world, including humans (Chapter 14).

2.6.2. Food

Natural ("organic") food is just a mixture of chemicals, but manmade chemicals are usually present also. Although pesticide residues in food are sufficiently important and controversial to warrant special attention (Special Topic 2), they are classified for regulatory purposes as unintentional or incidental food additives. This classification also includes metals acquired during food processing and plasticizers leached into food from packaging. However, hundreds of chemical substances are added intentionally to assist the processing, preservation, nutritional quality, and organoleptic properties such as flavor, odor, color, and texture; a glance at almost any food package reveals intentional additives. The U.S. Food and Drug Administration keeps close watch over possible toxic food additives, although many substances that have been added for centuries are classified as Generally Regarded as Safe (GRAS) and so are exempt from regulation.

About a dozen chemicals represent 95% of the total weight of additives used, including sucrose, citric acid, dextrose, salt, carbon dioxide, and sodium bicarbonate. Nevertheless, a few of the approved additives remain controversial. The flavor enhancer monosodium glutamate (MSG) can cause headache and flushing ("Chinese restaurant syndrome"), the preservative butylated hydroxyanisole (BHA) has produced adverse effects in rat tests, and the tests also suggest that sulfites, nitrates, and several azo dyes may be carcinogenic. The Delaney Clause of the 1958 Food Additives Amendment to the Pure Food and Drug Act states that "no additive shall be deemed safe if it is found to induce cancer when ingested by man or animal"; accordingly, the additive either causes "cancer" or it doesn't, without regard for dose, species, or tumor type (see Special Topic 9). For example, the very weak carcinogen safrol, used for many decades to give flavor to root beer, is now prohibited "to protect the public."

A brief survey of the author's kitchen produced the list of common food additives shown in Table 2.7. For further education, just look at the label on a soft drink

TABLE 2.7
Some Common Food Additives and Their Uses

Additive	Uses	Typical Foods[a]
Ascorbic acid	Preservative	Bread, cereal, soft drink
Aspartame	Sweetener	Soft drink
Butylated hydroxytoluene	Preservative	Cereal
Caffeine	Stimulant	Soft drink
β-Carotene	Color	Mayonnaise
Cellulose gum	Thickener	Ice cream
Disodium guanylate	Flavor	Pretzels
Ferrous gluconate	Color, flavor	Canned olives
Malic acid	Sequestrant[b]	Pretzels
Monosodium glutamate	Flavor	Noodles
Niacinamide	Nutrient	Cereal, bread
Phosphoric acid	Sequestrant[b]	Soft drink
Potassium sorbate	Preservative	Mayonnaise
Sodium aluminophosphate	Anticaking	Pancake mix
Vanillin	Flavor	Ice cream, pudding mix
Zinc oxide	Nutrient	Cereal

[a] From a brief tour of the author's kitchen.
[b] Binds metals that catalyze oxidations.

can and the toxicologist's view of food additives provided by Hayes and Campbell (1986).

2.7 | EVERYDAY LIFE

We are surrounded by chemicals every day. The workplace often is an important source of exposure: Roofers, construction workers, and automobile mechanics receive high levels of PAH; farm workers are exposed to pesticides; and anyone who cleans anything may be exposed to chlorinated solvents. Office workers contact photocopier toner, correction fluid, and rubber cement; hospital workers are surrounded by disinfectants; and everyone is exposed in their home to cleaners, detergents, preservatives, carbon monoxide, paints, and toxic plants including onions and pepper. No one is exempt.

One ordinarily assumes that environmental pollutants come only from business and industry, but actually each of us plays a part. The contents of aerosol cans, tobacco smoke, perfume and cologne, the tin fluoride in toothpaste, and the shampoo that washes down the drain are among our frequent contributions to environmental pollution. When you take an aspirin tablet, where do you think its constituents go? What happens to the oil drippings in your driveway? What about the cadmium-containing rubber that continuously wears off your shoes and tires? Every manufactured product eventually ends up in the environment, where its constituent molecules disperse. This dispersal is the subject of the next two chapters.

2.8 | REFERENCES

Adams, D. F., S. O. Farwell, E. Robinson, M. R. Pack, and W. I. Bamesberger. 1981. Biogenic sulfur source strengths. *Environ. Sci. Technol.* **15:** 1493–98.

Ames, B. N. 1983. Dietary carcinogens annd anticarcinogens. *Science* **221:** 1249–64.

Andreae, M. O., and H. Raemdonck. 1983. Dimethyl sulfide in surface ocean and marine atmosphere: a global view. *Science* **221:** 744–47.

Cal EPA. 1996. *Pesticide Use Report, Annual 1994,* California Environmental Protection Agency, Dept. of Pesticide Regulation, Sacramento, CA.

CDHS. 1986. *Organic Chemical Contamination of Large Public Water Systems in California,* California Department of Health Services, Berkeley, CA.

Corbett, J. R., K. Wright, and A. C. Baille. 1984. *The Biochemical Mode of Action of Pesticides,* 2nd Ed., Academic Press, New York, NY.

Cullis, C. F., and M. M. Hirschler. 1980. Atmospheric sulfur: Natural and manmade sources. *Atmos. Environ.* **14:** 1263–78.

Duce, R. A., V. A. Mohnen, P. R. Zimmerman, D. Grosjean, W. Cautreels, R. Chatfield, R. Jaenicke, J. A. Ogren, E. D. Pellizzari, and G. T. Wallace. 1983. Organic material in the global troposphere. *Rev. Geophys. Space Phys.* **21:** 921–52.

Eisenreich, S. J., B. B. Looney, and J. D. Thornton. 1981. Airborne organic contaminants in the Great Lakes ecosystem. *Environ. Sci. Technol.* **15:** 30–38.

Findlayson-Pitts, B. J., and Pitts, J. N. 1986. *Atmospheric Chemistry: Fundamentals and Experimental Techniques,* John Wiley & Sons, New York, NY.

Gribble, G. W. 1994. The natural production of chlorinated compounds. *Environ. Sci. Technol.* **28:** 310A–19A.

Gschwend, P. M., J. K. MacFarlane, and K. A. Newman. 1985. Volatile halogenated organic compounds released to seawater from temperate marine macroalgae. *Science* **227:** 1033–35.

Hayes, J. R., and T. C. Campbell. 1986. Food additives and contaminants, in *Casarett and Doull's Toxicology,* 3rd Edition (C. D. Klaassen, M. O. Amdur, and J. Doull, eds.), Macmillan, New York, NY.

Holland, H. D. 1978. *The Chemistry of the Atmosphere and Oceans,* John Wiley & Sons, New York, NY.

Honeycutt, R. C., J. P. Wargo, and I. L. Adler. 1976. Bound residues of nitrofen in cereal grain and straw, in *Bound and Conjugated Pesticide Residues* (D. D. Kaufman, G. G. Still, G. D. Paulson, and S. K. Bandal, eds.), American Chemical Society, Washington DC, pp. 170–72.

Johnson, R. D., D. D. Manske, and D. S. Podrebarac. 1981. Pesticide, metal, and other chemical residues in adult total diet samples. XII. August 1975–July 1976. *Pestic. Monit. J.* **15:** 54–69.

Kaufman, D. D., G. G. Still, G. D. Paulson, and S. K. Bandal. 1976. *Bound and Conjugated Pesticide Residues,* American Chemical Society, Washington, DC.

Kearney, P. C., R. G. Nash, and A. R. Isensee. 1969. Persistence of pesticide residues in soils, in *Chemical Fallout:* Current Research on Persistent Pesticides (M. W. Miller and G. G. Berg, eds.), Thomas, Springfield, IL pp. 54–67.

Keith, L. J. 1979. Priority pollutants. I. A perspective view. *Environ. Sci. Technol.* **13:** 416–23.

Khalil, M. A. K., and R. A. Rasmussen. 1983. Sources, sinks, and seasonal cycles in atmospheric methane. *J. Geophys. Res.* **88C:** 5131–44.

Matolcsy, G., M. Nadasy, and V. Andriska. 1988. *Pesticide Chemistry,* Elsevier Publishers, Amsterdam, the Netherlands.

NRC. 1985. *Oil in the Sea,* National Academy of Sciences, Washington, DC.

Rogge, W. F., L. M. Hilderman, M. A. Mazurek, G. R. Cass, and B.R.T. Simoneit. 1994. Sources of fine organic aerosol. 6. Cigarette smoke in the urban atmosphere. *Environ. Sci. Technol.* **28:** 1375–88.

Tomlin, C. 1994. *The Pesticide Manual,* 10th Edition, British Crop Protection Council, Farnham, U.K., and the Royal Society of Chemistry, Cambridge, U.K.

Turiel, I. 1985. *Indoor Air Quality and Human Health,* Stanford University Press, Stanford, CA.

USEPA. 1990. *National Survey of Pesticides in Drinking Water,* National Pesticide Survey, Phase 1 Report, U.S. Environmental Protection Agency, Washington, DC.

USEPA. 1991. *Toxics in the Community: The 1989 Toxics Release Inventory National Report,* U.S. Environmental Protection Agency, Washington, DC.

USITC. 1994. *Synthetic Organic Chemicals, 1992,* U.S. International Trade Commission, U.S. Department of Commerce, Washington, DC. Publ. 720.

Van Emon, J. M., and V. Lopez-Avila. 1992. Immunochemical methods for environmental analysis. *Anal. Chem.* **64:** 78A–88A.

Yih, R. Y., C. Swithenbank, and D. H. McRae. 1970. Transformations of the herbicide *N*-(1,1-dimethylpropynyl)-3,5-dichlorobenzamide in soil. *Weed Sci.* **18:** 604–7.

Special Topic 2. Pesticide Residues

Whenever a crop is treated with a pesticide, some of the formulation necessarily remains so that it can be effective. This spray *deposit* often is visible for a short time after application, but most of it is quickly lost until only an invisible *pesticide residue* remains. Figure 10.1 illustrates this biphasic process, known as *dissipation,* which continues as the residual chemicals are *weathered* by volatilization, sunlight, and mechanical removal until most resides in the waxy leaf cuticle, where it is diluted by absorption and plant growth and is said to be "aged."

Pesticide residues in food remain a subject of public concern. Due to extensive publicity, the perception persists of more residues than are actually there. The U.S. government (as well as other nations and the U.N.) establishes safe residue levels, called tolerances or maximum residue levels (MRL), for each commodity (Section 10.5). Among other means, residues are checked by annual "market-basket surveys" in which the components of a robust standard diet—that of a 17-year-old boy—are purchased at food markets around the country, composited, and analyzed for over 100 different pesticides. The results (Table 2.8) show that although residues may indeed be present, over 80% of the tested composites contain no detectable pesticide, and fewer than 1% contain levels above the tolerance. Food usually contains higher levels of toxic natural products than it does of pesticide residues (Ames, 1983).

Table 2.8 shows that parathion residues, for example, were found in only 2% of the test composites in 1981 and in fewer than 1%—three out of 926 composites—by 1989. The maximum amount was 0.006 mg in a kilogram (2.2 lbs) of food, less than 1/100 of the legal limit. Further, the residues represent what was found on the raw agricultural commodity, and much less remains after the food is peeled, cooked, or processed. The outline of pesticide residue analysis in Section 2.2 suggests the possibility of pesticide residue measurement down into the ng/kg (part per trillion) range or lower, far below any measurable toxicological significance for humans.

Although analysis measures individual pesticide chemicals, breakdown products often are

TABLE 2.8
Some Pesticide Residues in Foods, 1981 and 1989[a]

Pesticide[b]	% Occurrence		Maximum Residue (ppb)[e]	Tolerance (ppb) (Commodity)	
	1981[c]	1989[d]			
Dieldrin	18	6	86	100	(fruit)
Malathion	11	19	96	8,000	(grain)
Lindane	5	2	4	7,000	(fat)
Heptachlor epoxide	5	nd[f]	3	100	(milk fat)
Dichloran	4	4	163	5,000	(tomato)
Diazinon	3	6	4	500	(almonds)
DDT	3	16	10	5,000	(fat)
Parathion	2	1	6	1,000	(tomato)
PCNB (Quintozene)	1	2	3	none	
Captan	1	nd	40	25,000	(apple)

[a] Johnson et al., 1981.
[b] Out of about 30 detected.
[c] From 240 composites.
[d] From 926 composites.
[e] Reported in 1981.
[f] nd = none detected.

included in the results. For example, DDE and DDD often are lumped together with DDT (written as DDTR), and paraoxon is included with parathion. However, research with radiolabeled pesticides reveals that some residue may remain unextracted from the plant material despite the most drastic treatment. These **bound residues** have become incorporated into the structural material of the plant itself, often in the lignin component, or metabolized and reincorporated into other plant constituents (Kaufman et al., 1976). For example, wheat treated with the now-discontinued herbicide nitrofen contained 30% of the residue in the straw in the form of lignin, and 70% of that in the grain was incorporated into the glucose of the starch (Honeycutt et al., 1976).

"Residues" also are found in water, air, and soil, although perhaps this is not a strictly correct use of the term; a residue is something left over after a part has been removed, and pesticides seldom are applied intentionally to the abiotic environment. The pesticides detected in these environmental compartments are subject to the same dissipative forces that act on plant residues, to be discussed more thoroughly in later chapters, and usually are similar to those in food. Of legitimate concern is the accumulation of persistent residues in aquatic animals, especially those to be consumed as food. Few modern pesticides persist long enough to be detectable, but the long-discontinued chlorinated insecticides such as DDT still remain (Section 14.2).

If today's pesticide residues in food have little or no toxicological significance, why then the continuing controversy? It is partly because few among the general public understand that the risk is negligible; pesticides are always portrayed as "bad" and "dangerous" without regard to the context. The more philosophical aspect is that one has little or no choice about the natural substances in our food, but somebody *put* the pesticides there without our explicit permission. However, as long as people insist on cheap and blemish-free fruits and vegetables, and the demand for ever more sensitive chemical analysis continues, some pesticide residues inevitably will be detected in our food.

APPENDIX 2.1.
Toxic Inventory Release (TRI) Data, 1989[a]

Rank	Chemical	Total Releases (10^6 lbs)	% of Total Releases[b]					
			Air	Water	Land	UG[c]	PS[c]	Offsite
1	Ammonium sulfate[d]	750.6	0.1	9.1	2.0	**61.5**	26.8	0.5
2	Hydrochloric acid	495.6	12.3	0.6	0.9	**60.7**	5.7	19.9
3	Methanol	408.1	**48.9**	4.2	1.9	5.7	26.7	12.6
4	Ammonia	377.2	**64.8**	6.4	2.0	17.3	7.6	1.8
5	Toluene	322.5	**79.2**	0.1	0.1	0.2	1.0	19.5
6	Sulfuric acid	318.4	7.6	6.4	1.5	**47.0**	13.5	24.1
7	Acetone	255.5	**78.0**	0.4	0.1	1.8	5.2	14.6
8	Xylenes	185.4	**79.5**	0.1	0.3	<0.1	2.1	18.0
9	1,1,1-Trichloroethane	185.0	**91.1**	<0.1	<0.1	0.0	0.1	8.7
10	Zn compounds	164.8	2.8	0.5	59.9	0.1	1.2	35.4
11	2-Butanone	157.0	**81.3**	<0.1	0.1	0.1	0.5	17.9
12	Chlorine	141.4	**93.6**	1.7	0.2	0.3	2.0	2.2
13	Dichloromethane	130.4	**83.8**	0.2	<0.1	0.6	1.1	14.3
14	Mn compounds	119.8	1.4	0.7	**70.9**	0.8	5.5	20.7
15	Carbon disulfide	100.2	**99.6**	<0.1	0.0	<0.1	0.1	0.2
16	Phosphoric acid	98.6	1.8	27.5	**50.7**	0.1	11.7	8.1
17	Nitric acid	74.9	7.0	1.1	0.7	**41.4**	18.5	31.3
18	Ammonium nitrate[c]	73.3	4.7	12.1	10.3	**61.0**	9.7	2.2
19	Freon 113	67.8	**93.2**	<0.1	<0.1	0.0	0.1	6.6
20	Glycol ethers	65.7	**72.5**	0.2	0.2	0.6	13.2	13.3
21	Ethylene glycol	57.8	22.3	6.5	1.5	14.8	**28.7**	26.2
22	Zinc metal	57.5	6.1	0.2	34.7	0.0	0.2	**58.8**
Subtotal		4607.7[b]	46.9	3.6	10.8	11.5	8.3	16.2
Other Releases		1097.9						
Grand Total		5705.6	42.5	3.3	7.8	20.7	9.7	16.0

[a] USEPA (1991).
[b] Totals may not be exact, due to rounding-off.
[c] UG = underground, PS = public sewer.
[d] Aqueous solution.

APPENDIX 2.2.
EPA Priority Water Pollutants[a]

Hydrocarbons:
  Toluene (29.3)[b]; benzene (29.1); ethylbenzene (16.7); naphthalene (10.6).

Polycyclic aromatic hydrocarbons:
  Phenanthrene/anthracene (10.6); pyrene (7.8); fluoranthene (7.2); fluorene (5.7); chrysene (5.1); acenaphthylene (4.5); acenaphthene (4.2); benzo[a]pyrene (3.2); benzo[a]anthracene (2.3); benzo[k]fluoranthene (1.8); benzo[b]fluoranthene (1.6); indeno[1,2,3–c,d]pyrene (0.8); benzo[g,h]perylene (0.6); dibenzo [a,h]anthracene (0.2).

Halomethanes:
  Trichloromethane (chloroform) (40.2); dichloromethane (methylene chloride) (34.2); tetrachloromethane (carbon tetrachloride) (7.7); fluorotrichloromethane (Freon 11) (6.8); bromodichloromethane (4.3); chlorodibromomethane (2.5); chloromethane (methyl chloride) (1.9); tribromomethane (bromoform) (1.9); dichlorodifluoromethane (0.3); bromomethane (methylbromide) (0.1).

Haloethanes and -propanes:
Trichloroethylene (10.5); tetrachloroethane (10.2); Z-1,2-dichloroethylene (7.7); 1,1-dichloroethylene (7.7); 1,1,2,2-tetrachloroethane (4.2);1,2-dichloropropane (2.1); 1,1,2-trichloroethane (1.9); 1,2-dichloroethane (ethylene dichloride) (1.4); 1,3-dichloropropene (1.0); hexachloroethane (0.5); chloroethane (ethyl chloride) (0.4); chloroethylene (vinyl chloride) (0.2); hexachlorobutadiene (0.2).

Haloalicyclics:
β-BHC[c] (0.8); α-BHC (0.6); γ-BHC (0.5); aldrin (0.5); β-endosulfan (0.4); α-endosulfan (0.3); heptachlor (0.3); δ-BHC (0.2); camphechlor (toxaphene) (0.2); chlordane (0.2); endosulfan sulfate (0.2); endrin (0.2); endrin aldehyde (0.2); dieldrin (0.1); heptachlor epoxide (0.1); hexachlorocyclopentadiene (0.1).

Haloaromatics:
Dichlorobenzenes (6.0); hexachlorobenzene (1.1); 1,2,4-trichlorobenzene (1.0); 2-chloronaphthalene (0.9); Aroclors 1016, 1221, 1232, 1242,1248,1254, 1260 (0.5–0.9); 4,4'-DDT (0.2); 4,4'-DDD (0.1); 4,4'-DDE (0.04).

Phenols:
Phenol (26.1); pentachlorophenol (6.9); 2,4-dimethylphenol (5.2); 2,4,6-trichlorophenol (4.6); 2,4-dichlorophenol (3.3); 2-chlorophenol (2.3); 2-nitrophenol (2.3); 4-nitrophenol (2.2); 4-chloro-3-methylphenol (1.9); 2,4-dinitrophenol (1.6); 4,6-dinitro-2-methylphenol (DNOC) (1.1).

Phthalate esters:
bis(2-Ethylhexyl phthalate (41.9); dibutyl phthalate (18.9); butyl benzyl phthalate (8.5); diethyl phthalate (7.6); di-n-octyl phthalate (6.4); dimethyl phthalate (5.8).

Ethers and carbonyl compounds:
2-Chloroethyl vinyl ether (1.5); isophorone (1.5); acrolein (1.2); bis(chloroethyl) ether (1.1); bis(2-chloroethoxy)methane (0.4); bis(chloromethyl) ether (0.1); 4-chlorophenyl phenyl ether (0.1); 4-bromophenyl phenyl ether (0.04); 2,3,7,8-tetrachlorodibenzo-p-dioxin (TCDD)(unavailable).

Nitrogenous Compounds:
Acrylonitrile (2.7); nitrobenzene (1.8); 2,6-dinitrotoluene (1.5); N-nitrosodiphenylamine (1.2); 2,4-dinitrotoluene (1.1); 1,2-diphenylhydrazine (0.8); benzidine (0.2); N-nitrosodimethylamine (0.1); N-nitrosodi-n-propylamine (0.1).

Inorganic:
Cu (55.5); Zn (54.6); Cr (53.7); Pb (43.8); Ni (34.7); CN⁻ (33.4); Cd (30.7); Ag (22.9); As(19.9); Tl (19.2); Se (18.9); Sb (18.1); Hg (16.5); Be (14.1); asbestos (unavailable).

[a] Adapted from Keith (1979).
[b] Percentage of 2617 industrial wastewater samples in which this pollutant was detected.
[c] BHC is benzene hexachloride (1,2,3,4,5,6-hexachlorocyclohexane).

APPENDIX 2.3.
Organic Contaminants in California Drinking Water[a]

Chemical	Wells[b]	Chemical	Wells[b]
Tetrachloroethylene	199	Toluene	10
Trichloroethylene	188	Benzene	9
Dibromochloropropane	155	Freon 113	8
Chloroform	116	Xylenes	6
1,1-Dichloroethylene	63	1,2-Dichloropropane	4
1,1,1-Trichloroethane	63	1,1,2-Trichloroethane	4
Carbon tetrachloride	38	Chlorobenzene	3
Atrazine	37	Freon 11	3

APPENDIX 2.3. (*continued*)

Chemical	Wells[b]	Chemical	Wells[b]
1,2-Dichloroethane	17	C_3-C_4 alkylbenzenes	3
Atrazine	37	Freon 11	3
1,2-Dichloroethylene	29	1,2-Dichlorobenzene	2
Simazine	26	1,3-Dichlorobenzene	2
Bromodichloromethane	25	1,4-Dichlorobenzene	2
Dibromochloromethane	21	Ethylbenzene	2
1,2-Dichloroethane	17	C_3-C_4 alkylbenzenes	1
Bromoform	14	Freon 12	1
1,1-Dichloroethane	12	Vinyl chloride	1
Dichloromethane	11		

[a] Adapted from CDHS (1986).
[b] Number of contaminated wells out of 2947 analyzed.

APPENDIX 2.4.
Some Pesticides Used in California, 1991[a]

Rank	Pesticide	Use[b]	Class[c]	kg Applied[d]
1	Sulfur	F	Inorganic	25,224,715
3	Methyl bromide	S	OH	9,195,873
5	Metham sodium	S	Dithiocarbamate	2,697,310
6	1,3-Dichloropropene	N	OH	2,356,269
9	Glyphosate	H	Phosphonate	1,255,018
11	Chlorpyrifos	I	OP	1,120,354
12	Chloropicrin	S	OH	1,022,115
13	Propargite	I,N	Ester	939,296
14	Malathion	I	OP	861,866
15	Maneb	F	Dithiocarbamate	850,009
16	Molinate	H	Thiocarbamate	695,266
17	Trifluralin	H	Nitroaniline	648,372
18	Simazine	H	Triazine	537,757
19	Diazinon	I	OP	456,573
25	Carbaryl	I	Carbamate	334,794
34	Chlorthal dimethyl	H	Ester	319,558
36	2,4-D esters	H	Phenoxy ester	273,736
39	Endosulfan	I	OH	232,602
43	Aldicarb	I	Carbamate	215,139
47	Acephate	I	OP	184,772
50	Cyanazine	H	Triazine	174,165
53	Carbofuran	I	Carbamate	156,897
57	Dichloran	F	Nitroaniline	115,862
59	Benomyl	F	Benzimidazole	106,120
61	Fenvalerate	I	Pyrethroid	90,628

[a] Cal EPA, 1992.
[b] A = acaricide, F = fungicide, H = herbicide, I = insecticide, N = nematicide, S = soil fumigant.
[c] OH means organohalogen; OP means organophosphorus.
[d] Active ingredients only.

Environmental Chemodynamics **3**

3.1 ENVIRONMENTAL CHEMODYNAMICS

If you spill some fuel while filling your car's tank, you might notice that the liquid quickly "disappears" leaving a strong gasoline odor. The fuel evaporated—that is, it went from a liquid to a gaseous state—and the vapor moved far enough that your nose could detect it. Chemicals move about and change in the environment, and the study of that movement and transformation is *environmental chemodynamics*. In volatilizing or dissolving in water, a substance becomes diluted and less hazardous, while bioconcentration into a living organism or binding to soil concentrates it and can make it more hazardous.

For convenience, the physical environment can be divided into four "compartments:" the atmosphere (air), hydrosphere (water), lithosphere (soil), and biosphere (living organisms or biota) (Chapter 2). The atmosphere, of course, is mostly gaseous, natural water represents the liquid phase above 0°C, soil is largely a solid mixture of sand, silt, clay, and organic matter, and fat (lipid) is characteristic of the physically and chemically complex biota. However, these phases are seldom homogeneous, and soil consists of air, water, and biota in addition to the minerals, air contains liquid water (rain) and solid dust particles, and so on. Surfaces, the interfaces between the phases, have some important characteristics of their own.

Although released primarily into one compartment, chemicals quickly enter all adjacent compartments (Fig. 3.1). The change of the gasoline from liquid to vapor reflects a universal *escaping tendency,* a thermodynamic drive for molecules to "escape" from one state (phase) into another. This *fugacity* (from Latin *fugere,* to flee) is to molecular diffusion what temperature is to heat diffusion (see Special Topic 3), and just as heat diffuses spontaneously from a higher temperature to a lower one, chemicals diffuse from high to low fugacity. Even as pollutant molecules escape from water into some other solvent, those in the solvent already are escaping back; when the chemical's fugacity, f, in one phase equals that in the other, diffusion stops and the system is said to be at equilibrium.

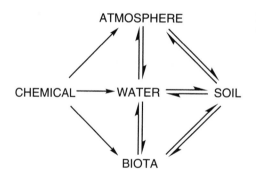

Figure 3.1. Intercompartment distribution of a chemical in the environment.

Fugacity dictates that the ***partitioning*** (distribution) of chemicals among environmental compartments or phases is inevitable. Theoretically, a chemical will escape from its original compartment until its concentration also is at equilibrium among all the adjacent compartments. While equilibrium seldom is reached in the real world, it often is close enough to permit general predictions of chemical distribution to be made. Partitioning takes place between adjacent, largely immiscible phases, such as between a solid, gas, or liquid and a liquid ***(dissolution),*** a solid or liquid and a gas ***(volatilization),*** a solution and a solid surface ***(adsorption),*** or a solution and an immiscible liquid ***(solvent partitioning and bioconcentration).*** Environmental partitioning will be discussed in this chapter and its consequences in the next, but additional detail is presented by Smith et al. (1988) and Schwarzenbach et al. (1993). The chemical transformations involved in environmental chemodynamics will be discussed in Chapters 5 and 6.

3.2 │ DISSOLUTION

One observes dissolution every day. Sugar dissolves in coffee, salt dissolves in hot water, varnish dissolves in paint thinner. The molecules or atoms of one become homogeneously dispersed among those of the other. The substance in the largest proportion is referred to as the ***solvent*** and the other as the ***solute,*** although sometimes each will be soluble in all proportions in the other *(e.g.,* alcohol and water). Dissolution in water is especially important for the environmental dissipation of chemicals, as almost all intercompartment transport starts from aqueous solution. *Every* substance is soluble in water to some extent, although its solubility may be so slight as to be listed as "insoluble" in a chemistry handbook: The maximum aqueous solubility of chalk (limestone) is 14 mg/L, that of metallic mercury 0.056 mg/L, and that of DDT 0.001 mg/L.

This maximum or saturation solubility, the "solubility" listed in handbooks and in Appendix 3.1, is the equilibrium point where molecules or atoms return to the surface of the undissolved chemical as rapidly as they leave. Molecular agitation and hence the solubility of most substances increases with increased temperature when dissolution is endothermic and decreases when it is exothermic; when a hot, saturated salt solution cools, crystals soon appear that are in equilibrium with the salt still in solution. The amount dissolved will always be the same at a given temperature, so solubility can be viewed as a ***partitioning*** of the salt between the solid state and solution.

But *why* does a salt crystal disappear when you put it in water? The electronegative oxygen of a water molecule strongly attracts the electrons it shares with the hydrogens, giving it a partial negative charge and each hydrogen a partial positive charge (Fig. 3.2). The positive and negative charges are separated by a distance, so water is said to be ***polar*** compared, say, to ***nonpolar*** hexane where the atoms are all similar and there is little charge separation. As a rule, the larger the proportion of oxygen or nitrogen in a molecule, the greater its polarity and the subsequent association with water—its solubility. Electrically, like dissolves like.

The charged sodium and chloride ions escape from the crystal surface by a combination of thermal agitation and the electrostatic attraction of adjacent water molecules. A similar attraction also causes the water molecules to form large aggregates ("polymer") connected by weak hydrogen bonds, each molecule surrounded by four other water molecules (Fig. 3.2). Dissolution of the salt requires the input of energy

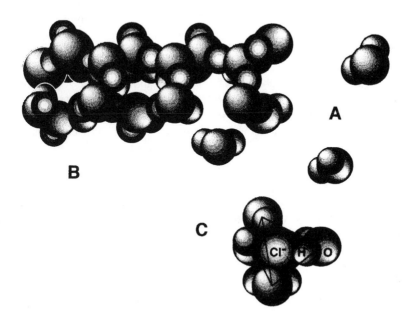

Figure 3.2. Schematic representation of water molecules (A), a segment of water polymer (B), and a hydrated chloride ion (C). Note the tetrahedral configuration of water molecules around the Cl. See Pauling (1952).

to overcome the cohesive forces that hold the crystal together and to break some of the water polymer's hydrogen bonds to make room for the new ions.

On the other hand, combination of the salt or other solute with water molecules (called **hydration,** Fig. 3.2) releases energy, and to the extent that this energy exceeds the total required to overcome cohesion, the salt can "dissolve." Water is the universal solvent because of its large electrostatic attraction for all polar species, although factors such as the symmetry, size, and crystal structure of the solute also affect solubility and make accurate prediction difficult.

The dissolution of pollutants in seawater provides a good example. Saturation solubilities of organic compounds in seawater generally are well below those in distilled water (Table 3.1). The reason is that hydration of the inorganic ions in seawater allows fewer openings for the solute to fit into the water polymer. Equation 3.1 shows that the difference between the log molar solubilities of an organic substance in seawater (S_{sw}) and distilled water (S) is related to the total electrolyte concentration (C_s) by the Setschenow constant, K_s.

$$\log S - \log S_{sw} = K_s C_s \tag{3.1}$$

The solubility of acids and bases is strongly affected by pH (see Section 7.2.1). Some proportion of an acidic compound will be ionized under alkaline conditions, and the negatively charged anionic form associates more readily with water and thus is more "soluble" than is the un-ionized form. Although only 32.37% of the weak acid pentachlorophenol (pK_a 4.82) is in ionic form at its natural pH of 4.50 in distilled water at 20°C (Table 3.2), virtually all of it is in ionic form in pH 8.2 seawater, and its solubility ranges from 18 mg/L at low pH to over 10^5 mg/L above pH 8. Organic bases behave similarly under acidic conditions.

Many substances of toxicological interest exhibit very low aqueous solubilities (Appendix 3.1), although this does not mean that the resulting concentrations are insignificant. The saturation solubility of DDT is only about 1 μg/L, yet the substance is readily bioconcentrated, volatilized, and toxic to aquatic invertebrates at this level. Dealing with such low concentrations requires a conceptual shift downward from the familiar units used by most chemists and biologists, the most common units of environmental concentration being parts per million (ppm, mg/L), parts per billion

TABLE 3.1

Solubility of Selected Compounds in Distilled Water (dw) and Seawater (sw) at 20°C[a]

Compound	S_{dw} (mg/L)	S_{sw}[b] (mg/L)
p-Toluidine	6643	5239
p-Nitrotoluene	288.0	251.0
Naphthalene	24.35	17.17
Phenanthrene	1.105	0.740
Pyrene	0.095	0.065

[a] Hashimoto et al., 1984.
[b] Total electrolyte 3.50%.

TABLE 3.2
Ionization of Pentachlorophenol in Water[a]

pH	% Ionized	pH	% Ionized
4.0	13.15	7.0	99.34
5.0	53.52	8.0	99.93
6.0	93.80	9.0	99.99

[a] Calculated by Eq. 7.5A.

(ppb, $pp10^9$, $\mu g/L$), and parts per trillion (ppt, $pp10^{12}$, ng/L). M is replaced by μM ($10^{-6}M$) or nM ($10^{-9}M$).

As concentrations below 1 ppm become increasingly relevant, the aqueous solubilities of substances previously considered to be "insoluble" must be reassessed. For example, barium sulfate and silver chloride, traditionally precipitated for quantitative analysis of their cations, are found to have saturation solubilities of several mg/L; over 30 mg/L of mothballs (naphthalene) dissolve in water at room temperature, and a saturated aqueous solution of benzene contains almost 1800 mg/L. The concentrations of pollutants in natural waters generally are far below such solubility limits.

How does one measure solubilities of less than 1 mg/L? Generally, an acetone solution of the organic compound is evaporated onto the surface of glass beads, which then are placed in a glass generating column and pure water allowed to flow slowly over them (May et al., 1978). After the water becomes saturated with solute, a measured volume of flow is collected directly onto a solid-phase extraction cartridge containing a nonionic adsorption resin to trap the solute, the resin is subsequently extracted by flushing an organic solvent through it, and the organic solution is analyzed. In this and other methods, contact of the saturated solution with glass surfaces must be minimized to avoid losses due to adsorption (Section 3.4). As with other physical properties, one should not be surprised to find that values from different sources may vary considerably, depending on the method of measurement.

3.3 | SOLVENT PARTITIONING

Solvent to Solvent

When an aqueous solution of a typical organic compound is brought into contact with a largely immiscible solvent such as hexane or ether (or fat), the solute molecules will escape into the new solvent until the fugacities are equal and equilibrium is established. This is the basis of the solvent extractions so widely used in organic chemistry laboratories, in the defatting of wool and soybeans, and in dry cleaning. The ratio of the compound's concentrations in the two phases is found always to be the same, and this constant is called the *partition coefficient,* K_p:

$$K_p = C_{\text{solvent}}/C_{\text{water}} \tag{3.2}$$

Each Solute(solvent?) has its own K_p

Bioconcentration is the transfer of an aqueous solute into a living organism to result in a higher concentration. Although people had known for a long time that those who eat too many carrots turn yellow from the fat-soluble carotenes, and that a cow that consumes wild onions produces onion-flavored milk, the recognition that synthetic environmental chemicals also could concentrate in animals came only with the advent of DDT. A fish living in water that contains 0.1 ppb of DDT typically might have more than 1000 ppb of the chemical in its body, a 10,000-fold increase in concentration. As late as 1970, this phenomenon was thought to be quite mysterious.

Actually, it is readily understood if one thinks of body fat as a "solvent." The solute (DDT) becomes distributed between water and fat until equilibrium is established (the fugacities are equal). As natural fat is hard to work with, triolein (glyceryl trioleate) has been used successfully as a surrogate (Chiou, 1985), but by far the most common substitute is 1-octanol (*n*-octyl alcohol). The partition coefficient of a substance between water and octanol is represented as K_{ow}. As most organic compounds are more soluble in fatty solvents than in water, K_{ow} values usually are >1 and often are so large that they are expressed as their $\log_{10}$ (Appendix 3.1). K_{ow} is measured directly by shaking the compound with a mixture of octanol and water and analyzing either phase, or it can be estimated from retention times on an HPLC (high pressure liquid chromatography) column (Veith et al., 1979).

The fat of a fish living in polluted water is in contact with the pollutants via the blood circulating through the gills (90%) and skin (10%). As with any other solvent, pollutant uptake initially is rapid (Fig. 3.3) but eventually slows as equilibrium is approached. If the pollutant level in the water increases, the equilibrium concentration in the fish also increases, but otherwise the uptake curves are similar. One observes that the ratio of the concentration in the fish to that in the water, the *bioconcentration factor* (**BCF**), is the same at all exposure levels (500 in this instance), although it varies with the chemical, species, and fat content:

$$BCF = C_{\text{organism}}/C_{\text{water}} \tag{3.3}$$

When the fish is returned to clean water, the distribution process is reversed (called *depuration*), and tissue concentrations decline as the chemical is released back into the water through the gills until a new equilibrium is established (Fig. 3.3). BCF is indeed a partition coefficient, as shown when its log values produce a linear relationship when plotted against the corresponding K_{ow} (Fig. 3.4). Although many regression equations, representing different conditions, have been reported for this relationship, that of Veith et al. (1980) is typical.

$K_{ow} = C_{fat}/C_{water}$ *(handwritten)*

$$\log BCF = 0.76 \log K_{ow} - 0.23, \qquad r^2\ 0.82 \tag{3.4}$$

overall for most con. (handwritten)

Mackay (1984) showed that when several outlying points due perhaps to questionable data are ignored, the statistical significance is improved (Eq. 3.5). A recent plot of the most reliable data (Fig. 3.4) confirmed the accuracy of this equation (Deanna Dowdy, personal communication, 1996).

$$\log BCF = \log K_{ow} - 1.32, \qquad r^2\ 0.95 \tag{3.5}$$

BCF always is less than $K_{ow,}$ largely due to differences in absorption rates, fat composition, and metabolism. At log BCF values above 5 and below 2, prediction be-

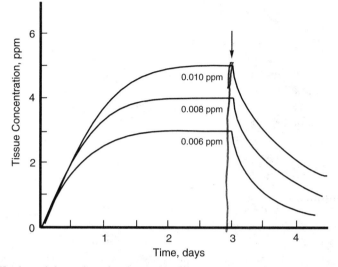

Figure 3.3. Uptake and depuration of a chemical in fish. Initial aqueous concentrations are shown, and the arrow marks return of the fish to clean water. BCF = 500.

comes less accurate due to slow establishment of equilibrium in the first case and to biodegradation and elimination in the second. Although the uptake over time, called **bioaccumulation,** may require months to complete with DDT (log K_{ow} 5.75), most pollutants equilibrate within less than a day. However, it is important to recognize that the bioconcentration and eventual equilibration are *required* by thermodynamics, regardless of the rates.

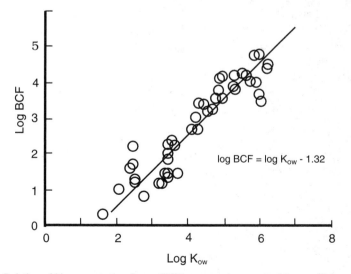

Figure 3.4. Relation of bioconcentration factor (BCF) to octanol–water partition coefficient (K_{ow}) for 44 selected chemicals. Data compiled by Deanna Dowdy.

Most uptake of chemicals by aquatic animals is directly from the water and relatively little from food (Table 3.3), and the much-studied DDT is an exception. The situation changes in terrestrial animals, for whom food normally is the principal exposure route. On land, several additional transfer steps are interposed between food and fat, and equilibrium often comes slowly. The larger and older the animal, the slower the equilibration and depuration; for example, equilibration of DDT in human fat may require years. However, BCFs still apply even in large terrestrial mammals, and Kenaga (1980) has shown that the dietary uptake of a range of chemicals into beef animals results in predictable bioconcentration, albeit with more variability than in fish.

$$\log \text{BCF} = 0.50 \log K_{ow} - 3.457, \quad r^2 \ 0.790 \qquad (3.6)$$

Again, intake is balanced by excretion and metabolism.

Comparison of pollutant residues in animals with the corresponding LD_{50} often presents an anomaly: Tissue concentrations can be far above the acutely toxic intake levels. This is because slow accumulation causes the chemical to be deposited in storage (depot) fat without ever reaching toxic levels in the blood. Likewise, a loss of fat, by itself, does not provide a toxic dose. However, if the contaminated organism is consumed by a predator, the delivered dose now may be lethal, and there are many accounts of this kind of "secondary poisoning." For example, earthworms accumulate levels of DDT that are toxic to robins.

One possible consequence of this nontoxic kind of bioconcentration could be *foodchain biomagnification.* When larger animals eat smaller ones, the transferred pollutants might build up at each step, and indeed, early work with DDT indicated that such was the case. For example, at Lake Michigan, Hickey et al. (1966) found that DDT residues increased at each trophic level, with an average of 0.014 ppm in bottom mud, 0.41 ppm in bottom-feeding crustaceans, 3–6 ppm in fish, and 2400 ppm in fish-eating gulls. However, most other studies have failed to show consistent biomagnification, and no wonder. At each step in an aquatic foodchain, the pollutant concentration in the fat equilibrates with that in the water according to the partition principal. As BCF is a constant, any new intake *must* be accompanied by an equal release back into the environment. On a fat basis, BCF actually can be mimicked with simply a plastic bag of triolein (Chiou, 1985), where even the rates of uptake and depuration approximate those in living organisms (Hawker and Connell, 1989).

TABLE 3.3
Pollutant Uptake from Water versus Food[a]

Chemical	Species	BCF	% from Diet
DEHP[b]	Bluegill	112	14
TCB[c]	Bluegill	182	6
Leptophos	Bluegill	773	1.2
Kepone	Sheepshead	7,400	<0.1
DDT	Fathead	133.000	27–62

[a]Macek et al., 1979.
[b]Di(2-ethylhexyl) phthalate.
[c]1,2,4-Trichlorobenzene.

Considering the fugacity principle, it should come as no surprise that bioconcentration has been observed in such diverse organisms as bacteria, earthworms, oysters, whales, and falcons. Plants absorb chemicals through their roots based on solubility in water rather than in fat, but leaves often are covered with a waxy cuticle that can concentrate chemicals in the usual way as shown by the uptake of DDT from soil onto the skin of a carrot. Most surface waters, too, are coated with a thin film of organic material, called the *surface microlayer,* which concentrates lipophilic pollutants by more than 10,000-fold (Gever et al., 1996). In fact, just about everything on earth seems to contain or be covered with a fatty layer that can concentrate pollutants, leading zoologist Robert Rudd to suggest the existence of an independent global *liposphere* through which lipophilic substances continually ebb and flow.

3.4 | VOLATILIZATION

Volatilization is perhaps the most important route by which chemicals dissipate. This fugacity is reflected in *vapor pressure,* the pressure exerted by the vapor of a substance on its own solid or liquid surface at equilibrium. Vapor pressure may have units of atmospheres (atm) or pascals (Pa), but here we will use the more familiar millimeters of mercury (also known as torr):

$$1 \text{ atm } = 760 \text{ torr } = 1.013 \times 10^5 \text{ Pa}; \quad 1 \text{ Pa } = 0.0075 \text{ torr} \qquad (3.7)$$

Just as every substance must be soluble in water, the fugacity principle dictates that every substance must have a vapor pressure at a given temperature. Liquid water turns to vapor (boils) at 100°C, where its vapor pressure equals the 760 torr mean atmospheric pressure, but we also expect it to evaporate eventually from an open cup even at room temperature. We hardly expect an iron bar to evaporate that way, yet its boiling point of 2735°C says that it, too, must exert some vapor pressure at room temperature. Even rocks evaporate (very slowly); quartz crystals vaporize at 2227°C, although their vapor pressure is still 10 torr at 1732°C. At the other end of the scale, the toxic gas, ammonia, has a vapor pressure of 7600 torr (10 atm) at 25°C. However, most substances of toxicological interest have vapor pressures of between 10^{-5} and 25 torr at room temperature (Appendix 3.1).

Vapor pressure increases with increasing temperature, sometimes quite rapidly (Fig. 3.5), as the surface molecules absorb energy. For example, the vapor pressure of methanol goes from 4 torr at a chilly morning temperature of 5°C to 500 torr at the 50°C temperature of a sidewalk on a hot day. Many solids vaporize without melting (they *sublime*); with rapid heating, arsenic trioxide melts at 313°C, but, heated slowly, it vaporizes visibly at 100°C and completely by 193°C. A solid deposit of a pesticide on a leaf surface will volatilize into, and equilibrate with, the very thin layer of still vapor just above the leaf surface.

At the usual environmental concentrations, volatilized pollutants behave quite like ideal gases, so the universal gas law, *PV = NRT,* applies. This is the basis of the most widely used method for measuring vapor pressures, where a slow stream of

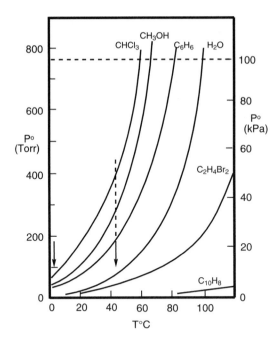

Figure 3.5. The effect of temperature on vapor pressures ($P°$) of some environmental pollutants. Arrows mark the range of moderate ambient temperatures (2–42°C).

inert gas is passed through the substance coated on the surface of glass beads or sand in a generating column similar to that mentioned for dissolution, a measured volume of saturated vapor is collected and analyzed, and the pressure P is calculated:

$$P = nRT/V = 244.4 \; (n/V) \; \text{atm} \tag{3.8}$$

where n/V is the vapor density in moles/L, the universal gas constant R is 0.082 L atm/°K/mole, and T is the temperature of the generating column in °K. Depending on the chemical, the method can measure vapor pressures down to about 10^{-8} torr.

Volatilization from water, that is, partition of a substance between aqueous solution and the atmosphere, is subject to **Henry's law.** This law states that, *at equilibrium,* the concentration of a chemical in the vapor state bears a constant relation to the concentration in aqueous solution:

$$H' = C_{\text{air}}/C_{\text{water}} = (n/V)\,(1/S) = P/RTS \tag{3.9}$$

Here H' is viewed as a partition coefficient and is dimensionless, so the concentration units do not matter. The term n/V is best seen as the molar concentration of the vapor and S as the saturation solubility in water at the same temperature. The necessity for equilibrium must be stressed: In a closed jar, bringing the whole system into equilibrium seems easy enough, but in a large lake or the ocean, equilibrium must be pictured as existing between thin, stagnant **boundary layers** that extend at most only a few millimeters above and below the actual surface (Fig. 4.3).

Converting the P of Eq. 3.9 into torr and S into mg/L (ppm), with T in °K and with the molecular weight M to convert moles to grams, provides a convenient calculation of the dimensionless H':

$$H' = 16.04PM/TS \qquad (3.10)$$

With RT constant at any given temperature, the remaining ratio of vapor pressure to solubility in Eq. 3.9 is seen to yield a different Henry's law constant, $H = P/S$, with units of atm m^3/mole. That is, at 25°C, $H = 0.024H'$. Although Henry's law constants often are calculated, they may also be measured directly (Mackay et al., 1979) by leading a slow, measured stream of air or nitrogen through an aqueous solution of the chemical at constant temperature and periodically measuring the aqueous concentration to determine the rate of solute loss from the "bubble chamber."

Typical values H' values are shown in Appendix 3.1. It may come as a surprise that although the vapor pressure of DDT or TCDD is very low, the H' and hence the volatility is relatively high (see Fig. 4.4). This is because aqueous solubilities also are very low, and so the ratio of P/S remains moderate. On the other hand, the low vapor pressure and relatively high solubility of malathion combine to reflect a low volatility. The **rate** of movement of a chemical into the atmosphere is controlled by the rates of transport, that is, through bulk water, across the aqueous and gaseous boundary layers, and into the open air (see Section 4.2 and Special Topic 4).

3.5 | ADSORPTION

Section 3.2 discussed the behavior of a salt solution in equilibrium with solid salt. Imagine now an aqueous solution in contact with a solid phase whose molecular structure is *different* from that of the solute. Some solute molecules still will be attracted to the surface and held there by weak forces, and eventually an equilibrium will be established between those in the water and those on the surface. Attachment to the surface is termed **adsorption,** while the corresponding detachment is **desorption.** Adsorption of most substances to a mineral surface such as glass is largely electrostatic and involves at most a few kcal/mole of binding energy, although some compounds may literally react with the surface to form strong bonds (**chemisorption**). The bound substance, or **sorbate,** may be also have been adsorbed from its gaseous state.

As a surface phenomenon, adsorption depends on both the chemical nature of the surface and its area. The lower the aqueous solubility of the solute, the greater is its binding potential. The adsorption of pollutants to the walls of a sample container is a frequent problem in water analysis, and Table 3.4 shows the adsorption of two common PAH congeners, phenanthrene and benzo[a]pyrene, to various materials that effectively remove them from aqueous solution. Some adsorbents have very large surface areas for reaction, for example the 200 m^2/g of heat-activated carbon, so the capacity to hold sorbate can be considerable.

Heat decreases sorption, so measurements must be conducted at constant temper-

TABLE 3.4
**Adsorption of PAH Congeners from Water onto Various
Materials (%)[a]**

	Phenanthrene		Benzo[a]pyrene	
Surface	1 h	13 h	1 h	13 h
Glass	8	—	53	82
Glass (silanized)	5	73	73	93
Platinum	35	87	57	93
Aluminum	5	76	67	95

[a] May et al., 1978.

ature. When the log of the equilibrium concentration of an adsorbed compound, in $\mu g/g$ or $\mu g/cm^2$, is plotted against the log concentration of the solution in $\mu g/L$ at constant temperature, the resulting graph is called an ***isotherm*** (Fig. 3.6). In 1909, the German chemist Freundlich proposed that such isotherms could be represented by:

$$C_s = K_F C_w^{1/n} \qquad (3.11)$$

where C_s and C_w are the observed concentrations on the solid and in solution, respectively, K_F is the ***Freundlich constant,*** and n has some empirical value determined by the adsorbent. When n is close to unity, as is usual, the proportionality constant becomes the ***distribution coefficient,*** K_d, and Equation 3.11 assumes the familiar form of a partition coefficient:

$$K_d = C_s/C_w \qquad (3.12)$$

To measure a Freundlich constant, an aqueous solution of the chemical is shaken with a measured weight of adsorbent until equilibrium is established, the solution is

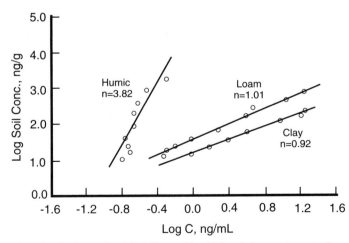

Figure 3.6. Adsorption isotherms for 2,2',4 4'-tetrachlorobiphenyl from water onto three types of soil. Adapted from Haque, 1975.

analyzed, and the ratio of adsorbed to dissolved substance is calculated at several concentrations and plotted.

In the environment, soil is the most widespread adsorbent. Soils represent a heterogeneous mixture of clay minerals, sand and silt, organic matter, and water (Section 2.5). Although the mineral fraction does adsorb pollutants to some degree, the organic fraction now is recognized to be a more important adsorbent. This fraction is composed of many kinds of low-molecular-weight organic molecules as well as organic polymers such as humic acids that give soil its overall negative charge. The base-insoluble polymeric material is called *kerogen.* Similar substances are found suspended and dissolved in both fresh water and sea water.

Not surprisingly, it is with the organic fraction that neutral pollutants associate. Rather than strictly physical adsorption, this binding often may be more a dissolution, that is, a mixing into the organic phase. When K_F is corrected to represent all the soil as if composed of organic carbon (as distinguished from inorganic carbon like that in carbonate), the organic-carbon distribution coefficient K_{oc} provides a relatively accurate and reproducible estimate of pollutant binding to soil (Appendix 3.1):

$$\mathbf{K_{oc}} = (K_F/\% \text{ organic carbon}) \times 100 \qquad (3.13)$$

Organic carbon is measured by combusting a small sample of the soil, collecting and weighing the carbon dioxide produced, and multiplying by 0.27 (the proportion of C in CO_2). The analogy to dissolution is strengthened by the often near-unity of K_{oc} and K_{ow} (Section 16.2):

$$\log \mathbf{K_{oc}} = 0.98 \log K_{ow} - 0.0002 \qquad (3.14)$$

Although there are now a number of sources of data for the main environmental partition coefficients representing solubility (S), solvent partitioning (K_{ow}), volatilization (H'), and soil sorption (K_{oc}), their values are surprisingly divergent. For example, Montgomey and Welkom (1990) list 12 values for the log K_{oc} of lindane, ranging from 3.20 to 3.89. That is, K_{oc} varies from 1584 to 7760! The discrepancies are due primarily to different methods of measurement, different soil type, and inattention to temperature. The numbers presented in Appendix 3.1 reflect the reliability of the sources, analogy to similar compounds, and any concensus among published figures, but readers should not be surprised to run across others that seem equally valid.

3.6 | SIGNIFICANCE

Comparison of Eqs. 3.3, 3.9, 3.12, and the concept of saturation solubility reveals that the concentration of a chemical in one environmental compartment bears a constant relation to that in any adjacent compartment *at equilibrium.* It is important to understand that all these partition equilibria exist simultaneously; a chemical released into one compartment must soon appear in all the rest. Remembering the escaping tendency lets one no longer be surprised when DDT is found in arctic ice, tropical fish, California fruit, and Antarctic penguins.

3.7 | REFERENCES

Chiou, C. T. 1985. Partition coefficients of organic compounds in lipid-water systems and correlations with fish bioconcentration factors. *Environ. Sci. Technol.* **19:** 57–62.

Gever, G. R., S. A. Mabury, and D. G. Crosby. 1996. Rice field surface microlayer: Collection, composition, and pesticide enrichment. *Environ. Toxicol. Chem.* **15:** 1676–82.

Hansch, C., A. Leo, and D. Hoekman. 1995. *Exploring QSAR: Hydrophobic, Electronic, and Steric Constants,* American Chemical Society, Washington, DC.

Haque, R. 1975. Role of adsorption in studying the dynamics of pesticides in a soil environment, in *Environmental Dynamics of Pesticides* (R. Haque and V.H. Freed, eds.), Plenum Press, New York, NY, pp. 97–114.

Hashimoto, Y., K. Tokura, H. Kishi, and W. M. J. Strachan. 1984. Prediction of seawater solubility of aromatic compounds. *Chemosphere* **13:** 881–88.

Hawker, D. W., and D. W. Connell. 1989. A simple octanol/water partition system for bioconcentration investigations. *Environ. Sci. Technol.* **23:** 961–65.

Hickey, J. J., J. A. Keith, and F. B. Coon. 1966. Exploration of pesticides in a Lake Michigan ecosystem. *J. Appl. Ecol.* **3** suppl.: 141–54.

Kenaga, E. E. 1980. Correlation of bioconcentration factors of chemicals in aquatic and terrestrial organisms with their physical and chemical properties. *Environ. Sci. Technol.* **14:** 553–56.

Macek, K. J., S. R. Petrocelli, and B. H. Sleight III. 1979. Considerations in assessing the potential for, and significance of, biomagnification of chemical residues in aquatic foodchains, in *Aquatic Toxicology* (L. L. Marking and R. A. Kimerle, eds.), ASTM, Philadelphia, PA. STP 667, pp. 251–68.

Mackay, D. 1979. Finding fugacity feasible. *Environ. Sci. Technol.* **13:** 1218–23.

Mackay, D. 1984. Correlation of bioconcentration factors. *Environ. Sci. Technol.* **16:**274–78.

Mackay, D., and S. Paterson. 1981. Calculating fugacity. *Environ. Sci. Technol.* **15:**1006–14.

Mackay, D., and S. Paterson. 1982. Fugacity revisited. *Environ. Sci. Technol.* **16:** 654A–60A.

Mackay, D., W. Y. Shiu, and R. P. Sutherland. 1979. Determination of air-water Henry's Law constants for hydrophobic pollutants. *Environ. Sci. Technol.* **13:** 333–37.

May, W. E., S. P. Wasik, and D. H.Freeman.1978. Determination of the solubility behavi- or of some polycyclic aromatic hydrocarbons in water. *Anal. Chem.* **50:** 997–1000.

Montgomery, J. H., and L. M. Welkom. 1990. *Groundwater Chemicals Desk Reference,* Lewis Publ., Chelsea, MA.

Pauling, L. 1952. *College Chemistry,* W. H. Freeman and Co., San Francisco, CA.

Schwarzenbach, R. P., P. M. Gschwend, and D. M. Imboden. 1993. *EnvironmentalOrganic Chemistry,* John Wiley and Sons, New York, NY.

Smith, J. A., P. J. Witkowski, and C. T. Chiou. 1988. Partition of nonionic organic compounds in aquatic systems. *Rev. Environ. Contam. Toxicol.* **103:** 127–51.

Veith, G. D., N. M. Austin, and R. T. Morris. 1979. A rapid method for estimating log *P* for organic chemicals. *Water Res.* **13:** 43–47.

Veith, G. D., K. J. Macek, S. R. Petrocelli, and J. Carrol. 1980. An evaluation of using partition coefficients and water solubility to estimate bioconcentration factors for organic chemicals in fish, in *Aquatic Toxicology* (J. G. Eaton, P. R. Parrish, and A. C. Hendricks, eds.), ASTM, Philadelphia, PA. STP 707, pp. 116–29.

Special Topic 3. The Great Escape

If one immerses a hot spoon in a small glass of cool water, the temperature of the spoon falls while that of the water rises. Within a short time, the two temperatures are identical, and spoon and water are in **thermal equilibrium.** The excess thermal energy in the spoon "escaped" (diffused) into the water until it equalled that removed from the water, running down a thermal gradient from higher to lower concentration of heat (temperature).

If several spoonful of rock salt are now placed in the water, most of it dissolves. That is, it escapes from solid to dissolved form by diffusion. However, at some point, escape back into the solid state equals that into solution, an equilibrium is established, and we would say that the solution was "saturated." This tendency for matter to escape its confining phase is called *fugacity.* Fugacity has the units of pressure and is to the diffusion of mass what tempera- ture is to diffusion of heat. Like temperature, it flows down a gradient from high to low. When the fugacities of a substance in two different phases become equal, a state of equilibrium exists.

Vapor pressure offers a close approximation to a chemical's fugacity. For dilute aqueous solutions, Henry's law states that the fugacity f' of a solute is proportional to the mole fraction N', that is, its proportion of the total moles of solute and solvent combined. At the low concen- trations usual among environmental pollutants, this total is very close to the number of moles **n** of solvent (water) alone and approximates both molality (moles per 1000 g) and molarity (moles per L) with the appropriate change of k' to k'' (1000 g of water at 25°C is 1.003 L).

$$f' = kN' = kn'/(\mathbf{n} + n') = k'n'/\mathbf{n} \text{ (where } \mathbf{n} \gg n') = k''C \text{ moles/L} \qquad (3.15)$$

If N_1' is the mole fraction of the substance in a solvent such as water, and N_2' is its mole fraction in a virtually immiscible one such as 1-octanol in contact with the solution, the fugaci- ties will be equal at equilibrium (no net change in either concentration):

$$f_1'/f_2' = k_1N_1'/k_2N_2' = (k_1''C_1)/(k_2''C_2) = k_3(C_1/C_2) = 1 \qquad (3.16)$$

Therefore, $C_1/C_2 = k_4$, the familiar Nernst equation. Where the solvents are 1-octanol and water, k_4 is the octanol–water partition coefficient K_{ow}, and where octanol is replaced by lipid, k_4 is the bioconcentration factor or BCF.

The concept of fugacity was first proposed in 1901 by UC-Berkeley professor G. N. Lewis, but its application to environmental chemodynamics by Canadian professor Donald Mackay came only after another 80 years (Mackay, 1979). Mackay also proposed a new term, the *fugacity capacity* (Z), corresponding to the heat capacity of a substance. Just as heat capac- ity relates temperature to heat content, Z relates fugacity to chemical concentration and quanti- fies the capacity of a particular phase for fugacity; that is, $C_1 = Zf_1$. As explained by Mackay, environmental pollutants tend to accumulate where Z is high and C can also become high before the substance escapes.

The particular utility of fugacity capacity is that it simplifies the estimation and evaluation of the relative equilibrium concentrations of a substance in a complex set of environmental phases. In the vapor phase (generally the atmosphere), f for any substance is its partial pressure (Eq. 3.17), and Z often is about 40 moles/m^3. In the liquid phase or solution, f (that is, P) and C are related by Henry's law, and $Z = 1/H$. Where V_B is the fraction of total environmental volume occupied by biota that contains **y** percent lipid (represented here by octanol), Z is provided by Eq. 3.18, and where a sorbent concentration in the environment is **a** g/m^3, Z for an adsorbed substance is given by Equation 3.19.

$$Z = C/f = n/PV = 1/RT = 40.8 \text{ mole/atm m}^3 \text{ at 298 K} \qquad (3.17)$$

$$Z = 100\mathbf{y}V_BK_{ow}/H \qquad (3.18)$$

$$Z = 10^{-6}aK_d/H \qquad (3.19)$$

Using these relationships, Z can be calculated for the various environmental compartments as discussed in Section 16.4, together with the proportion of chemical to be expected in each at equilibrium. However, values may also be estimated for more realistic conditions where environmental transport and degradation occur (Mackay and Paterson, 1981, 1982).

It may seem startling at first to realize that *every* material object accomodates escaping tendency and fugacity. As a chemical or mixture of chemicals, a rock can be expected to have a finite solubility and vapor pressure and so, to some degree, dissolve and volatilize. Fortunately for our sanity, these processes usually are extremely slow. Your spoon may occasionally escape from you, but it is not likely to evaporate before your very eyes.

APPENDIX 3.1.
Physical Properties of Selected Chemicals

Chemical	Formula	MW	$\log P_0^a$ Torr	$\log P_0^a$ Pa	S (mg/L)^a	$\log H'^b$	$\log K_{ow}^c$	$\log K_{oc}^a$
Hydrogen cyanide	CHN	27.0	6.20E2	8.27E4	∞^d	—	-0.25	—
Chloroform	$CHCl_3$	119.4	1.60E2	2.13E4	8.22E3	-0.90	1.97	1.64
Methyl bromide	CH_3Br	94.9	1.70E3	2.27E5	1.34E4	-0.38	1.19	1.92
Methanol	CH_4O	32.0	0.97E2	1.29E4	∞	—	-0.77	—
Trichloroethylene	C_2HCl_3	131.4	0.55E2	7.33E3	1.10E3	-0.44	2.61	2.03
HCB	C_6Cl_6	284.8	1.09E-5	1.45E-3	3.50E-2	-2.31	5.73	3.59
Benzene	C_6H_6	78.1	0.75E2	1.00E4	1.88E3	-0.77	2.13	1.92
Lindane	$C_6H_6Cl_6$	290.9	3.23E-5	4.31E-3	7.52^e	-4.16	3.72	2.96
Naphthalene	$C_{10}H_8$	128.2	0.23	0.37E2	0.31E2	-1.28	3.30	3.11
Nicotine	$C_{10}H_{14}N_2$	162.2	4.24E-2^e	5.65^e	∞	—	1.17	—
Parathion	$C_{10}H_{14}NO_5PS$	291.3	6.68E-6	8.91E-4	0.11E2	-5.01	3.83	3.68
Malathion	$C_{10}H_{19}O_6PS_2$	330.4	3.98E-5	5.3^e	1.45E2^e	-5.31	2.36	2.46
2,2',4,4',5,5'-PCB	$C_{12}H_4Cl_6$	360.9	8.14E-7^e	1.09E-4^e	5.46E-3	-2.54	6.57	6.08
TCDD	$C_{12}H_4Cl_4O$	322.0	1.5E-9^e	2.0E-7^e	8.0E-6^e	-2.49	6.53	6.60
DDE	$C_{14}H_8Cl_4$	318.0	6.49E-6^e	8.65E-4	1.00E-2	-1.96	6.96	6.00
DDT	$C_{14}H_9Cl_5$	354.5	1.88E-7	2.51E-5	1.2E-3	-2.52	6.91	5.77
Dibutyl phthalate	$C_{16}H_{22}O_4$	278.3	7.28E-5^e	9.7E-3^e	0.11E2	-3.91	4.72	3.14
Benzo[a]pyrene	$C_{20}H_{12}$	252.3	5.49E-9	7.32E-7	3.8E-3^e	-4.71	5.97	5.60

^{a}Data selected from Montgomery and Welkom, 1990, all at 20°C except where noted.
^{b}Calculated.
^{c}Data from Hansch et al., 1995.
^{d}Miscible in all proportions.
^{e}25°C, except P_0 for DDE and malathion at 30°C.

CHAPTER

Environmental 4
Transport

4.1 | DISSIPATION OF CHEMICALS

Chemicals move in the environment. Chapter 3 showed that the release of a substance into an environmental compartment results in its rapid redistribution into all adjacent compartments (Fig. 3.1), and as dictated by the fugacity principle, the molecules continue to diffuse until chemical potentials are equal between phases and equilibrium is attained.

However, they generally continue to move. Spilled gasoline volatilizes into the atmosphere and is carried away by the wind so that equilibrium never is reached. Similarly, application of molinate herbicide (Ordram®) to a flooded field immediately results in residues in the air above the field, in bottom sediment, and in fish that live in the water. Water samples collected at susequent intervals show a steady loss of the herbicide over time as it moves between compartments and then away from the field (Fig. 4.1A).

Such a plot of concentration versus time represents a "dissipation curve" that usually reflects *first-order kinetics.* That is, the rate of molinate loss at any given time **t** is proportional to the remaining concentration **C**. If C_o was the starting concentration, then

$$\ln (C_o/C) = k_1 t \qquad (4.1)$$

and a plot of ln (**C/Co**) against time yields a straight line whose slope is the first-order rate constant, k_1 (Fig. 4.1B). A characteristic of first-order plots is that the time required for half of the remaining chemical to dissipate at any point, known as the half-life or $t_{1/2}$, remains constant. In our example, half of the molinate dissipates from the water into air, sediment, and biota within 48 h, another 25% is lost during the next 48 h, there is only 13% left after 144 h, and so on, as described by Eq. 4.2. A quick check shows that molinate dissipation indeed is close to first order.

$$t_{1/2} = \ln 2/k_1 = 0.693/k_1 \qquad (4.2)$$

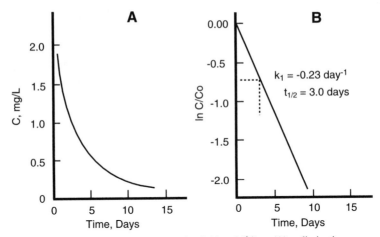

Figure 4.1. Dissipation rate of molinate from a rice field at 26°C as (A) a dissipation curve and (B) as a first-order plot. C_0 is the initial concentration and C the concentration at time t (see Soderquist et al., 1977).

However, as environmental dissipation of a chemical is much more complex than a simple unimolecular reaction, such rates often are referred to as ***pseudo first order.***

As the molinate molecules leave the water surface, they are swept away by air currents. Moving air, water, and soil continually displace the intercompartmental equilibria. As seen in Fig. 4.2, chemicals flow with surface water, penetrate into soil and groundwater, and move into and through the atmosphere. Their bulk movement in air and water is called ***advection*** and is characterized as ***transport.*** This chapter discusses the various routes and factors involved in transport, while dissipation by chemical transformations will be described in Chapters 5 and 6.

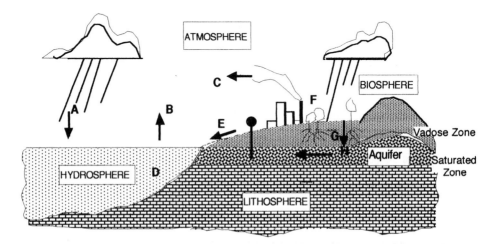

Figure 4.2. Modes of chemical transport: Wet and dry deposition (A), volatility from water (B), advection of airborne particles and vapor (C), movement in solution and on suspended particles (D), runoff (E), volatilization from plants and soil (F), percolation through soil (G), and movement in groundwater (H).

4.2 | TRANSPORT IN AND FROM SURFACE WATER

4.2.1. Transport in Solution

As you watch the water in a river, ocean, or just in an irrigation ditch, you become aware of its movement. Any chemical dissolved in that water, or attached to particles suspended in it, will also be moved—transported. Any or all of several mechanisms are involved:

- **Advection** is bulk movement by currents. Advection rate is determined by concentration C and water velocity v_w.

$$J = Cv_w \qquad \frac{mg}{L} \cdot \frac{m}{s} = mg/s \qquad (4.3)$$

The *flux density* J is the quantity of chemical transported through a surface of unit area in unit time. That is, at 10 mg/L (0.01 mg/cm^3) in a stream moving at 10 cm/s, the *flux* of the chemical will be 0.01 mg/cm^3 x 10 cm/s, or 0.1 mg/cm^2/s;

- **Molecular diffusion** is due to the random thermal movement of molecules. It obeys *Fick's law*

$$D_m = -J \frac{dz}{dC}$$

$$J = -D_m \frac{dC}{dz} \qquad \frac{\Delta C}{\Delta z} \qquad \frac{|c_r c_0|}{|z - z_0|} \qquad (4.4)$$

where D_m is the molecular diffusion coefficient in cm^2/s, C is concentration, and z is the distance through which the molecules have moved. Therefore, dC/dz represents the change of concentration with distance. This expression of Fick's law holds for diffusion in one dimension only, but it can be readily extended to the more realistic three-dimensional situation (see Hemond and Fechner, 1994).

For many compounds, $D_m = 2.7 \times 10^{-4}/M^{0.71}$ cm^2/s, where M is the molecular weight (Hayduk and Laudie, 1974). As D_m is inversely proportional to the square root of molecular weight, it also may be calculated by Eq. 4.5 when a value for a similar compound already is known. D_m for ethylbenzene (e) is 0.81×10^{-5} at 20°C, and $M = 106.2$; with this, D_m for n-butylbenzene (b), $M = 134.2$, may be calculated to within $\pm 10\%$.

$$D_{mb}/D_{me} = M_e^{0.5}/M_b^{0.5} \qquad (4.5)$$

$$D_{mb} = (0.81 \times 10^{-5})(10.3/11.6) = 0.72 \times 10^{-5}$$

- **Turbulent diffusion** is what one generally refers to as "mixing," as when sugar is stirred into coffee. It, too, obeys Fick's law but is far more rapid than molecular diffusion. In still water, D_m usually is about 10^{-5} cm^2/s, whereas the highly variable D_t often can be thousands of times greater in turbulent water;

- **Dispersion** is typical of mixing in groundwater, where the low velocity of advection precludes turbulence but the water's movement around solid particles results in transport akin to turbulent diffusion.

Of course, both advection and diffusion normally operate simultaneously. The rate dC/dt at which the concentration changes as the solution flows past some fixed point will be represented by a combination of Eqs. 4.3 and 4.4:

$$\frac{dC}{dt} = -v_w \frac{dC}{dx} + \frac{d(D\ dC/dx)}{dx} + s \qquad (4.6)$$

The factor s accounts for any loss of chemical by reaction or intercompartmental movement.

4.2.2. Transport between Water and Air

The previous chapter introduced Henry's law and the partition coefficient H' which relates a compound's concentration in the aqueous boundary layer to that in the atmospheric boundary layer above. However, Henry's law describes an equilibrium situation and says nothing about volatilization *rate*. In the dynamic real world, a molecule moving from water into air encounters resistance by other molecules in both phases (Fig. 4.3). In addition, it must cross the aqueous boundary layer at a rate of k_1 cm/s, escape the surface, and move through the atmospheric boundary layer at k_g cm/s (k_1 and k_g are called the liquid-phase and gas-phase exchange coefficients, respectively). The rate at which it passes from bulk water to bulk air is represented by the ***mass-transfer coefficient***, K_L. The three coefficients are related by Eq. 4.7, their reciprocals being best thought of as resistances. The flux, J, from water into air is related to K_L by Eq. 4.8, where C_a and C_w are concentrations in air and water, respectively.

$$1/K_L = 1/k_1 + 1/H'k_g \text{ (h/cm)} \qquad (4.7)$$

$$J = K_L(C_w - C_a/H') \text{ (moles/cm}^2/\text{s)} \qquad (4.8)$$

As chemicals volatilize from water at a first-order rate (excluding advection), their half-life can provide important information. If the depth of the water is Z cm, the concentration C at any time t is related to the initial concentration C_o by Eq. 4.9, and to half-life by Eq. 4.10.

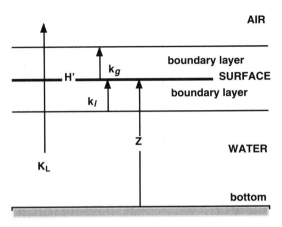

Figure 4.3. Transport of a chemical from water of depth Z into the atmosphere. K_L = total mass transfer coefficient (rate), k_1 = liquid-phase mass transfer coefficient, k_g = gas-phase mass transfer coefficient, H' = nondimensional Henry's law constant.

$$1/K_L = 1/k_1 + 1/H'k_g$$

$$C = C_o e^{-K_L t/Z} (g/mL) \qquad (4.9)$$

$$t_{1/2} = 0.69 Z/K_L (h) \qquad (4.10)$$

Wind speed, water motion, and temperature all affect volatilization (see Schwarzenbach et al., 1993, and Special Topic 4). Although nomograms to account for these factors are provided by Lyman et al. (1990), Fig. 4.4 offers a convenient and practical way to estimate volatilization rates from H' alone. If H' is $<10^{-5}$, the chemical is less volatile than water (H' 2.5×10^{-5} at 25°C) and will become more concentrated as the water evaporates. When $H = 10^{-5}$ to 5×10^{-4}, k_g dominates transport, and volatilization is slow. Between H' 5×10^{-4} and 5×10^{-2}, both k_g and k_l are influential, but liquid phase resistance is in control above 5×10^{-2}, and volatilization will be rapid (Lyman et al., 1990).

4.2.3. Transport on Particles

Chemicals also are transported on or in suspended particulate matter. They may travel for many miles adsorbed to particle surfaces or absorbed into a porous matrix. The particles come in a wide range of sizes, from clays with diameters of <1 μm to silt and fine sand with diameters of 1–200 μm and finally to coarse sand with diameters up to 1 mm or more. As the velocity at which particles settle, v_s varies with the square of their radius r

$$v_s = 0.22 g r^2 (d_p/d_w - 1)/\eta_k = 344 r^2 \qquad (4.11)$$

only the smaller ones travel very far. Particle density, d_p, averages 2.6 g/cm^3; the density of water, d_w, is near 1.0 g/cm^3 at ambient temperatures; water's kinematic

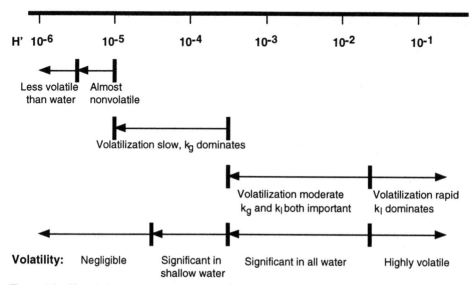

Figure 4.4. Henry's law constant as a measure of the volatility of chemicals from water. Adapted from Lyman et al., 1990.

viscosity η_k (absolute viscosity divided by the density at the same temperature) is 0.0101 cm^2/s at 20°C; and gravitational acceleration g is near 980 cm/s^2. Thus a sand particle 200 μm in diameter would fall through 1 m of 20°C water in 29 s, while a clay particle 1.0 μm in diameter would take 1.17 million seconds (13.5 days) to fall the same distance. In fact, with a radius that small, turbulence and even Brownian movement might keep the clay suspended indefinitely.

Viscosity is important to both water movement and solute diffusion as well as to the settling and dispersion of water-borne particles. It represents resistance to fluid deformation (flow), and its measure is the poise, 1 dyne-s/cm^2 and also 1 g/cm/s. The centipoise (cp) is the more common unit and equals 0.01 poise or 1 millipascal second (mPa s). The viscosity of aqueous solutions of both organic and inorganic compounds increases with increasing molecular weight and polarity and decreases with temperature. The absolute viscosity of pure water is 0.89 cp at 25°C (1.00 cp at 20°C), while that of 35‰ seawater is 0.96 cp, meaning that seawater is "thicker" and that suspended particles will sink even more slowly than they do in fresh water.

Fish readily bioconcentrate hydrophobic pollutants. Does that mean that fish migration provides a route for the environmental transport of chemicals such as DDT? It does, but this route is considered to be insignificant compared to other pathways.

4.3 | TRANSPORT IN SOIL AND GROUNDWATER

4.3.1. Transport Through Soil

The seemingly solid earth beneath us, loosely called "soil," is composed of gravel, sand, silt, clay and organic matter and is highly heterogeneous. In its upper layers, known as the unsaturated or *vadose* zone, the *interstices* or *pores* (spaces) between particles represent 20–40% of the total soil volume and normally are filled partly with water and partly with air. Below this lies the *saturated zone,* generally an aquifer composed of sand and gravel whose interstices are completely filled with water. Still deeper is the aquiclude or basal rock. Chemicals move through the vadose zone by at least three mechanisms: In solution, as vapor, and adsorbed to solid particles.

Downward and lateral percolation of aqueous solutions, probably the most common mechanism of transport, are influenced by soil porosity, adsorption, abiotic and biological reactions, and, of course, by gravity. Often forced along by the pressure of rain or irrigation water, the soil-filtered solution is in intimate contact with adsorbing surfaces, and a form of chromatography results that separates solutes according to their adsorption coefficients. The relative distance through which a substance moves, the *leaching* **distance,** is inversely proportional to its soil distribution coefficient K_d (Fig. 4.5). Hydrophobic chemicals such as DDT, which have a very high K_d or K_{oc}, remain almost at the soil surface, while others migrate proportionately farther. Due primarily to the carboxyl groups of its organic fraction, soil has an overall negative

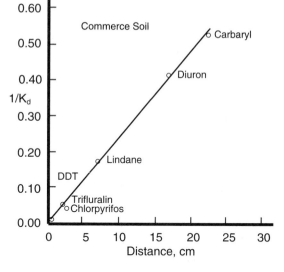

Figure 4.5. Relation of leaching distance to soil distribution coefficient K_d for several pesticides. Adapted from McCall et al., 1981.

charge; anionic substances such as the acidic herbicides are repelled and travel almost unimpeded, while cations such as paraquat are bound tightly to particles.

The relative ***mobility*** of a chemical often is measured in the laboratory by allowing a solution of it to be washed through a vertical column of the soil of interest and either detecting how far it has travelled when the water first drips out or measuring the length of time required for it to emerge relative to another substance of known mobility. However, a more convenient method is to coat a thin-layer chromatography plate with the soil, apply a spot of the chemical, allow water to be pulled up across the surface by capillary action, and detect how far the spot has migrated (Helling and Turner, 1968; Helling, 1971).

Volatile chemicals also migrate in vapor form in the soil interstices. Those such as methyl bromide and ethylene dibromide, which exhibit a high vapor pressure and relatively weak adsorption, may prefer this route. Even in the vadose zone, water competes for sorption sites, and absorbed chemicals can be displaced by the large molar excess of water resulting from rain or irrigation. For example, analysis for soil-incorporated trifluralin in the atmosphere above a treated field showed only low levels, but immediately after a brief rain, concentrations increased 500–fold as the herbicide was forced out of the soil (Soderquist et al., 1975). Upward movement of adsorbed chemicals also occurs as water is wicked up to the soil surface from below.

Chemicals adsorbed to soil particles are transported when the particles move. While particle movement in intact soil is limited, it is common during the tillage of fields, construction activities, and especially erosion and surface runoff. However, contrary to popular opinion, very little pesticide—less than 1%—ordinarily leaves a treated field in this way. Runoff depends on climate, field management, and watershed characteristics. Very little soil is lost to irrigation water, and rainfall causes runoff only when it exceeds the soil's ability to absorb it, that is, the ***field capacity.***

Rainfall intensity, or volume per unit time, is the key element; high intensity can lead to heavy storm runoff and extensive transport of chemicals. Urban runoff is the main source of aquatic pollutants, including petroleum hydrocarbons, fatty acids, fecal sterols, phthalate plasticizers (Eganhouse et al., 1981), metals, and others. The short Los Angeles River in Southern California, which drains only part of the city, dumps 1% of the *world's* marine oil pollution into the Pacific Ocean each year.

4.3.2. Transport through Groundwater

Just a few years ago, the penetration of chemicals through the soil surface and vadose zone into groundwater was considered unlikely. However, it finally has happened, although the migration often was very slow. Among the first signs of trouble was the detection of the highly volatile and toxic insecticide aldicarb in shallow groundwater, where it probably had penetrated primarily as vapor through sandy soil. Pesticides and other chemicals now are routinely detected in groundwater from wells (Appendix 2.3), but generally only those exhibiting high H' and low K_{oc}. Others occasionally enter via leaks around well casings or through **macropores** (deep cracks in the ground).

The transport of a chemical in groundwater occurs mostly through diffusion and advection in an **aquifer** that may lie hundreds of feet below the surface to only a few feet down. Where the aquifer reaches the surface, a spring occurs. Water flows through it due to hydrostatic pressure or "head," but where the pressure is low (a low **hydraulic gradient**), the movement may be slow. The ease with which water moves is experimentally measured as the **hydraulic conductivity,** K_c, of the porous layer and may be as high as 1 cm/s in a coarse gravel or less than 10^{-3} cm/s in fine sand. The one-dimensional **specific discharge, d,** represents the rate at which water flows through a perpendicular unit area in unit time, as described by Darcy's law, where dh/dx is the hydraulic gradient or rate of change in hydrostatic pressure over a distance z:

$$\mathbf{d} = - K_c \frac{dh}{dz} \tag{4.12}$$

For example, with a gradient of 1 m over a distance of 100 m, and a hydraulic conductivity of 0.01 cm/s, the specific discharge will be 1×10^{-4} cm/s. The minus sign indicates the direction of flow.

A relatively steep and porous aquifer could generate a specific discharge of >5 cm/s. The **porosity (n)** of the aquifer, which determines the advective movement of a chemical in groundwater, is defined as the proportion of interstices (void space between particles) in a unit volume of porous medium. The rate, or **seepage velocity,** at which the chemical moves past a fixed point in the aquifer is defined as $v_s = \mathbf{d}/n$. For $n = 0.1$, where the void space makes up only 10% of the aquifer volume, the v_s of trichloroethylene would be 1×10^{-5} cm/s when $\mathbf{d} = 1 \times 10^{-4}$ cm/s, and the TCE should move only 6 cm in a week! For more discussion of this complex subject, see Hemond and Fechner (1994).

4.4 | ATMOSPHERIC TRANSPORT

4.4.1. Volatilization

Volatile chemicals reach the atmosphere by diffusion from water, soil and other surfaces, and even from the respiration of plants and animals. Like water, the atmosphere is a fluid, and diffusion is governed by Fick's law (Eq. 4.4). The rate of this dissipation is of considerable interest to both manufacturers and regulators. The volatilization of a substance from a leaf surface starts with diffusion into the static air boundary layer above it, with a flux J proportional to the concentration in the layer (n/V) and the gaseous diffusion rate (D_v) and inversely proportional to the layer's thickness l

$$J = k_1(n/V)D_v/l \qquad (4.13)$$

D_v is inversely proportional to the square root of the molecular weight $(M^{0.5})$, molar concentration is converted to the more convenient g/L by multiplying by M, and $n/V = P/RT$ (P is the vapor pressure):

$$J = k_2 \, (n/V)(1/M^{0.5})(M)/l = k_3 \, PM^{0.5}/lRT \qquad (4.14)$$

$$= 0.0029PM^{-2} \text{ mg/cm}^2/\text{min}$$

Combining the constants k_2, R, T, and l with the appropriate conversion factors results in $k_3 = 2.90 \times 10^{-3}$ (about 1/345). The volatilization rates of several common chemicals calculated in this way (Table 4.1) correspond fairly well with the few that have actually been measured in the field. For example, the calculated rate for DDT is 1.0×10^{-8} mg/cm^2/min, while the measured value is 1.6×10^{-8} mg/cm^2/min (about 0.001 lb/acre/day).

4.4.2. Advection

Once beyond the surface, the vapor is quickly borne away by air currents and can be transported over great distances. Sophisticated methods are available to

TABLE 4.1
Predicted Volatilization Rates of Some Pesticides[a]

Pesticide	Vapor Pressure (Torr), 20°C	Mol. Wt.	Calculated Flux	
			(Mg/cm^2/min)	(Lbs/acre/day)[b]
EPTC	0.034 (25°C)	189	1.35×10^{-3}	167
Dichlobenil	5.48×10^{-7}	172	2.08×10^{-5}	2.6
Lindane	4.20×10^{-5}	291	2.08×10^{-6}	0.26
Parathion	6.68×10^{-6}	291	3.30×10^{-7}	0.04
DDT	1.88×10^{-7}	355	1.03×10^{-8}	0.0013
Simazine	6.08×10^{-9}	202	2.50×10^{-10}	3.1×10^{-5}

[a] See Hartley, 1969.
[b] mg/cm^2/min $\times 1.24 \times 10^5$ = lbs/acre/day.

sample and quantitatively measure airborne chemicals, but one's nose often is a sensitive and reliable detector. The molinate herbicide mentioned earlier is clearly detectable by odor at a distance of over 20 miles during field applications, and dimethyl sulfide released by a paper mill likewise can be smelled for many miles. More serious, airborne sulfur dioxide and nitrogen oxides travel great distances only to be converted eventually into "acid rain" (sulfuric and nitric acids), create major damage to faraway forests and lakes, and strain relations between nations such as the United States and Canada (Section 2.3).

Chemicals also are transported long distance as particles or on the surface of airborne dust. On 25 January 1965, a dust storm was generated in the high plains and agricultural lands of eastern New Mexico and western Texas where pesticides had been applied. The cloud was followed as it moved northeast until it arrived over Cincinnatti, Ohio, the next day. The city's air had been cleared by an earlier rain, and a brief noon shower deposited reddish mud which was collected and analyzed. Substantial amounts (200–600 μg/kg) of DDT, DDE, chlordane, and ronnel were measured, along with smaller amounts of several other pesticides (Cohen and Pinkerton, 1966).

Pesticides also move in the atmosphere as *spray drift* generated as droplets or particles of formulation fall toward their target. Most aerially applied pesticides have been diluted with water, which evaporates as the droplets descend. A droplet with an initial diameter of 200 μm lasts about 56 s and falls 21 m, while one of 50 μm diameter survives only 3.5 s and falls 3 cm before becoming an airborne particle (Akesson and Yates, 1964). Such particles may travel for a considerable distance, as when application of 2,4-D herbicide on a breezy day in Oregon's Willamette Valley resulted in a band of plant damage for 100 miles downwind.

4.4.3. Deposition

"What goes up must come down." To a large degree, this is true of both volatilized and particulate chemicals in the atmosphere. Eventually, dust settles and rain washes both vapor and particles back to earth, often far from the source as in the Cincinnatti incident. During the trip, both solid and liquid particles absorb vaporized chemicals. The mystery of how pesticide residues appeared on the leaves of vegetables in California's Central Valley where these chemicals had never been sprayed was solved by collection and analysis of fog water. The chemicals proved to be volatilizing from treated orchards miles away and moving in fog droplets, which eventually settled onto the leaves (Glotfelty et al., 1990).

The deposition of atmospheric vapor and dust by absorption and physical entrainment into raindrops is an important transport route over much of the world. The same transport rules apply to raindrops as to deeper water bodies, but (1) the water boundary layer is much thinner and transport is faster, and (2) the air boundary layer is very thin and requires less than 10 s to reach equilibrium as the drop falls. Henry's law still applies. All but the largest drops may remain airborne for many minutes, certainly long enough for dissolution and equilibration to be achieved. Analysis of rain from over the Great Lakes (Table 4.2) reveals the same spectrum of pollutants as found in the surrounding air.

TABLE 4.2
Pollutant Deposition in Rain[a]

Pollutant	Concentration[b] in Air (ng/m^3)	Concentration[b] in Rain (ng/L)
PCB, total	1.0	30
Chlorinated pesticides		
DDT, total	0.03	5
BHC, total	2.3	20
Dieldrin	0.05	2
HCB	0.2	2
Methoxychlor	1	8
PAH, total	20	100
Phthalate esters	4	12

[a] Data from Eisenreich et al., 1981.
[b] Average concentrations.

Dry deposition also is very important, especially in the more arid parts of the world. Even in the damp English climate, the dry deposition of SO_2 is about twice that of wet (rain) deposition. The deposition rate or flux J is related to atmospheric concentration, C_a, by the *deposition velocity* v_g (Eq. 4.15). The typical v_g values shown in Table 4.3 are on the order of centimeters per second.

$$J = v_g C_a \quad (g/cm^2/s) \tag{4.15}$$

This **fallout** has received its greatest recognition in the deposition of radioactive particles. Atmospheric testing of nuclear weapons resulted in local deposition of the heavier particles for a day or so, tropospheric deposition during the following month, and worldwide stratospheric transport and deposition for many years. What typically followed was not the mass radiation poisoning depicted in the movies, but rather a slow deposition onto grass, eating of the grass by cows, transfer of the radioactivity into milk, and uptake of radioactive ^{131}I into one's thyroid gland in place of natural

TABLE 4.3
Typical Deposition Velocities[a]

Chemical	Surface	v_g(cm/s)
NO_2	Soil, concrete	0.30–0.80
	Alfalfa	1.90
O_3	Soil, grass, water	0.47–0.55
	Corn leaves	0.20–0.84
PAN[b]	Soil, grass	0.14–0.30
	Alfalfa leaves	0.63
S (solid)	Grass	0.02–0.42
	Pine forest	0.48–0.90

[a] Adapted from Finlayson-Pitts and Pitts, 1986.
[b] Peroxyacetyl nitrate.

iodine and ^{90}Sr into bone in place of calcium. This uptake and deposition of radioactivity was the most extensive and serious in children.

Deposition of airborne vapor into water takes place by the reverse of volatilization, except that atmospheric concentrations usually are lower than those in water. Atmospheric chemicals diffuse into the gaseous boundary layer from the outside and then cross the surface into the static aqueous layer (Eq. 4.16), equilibrium being approached from the direction opposite to that of Eq. 4.8.

$$J = K_L(C_a - H'C_l) \tag{4.16}$$

4.5 | COMMERCIAL CONVEYANCE

The conveyance of processed chemicals and their raw materials by truck, rail, and ship across the nation and the world, often also referred to as "transport," has become increasingly commonplace in recent decades (Table 4.4). Petroleum and petroleum products form a major part of the hazardous materials transported by truck, with mixed loads second and chemicals third. Chemicals were the principal hazardous cargo carried by rail, and petroleum, fertilizer, and chemicals such as sodium hydroxide, benzene, and toluene were those most transported by ship (U.S. Congress, 1986). Although accident statistics vary from source to source, those provided for Table 4.4 by the Congressional Office of Technology Assessment seem reliable.

An average of 15,000 annual accidents involving hazardous materials provides the potential for a lot of release and dispersal of toxic chemicals, and this figure is for the United States only. Notorious examples include the massive Alaskan oil release by the *Exxon Valdez,* a 1985 barge accident in the New Orleans Ship Canal that spilled tons of pentachlorophenol and shut down the port for days, and the 1991 derailment of a railroad tankcar whose thousands of gallons of the fumigant Vapam® spilled into California's Sacramento River and temporarily destroyed all riparian life.

TABLE 4.4
Transportation of Hazardous Materials in the United States[a]

Carrier	Number	Pounds (Billions)	Pound-miles (Trillions)	Accidents (Average)[b]
Truck	337,000 dry	1,854	187.2	13,751
	130,000 tank trucks			
Rail	115,600 tank cars	146	106	1,148
Waterborne	4,909 tank barges	1,098	1,273	16
Air	3,772 planes	0.57	0.218	113
Total	**591,281**	**3,099**	**1,556.4**	**15,066**[c]

[a] Data adapted from U.S. Congress, 1986.
[b] Annual average for 1976–1984.
[c] Includes 38 in miscellaneous modes.

4.6 | GLOBAL TRANSPORT

The global transport of pollutants, discussed by Ballschmiter (1992), takes place in the oceans and, especially, the atmosphere (Fig. 4.6). Wind is the primary force behind tropospheric mixing, as diffusion is much too slow. Average wind speed is about 5.5 m/s (12.5 mph), but both the northern and southern jet streams can reach 130 m/s (300 mph). The entire air mass over an industrial city such as Birmingham, England (point B on Fig. 4.6A), makes its easterly trip around the world in about a month, spreading pollution as it goes.

The global scale of pollutant transport is demonstrated by the worldwide circulation of volcanic dust from major volcanic eruptions such as those of Krakatoa in Indonesia and Mt. Pinatubo in the Philippines. Radioactivity from atmospheric bomb tests and the meltdown at Chernobyl also was soon detectable throughout the world. Although the closest test had been thousands of miles away, the ^{90}Sr and ^{137}Cs content of milk in New York state increased sharply for two years following U.S. and Soviet nuclear tests in 1956–1958 and 1962, and it dropped back during the intervening moratorium. A French nuclear test at Moruroa Atoll in the South Pacific (21°S latitude, point M on Fig. 4.6A) produced a spike of ^{131}I and ^{140}Ba detected 22 days later in Arkansas at 34°N latitude (point A), meaning that the radioactivity had somehow crossed the equator.

There is less evidence for global pollutant movement via the oceans. Certainly, surface currents (Fig. 4.6B) can be expected to move dissolved chemicals just as they do lost sailor's bottles and fishermen's floats. Chemicals are exchanged continuously between the atmosphere and the sea surface, and so the troposphere contains detectable levels of persistent chlorinated hydrocarbons even at remote locations. Zell and Ballschmiter (1980) reported that air samples from one of the world's most isolated islands, Enewetak atoll in the North Pacific, contained hexachlorobenzene and toxaphene, which could only have originated thousands of miles to the east. However, the ocean surface tends to circulate within confined gyres (Fig. 4.6B), generally at rates of 0.5–1.5 m/s, and communication between gyres occurs only at great depths and at rates of 0.1–0.2 m/s.

For many years, the source of chlorinated insecticide residues in Arctic and Antarctic ice was a mystery. It is now explained as the result of a fractional distillation or, perhaps more accurately, a form of global chromatography. Chlorinated hydrocarbons originating primarily in developing nations of the warm tropics volatilize into the atmosphere only to be absorbed into the soil and water downwind toward the poles. There, they are revolatilized and so transported stepwise by prevailing air and water currents until they are eventually condensed out at the frigid poles. The only requirements are that each chemical be volatile enough to evaporate from the surface of water and soil and remain stable long enough to survive the trip.

The transport of chemicals in air and water is not the last word on their fate. They are transformed to other and often more reactive forms. Oxidation is the main route of reactive loss for most, although many halogenated hydrocarbons are slow to change. Concentrations in the atmosphere, normally the final repository for most

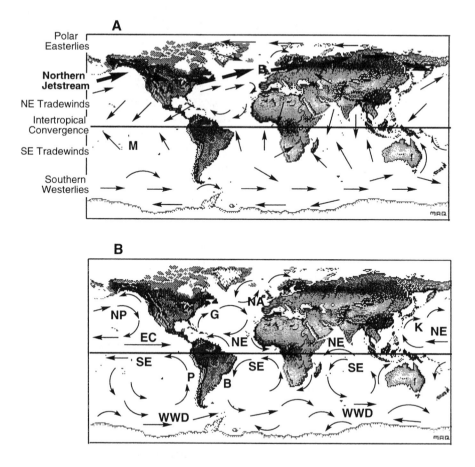

Currents: B = Brazil, EC = Equatorial countercurrent, G = Gulf Stream, K = Kuroshio, NA = North Atlantic, NE = North equatorial, NP = North Pacific, P = Peru, SE = South equatorial, WWD = West wind drift.

Figure 4.6. Prevailing winds (A) and ocean currents (B). A = Arkansas, B = Birmingham, UK, and M = Moruroa atoll. Heavy arrows show the transport route of airborne pollutants in the Northern Hemisphere. World Map Graphic© Apple Computer, Inc., 1991. All rights reserved, used with permission.

chemicals, remain low but often increase near the poles, and reaction rates between pollutants and natural reagents can become appreciable. In the case of stratospheric ozone, the results may be catestrophic. Chapter 5 describes many of these reactions, and Special Topic 5 concerns how atmospheric chemicals interact with earth's protective ozone layer.

4.7 | REFERENCES

Akesson, N. B., and W. E. Yates. 1964. Problems relating to application of agricultural chemicals and resulting drift residues. *Ann. Rev. Entomol.* **9:** 285–318.

Ballschmiter, K. 1992. Transport and fate of organic compounds in the global environment. *Angew. Chem. Int. Ed. Engl.* **31:** 487–515.

Broecker, H. C., J. Petermann, and W. Seims. 1978. The influence of wind on CO_2 exchange in a wind-wave tunnel, including the effects of monolayers. *J. Mar. Res.* **36:** 595–610.

Cohen, J. M., and C. Pinkerton. 1966. Widespread translocation of pesticides by air transport and rain-out, in *Organic Pesticides in the Environment* (A. A. Rosen and H. F. Kraybill, eds.), American Chemical Society, Washington, DC, pp. 163–76.

Downing, A. L., and G. A. Truesdale. 1955. Some factors affecting the rate of solution of oxygen in water. *J. Appl. Chem.* **5:** 570–81.

Eganhouse, R. P., and I. R. Kaplan. 1981. Extractable organic matter in urban stormwater runoff. 1. Transport dynamics and mass emission rates. *Environ. Sci. Technol.* **15:** 310–15.

Eisenreich, S. J, B. B. Looney, and J. D. Thornton. 1981. Airborne organic contaminants in the Great Lakes ecosystem. *Environ. Sci. Technol.* **15:** 30–38.

Finlayson-Pitts, B. J., and J. N. Pitts. 1986. *Atmospheric Chemistry: Fundamentals and Experimental Techniques,* John Wiley and Sons, New York, NY.

Gever, G. R., S. A. Mabury, and D. G. Crosby. 1996. Rice field surface microlayer: Collection, composition, and pesticide enrichment. *Environ.Toxicol. Chem.* **15:** 1676–82.

Glotfelty, D. E., M. S. Majewski, and J. N. Seiber. 1990. Distribution of several organophosphorus insecticides and their oxygen analogs in a foggy atmosphere. *Environ. Sci. Technol.* **24:** 353–57.

Hardy, J. T. 1982. The sea-surface microlayer: Biology, chemistry, and anthropogenic enrichment. *Prog. Oceanogr.* **11:** 307–28.

Hartley, G. S. 1969. Evaporation of pesticides, in *Pesticidal Formulations Research* (J. W. Van Valkenberg, ed.), American Chemical Society, Washington, DC. Advances in Chemistry No. 86, pp. 115–34.

Hayduk, W., and H. Laudie. 1974. Prediction of diffusion coefficients for nonelectrolytes in dilute aqueous solutions. *JAIChE* **20:** 611–15.

Helling, C. S. 1971. Pesticide mobility in soils. II. Applications of soil thin-layer chromatography. *Soil Sci. Soc. Amer. Proc.* **35:** 737–43.

Helling, C. S., and B. C. Turner. 1968. Pesticide mobility: Determination by soil thin-layer chromatography. *Science* **162:** 562–63.

Hemond, H. F., and E. J. Fechner. 1994. *Chemical Fate and Transport in the Environment,* Academic Press, New York, NY.

Lyman, W. J., W. F. Reehl, and D. H. Rosenblatt. 1990. *Handbook of Chemical Property Estimation Methods,* American Chemical Society, Washington, DC.

McCall, P. J., D. A. Laskowski, R. L. Swann, and H.J. Dishburger. 1981. Measurement of sorption coefficients of organic chemicals and their use in environmental fate analysis, in *Test Protocols for Environmental Fate and Movement of Toxicants,* Association of Official Analytical Chemists, Washington, DC.

Schwarzenbach, R. P., P. M. Gschwend, and D. M. Imboden. 1993. *Environmental Organic Chemistry,* John Wiley and Sons, New York, NY.

Soderquist, C. J., J. B. Bowers, and D. G. Crosby. 1977. Dissipation of molinate in a rice field. *J. Agr. Food Chem.* **25:** 940–46.

Soderquist, C. J., D. G. Crosby, K. W. Moilanen, J. N. Seiber, and J. E. Woodrow. 1975. Occurrence of trifluralin and its photoproducts in air. *J. Agr. Food Chem.* **23:** 304–9.

Southworth, G. R. 1979. The role of volatilization in removing polycyclic aromatic hydrocarbons from aquatic environments. *Bull. Environ. Contam. Toxicol.* **21:** 507–14.

U.S. Congress. 1986. Transportation of Hazardous Material, OTA SET-304, U.S. Government Printing Office, Washington, DC.

Zell, M., and K. Ballschmiter. 1980. *Fresenius Z. Anal. Chem.* **304:** 337–49.

Special Topic 4. Water, Wind, and Waves

As one watches the early morning stillness of a pond, the boundary layers illustrated in Fig. 4.3 and their equilibria are easy to visualize. Suddenly, a breeze ripples the water: What happens now to "equilibrium"? And what would happen in crashing ocean waves? We observe that chemicals such as water and gasoline evaporate most rapidly when the wind is blowing and that evaporation is faster when the water is turbulent.

At wind speeds up to 3 m/s (7 mph), a nice breeze, the water surface remains largely undisturbed, k_l is 1–3 cm/h, and mass transfer is dominated by water currents alone. In turbulent water, k_l is a function of current velocity v_c, depth Z, and the molecular diffusion coefficient D_m

$$k_l = (D_m v_c/Z)^{1/2} \text{ cm/h} \tag{4.17}$$

However, as wind speeds rise to 3–10 m/s (7–20 mph), k_l increases from 3 to 30 cm/h. Ripples appear before the breeze reaches 6 m/s (13 mph) and, above that, become wavelets. At speeds above 10 m/s, the waves break and the surface area upon which volatilization rate depends expands sharply; k_l can reach 70 cm/h. Spray and fog approach the ultimate in surface, as the 100 mm^2 surface of 1 cm^3 of quiet lake become 30,000 mm^2 when converted to droplets of 200 μm diameter.

For most compounds of environmental interest, k_l conforms to Eq. 4.18 at wind speeds below 1.9 m/s (4 mph) and to Eq. 4.19 for speeds between 1.9 and 5 m/s (4–11 mph).

$$k_l = 133 \, (v_c^{0.97}/Z^{0.67})/M^{0.5} \text{ cm/hr} \tag{4.18}$$

$$k_l = 133 \, [(v_c^{0.97}/Z^{0.67})/M^{0.5}]e^{0.53(v_w - 1.9)} \text{ cm/h} \tag{4.19}$$

Current velocity is represented by v_c (m/s), wind velocity by v_w (m/s), and water depth in meters by Z (Lyman et al., 1990). In winds of <10 m/s (20 mph), K_l normally will lie between 2 and 20 cm/h (Schwarzenbach et al., 1993). Values for k_g can be estimated by Eq. 4.20 (Southworth, 1979), and the total mass transfer coefficient K_L and flux J may now be calculated by Eq. 4.21 and 4.22, respectively, where C_w is the concentration in water and C_a that in air. Equation 4.8 (Section 4.2.2) shows that where C_a/H' is small, K_L will approximate C_w. The equations hold true for any body of water, from a raindrop to the Pacific Ocean.

$$k_g = 4.83 \times 10^3 \, (V_w + V_c)/\sqrt{M} \text{ cm/h} \tag{4.20}$$

$$K_L = H'k_g k_l/(H'k_g + k_l) \text{ cm/h} \tag{4.21}$$

$$J = K_L(C_w - C_a/H') \text{ moles/cm}^2/\text{h} \tag{4.22}$$

Many water bodies are covered with a very thin (<1 μm), natural film of organic material known as the **surface microlayer**. The marine film consists of a hydrophobic layer of hydrocarbons, fatty acids, and esters below which is found a hydrophilic polysaccharide–protein zone in equilibrium with the underlying water, most of it derived from decomposed plants and animals. This microlayer concentrates hydrophobic contaminants from both water and atmosphere, and PAH congeners have been detected at levels up to 10^6 times their level in the water column, PCBs at 1000–fold greater, and DDT at 2600–fold greater concentrations (Hardy, 1982). Breaking waves throw minute microlayer-coated water droplets high into the marine atmosphere, where they eventually evaporate to organic-coated salt crystals amounting to 10^{12} kg per year including the 5×10^{10} kg of film. These particles are responsible for the coastal haze so often seen during visits to the beach.

Fresh water now is known to be covered with a similar film, again consisting of long-chain fatty acids, esters, and a protein–carbohydrate matrix and capable of concentrating chemicals such as pesticides (Gever et al., 1996). While artificial films sometimes are used intentionally to reduce evaporation from freshwater reservoirs, knowedge of the influence of surface microlayers on the transport of hydrophobic chemicals from water into the atmosphere remains very limited. However, experiments with the flux of gases such as oxygen and carbon dioxide show that transport across yet another boundary layer reduces the flux by as much as 80% (Downing and Truesdale, 1955; Broecker et al., 1978).

5.1 | TRANSFORMATIONS

Evidence of environmental chemical change is everywhere. Colored fabrics fade, rubber cracks, paint peels, skin tans. Some of these transformations are brought about by living organisms (biotransformations, Chapter 6), but most are purely chemical *(abiotic)*. Powerful environmental reagents, including oxygen and other oxidants, reductants, and water, participate in most of them. Some are driven by ultraviolet (UV) energy from sunlight, while others operate continuously, day and night.

Most environmental reactions appear to follow *first-order kinetics* (Section 4.1), the rate depending only on the concentration of reactant remaining at any given time, and the half-life independent of starting concentration:

$$\ln(C_0/C) = k_1 t$$
$$t_{1/2} = 0.693/k_1 \tag{5.1}$$

If the concentration of another reactant also affects the rate, the reaction would be called *second order.* However, in most environmental reactions, the relative concentration of the second reagent is so large as to remain virtually constant (e.g., atmospheric oxygen), and the rates are best termed pseudo first order or second order regardless of their actual complexity. Photochemical reactions, driven by light energy, are distinguished from the purely thermal *"dark reactions."*

Reactions usually take place in several integrated steps which, overall, represent a reaction *mechanism* (March, 1992).

$$A + B \rightleftharpoons [AB] \longrightarrow C + D \tag{5.2}$$

These steps often involve a transition state or activated complex *(AB)* whose formation is rate controlling and requires the most energy. This *activation energy* (E_a) can be substantial and is supplied by heat or light. Because of this, most environmental reactions fall under kinetic rather than thermodynamic control and even substances as stable as PCBs react with oxygen when heated to 300°C. Reaction rates depend on temperature; where E_a is moderate, say 50 kJ/mole (12 kcal/mole), the rate doubles for each 10°C rise in temperature, but it triples at an E_a of 80 kJ/mole and quadruples at 100 kJ/mole. The activation energy of hydrolysis of esters lies between 40 and 80 kJ/mole, so one can expect the reaction to proceed 2–3 times faster at 30°C than it did at 20°C (Schwarzenbach et al., 1993).

For any type of substance, reactivity is governed by stereochemistry, bond stengths (Appendix 5.1), and the electronic configuration of functional groups. A measure of this configuration is the Hammett σ and ρ constants (Section 16.3.2), in which, for a given set of conditions, electron-withdrawing (*m*-directing) substituents stabilize a benzene ring while electron-releasing (*p*-directing) groups increase its reactivity. For example, the chlorine substituents stabilize a PCB, while the hydroxyl group accounts for the rapid environmental degradation of phenol.

5.2 | PHOTOCHEMISTRY

5.2.1. Radiant Energy

Chemical reactions require energy, and light can provide it. The wave nature of light is represented by the electromagnetic spectrum (Fig. 5.1), whose wavelengths—the distance from one wave crest to the next—are measured in nanometers (10^{-9} m, 10^{-7} cm), and whose frequency, or the number of wave cycles per second, is in hertz (cycles/second). Frequency is familiar to radio listeners, as a typical FM station broadcasts at 90 megahertz (9×10^7 waves/second). Wavelength λ is related to frequency ν by the speed of light, c, so that $c = \lambda\nu$.

However, in chemical reactions, light behaves as particles, called ***photons,*** whose energy E is related to wavelength λ (Equation 5.3):

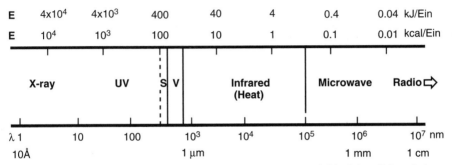

Figure 5.1. The electromagnetic spectrum. S = sunlight UV, V = visible light, all boundaries are indistinct. γ-Rays lie below 0.1 nm, the FM broadcast band at about 3×10^9 nm (100 MHz). E = energy.

$$E = \frac{1.196 \times 10^5}{\lambda} \text{ kJ/Einstein} = \frac{2.859 \times 10^4}{\lambda} \text{ kcal/mole}$$

$$(5.3)$$

where an einstein is Avogadro's number (6.023×10^{23}) of photons. **Ultraviolet (UV)** radiation is energetic enough to break chemical bonds (Appendix 5.1), provided it can be absorbed as shown by a compound's UV absorption spectrum. To record a spectrum, the absorbed energy is measured with a UV spectrophotometer and plotted at each desired wave-length in **absorbance units,** A, calculated by Eq. 5.4. C is molar concentration, l the cm of light path through the substance or solution, ϵ the compound's characteristic molar absorptivity or "extinction coefficient," and I_0/I the ratio of light intensity before and after traversing l.

$$A = \epsilon C l = \log I_0/I$$

$$(5.4)$$

The UV spectrum of the herbicide 2,4-D (Fig. 5.2) shows typical absorption maxima at 230 nm (ϵ 9,559) and 287 nm (ϵ 2,203).

To absorb UV energy, a substance must possess electronic unsaturation, usually as delocalized πelectrons. Benzene rings absorb UV strongly, most olefins absorb only weakly, and saturated compounds such as alkanes do not absorb at all. Long chains of alternating double bonds, including aromatic rings, absorb light far into the visible region and so exhibit "color" and very high molar absorptivities. While 2,4-D contains only a single benzene ring, its ϵ values are relatively high, and it undergoes photochemical reactions even in dilute aqueous solution.

5.2.2. Photochemical Reactions

Absorption of a photon provides energy (Eq. 5.3) that boosts an electron into a higher-energy state, which then quickly decays. The released energy appears as heat, fluorescence, or an activated chemical bond ready for reaction. Clearly, not every absorbed photon results in a chemical transformation. The efficiency of the

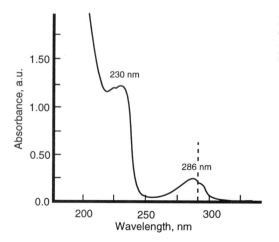

Figure 5.2. The UV absorption spectrum of 2,4-dichlorophenoxyacetic acid (2,4-D). $C = 1.36 \times 10^{-4}$ M, $l = 1.00$ cm. Dashed line shows sunlight UV cutoff.

photochemical reaction—the fraction of absorbed photons that causes a molecular change—is called the **quantum yield (ϕ)** and is usually less than unity. A ϕ of 0.21 means that 21% of energy-absorbing molecules undergo reaction. However, if that reaction results in an unpaired electron (a free radical), a single photon can lead to many molecules of product in a **chain reaction** (see Fig. 5.4), and ϕ can be much greater than 1. The physical basis of photochemistry is reviewed well by Finlayson-Pitts and Pitts (1986) and by March (1992).

Photochemical reactions were first investigated scientifically by Italian chemist G. Ciamician in the late 1800s. Most represent **direct photoreactions,** where UV energy is absorbed by the chemical's extended π-electron system **(chromophore)** and dissipated by bond-breaking and subsequent reactions. Environmental transformations may also involve **indirect photoreactions** in which a reactive species is generated photochemically from some UV-absorbing molecule and subsequently attacks the compound of interest. Certain chemicals, such as chlorophyll, can bring about **sensitized** reactions by absorbing UV energy and passing it by physical contact to another molecule, which then reacts. Examples of these reaction types will appear later.

Photochemical reactions are important. They allow photosynthesis in green plants, cause skin damage from UV exposure, and often degrade environmental pollutants. Both UV intensity and quantum yield can be measured by a chemical actinometer, in which a dilute solution of photoreactive substance of known ϕ_A is exposed to UV radiation and the first-order rate constant, k_{1A}, of its loss determined in a transparent vessel of known volume V and illuminated surface area A. The UV intensity I_0 (usually expressed in $\mu W/cm^2$) is then determined by Eq. 5.5, and, under the same conditions of I_0, A, and V, the quantum yield ϕ_B of another reaction can be calculated by Eq. 5.6 (Calvert and Pitts, 1966).

$$I_0 = \frac{k_{1A}\Phi_A}{2.303(A/V)\varepsilon_A} \tag{5.5}$$

$$\Phi_B = \frac{k_{1B}\varepsilon_A}{k_{1A}\varepsilon_B}\Phi_A \tag{5.6}$$

Suitable actinometer reactions include photoisomerization of cis- to trans-stilbene, decomposition of potassium ferrioxalate, and reaction of pyridine with p-nitroanisole (Dulin and Mill, 1982).

5.2.3. Solar Energy

The energy of most interest to environmental chemists and toxicologists is the UV component of sunlight. Below about 290 nm, solar radiation is absorbed by stratospheric ozone before it reaches the earth's surface, while above 400 nm it does not provide enough energy to break most chemical bonds (Fig. 5.3, Appendix 5.1). Solar radiation is very intense at the outer edge of the atmosphere, and UV energy at

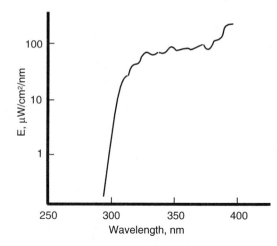

Figure 5.3. The intensity of sunlight ultraviolet (UV) radiation at the earth's surface.

the earth's surface often exceeds 20 watts/m^2 on a clear summer day (Finlayson-Pitts and Pitts, 1986).

Solar UV is attenuated by the dilute, gaseous layer of natural ozone that encircles the globe between about 32 and 48 km above sea level (Special Topic 5). Below that, the atmosphere is relatively transparent except for airborne particles that reflect and scatter light. Water vapor and clouds have relatively little influence on UV intensity, as anyone who has been sunburned on a cloudy day will recognize, and contrary to popular belief, most natural waters also are quite transparent to UV and easily transmit 300 nm radiation to depths of over 10 m (Baker and Smith, 1982). Like a dirty atmosphere, muddy water attenuates UV transmission, and UV radiation penetrates less than 1 mm below the surface of soil or sand.

Perhaps surprisingly, there are relatively few general types of abiotic environmental reactions, photochemical or not. Each environmental compartment is characterized by a particular reaction type: *Oxidation* predominates in the oxygen-rich atmosphere, *hydrolysis* is most important in water, and soil and sediments are conducive to *reduction.* These assignments are not exclusive—soil contains water and air, the atmosphere contains water vapor and suspended dust particles, and so on—but they do afford a useful first approximation. Each type will be discussed individually in the rest of this chapter, although all usually operate simultaneously.

5.3 | OXIDATION

Oxidation is the environmental reaction encountered most frequently. By the strict chemical definition, it simply means a loss of electrons. When iron (Fe°) rusts, it loses electrons and goes to its trivalent ferric form, Fe_2O_3. The oxygen gains them and is "reduced." The ease with which electrons are transferred is measured by the redox potential (Section 5.4).

5.3.1. Radical Oxidations

"Oxidation" also commonly indicates the introduction of an oxygen atom into a molecule, that is, *oxygenation.* As you light a propane gas stove, see its blue flame, and feel the heat, you are witnessing the most important and pervasive of all environmental reactions, a free radical oxidation. The stoichiometry shown in Eq. 5.7 scarcely begins to reveal the chemistry involved.

$$C_3H_8 + 5O_2 \longrightarrow 3CO_2 + 4H_2O + 531 \text{ kcal} \tag{5.7}$$

Figure 5.4 gives a better picture, although still incomplete. A free radical, **•R,** initiates the reaction by abstracting an hydrogen atom from a propane molecule. It approaches so close that the hydrogen transfers to it, leaving behind a carbon radical. A series of very fast, largely exothermic reactions follow that cleave off carbons one at a time. Almost any of the new radicals can serve as **•R** to propagate the chain, and the water and carbon dioxide produced appear at different stages of the sequence. The H abstraction occurs principally on the second carbon, as its C–H bond strength is lowest (Appendix 5.1), and a tertiary H would be removed even more readily.

What the flame provides in concentrated form has its dilute counterpart in the atmosphere, where the principal atmospheric reagents are oxidants (Table 5.1). The triplet oxygen ($^3O_2$) we breathe comprises about 21% of the atmosphere by volume. As a diradical, it is the key factor in the burning of propane, but the parallel spins of

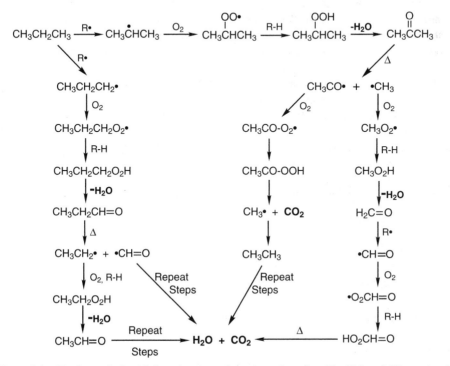

Figure 5.4. The free radical oxidation of propane. Δ = thermal reaction. The H_2O and CO_2 produced are shown in bold type.

TABLE 5.1
Atmospheric Oxidants[a]

		Concentration (Molecules/L)	
Oxidant	Formula	Clean Air	Smog
Triplet oxygen (^3S)	$\cdot O{-}O\cdot$	5×10^{21}	5×10^{21}
Singlet oxygen (^1D)	$O{=}O$	6×10^{11}	1×10^{13}
Oxygen atoms (^3P)	$\cdot O\cdot$	7×10^7	5×10^8
Ozone	$\cdot O{-}O{-}O\cdot$	7×10^{14}	1×10^{16}
Hydroxyl	$\cdot OH$	2×10^9	5×10^{10}
Nitrogen dioxide	$\cdot NO_2$	9×10^{13}	5×10^{15}

[a] Crosby, 1983.

its unpaired electrons reduces its reactivity, and it cannot initiate the chain. Other important oxidants include the highly reactive singlet molecular oxygen ($^1O_2$) that contains an oxygen–oxygen double bond, atomic oxygen in its ground state triplet (^3P) and excited singlet (^1D) states, and ozone, O_3, an allotropic form of oxygen. Hydroxyl radicals are the most reactive atmospheric oxidants of all.

5.3.2. Atmospheric Oxidants

The natural generation of most oxidants requires sunlight. The process starts with the oxidation of atmospheric nitrogen to nitric oxide (Eq. 5.8). Lightning provides one energy source for this reaction, but 80% of natural NO is derived from the bacterial oxidation of ammonia in soil and water. Itself a stable free radical, NO is rapidly oxidized to nitrogen dioxide by ozone or peroxyl radicals, and the stage is set for oxidant generation.

Sunlight UV now decomposes NO_2 to return NO to the pool and generates ^3P oxygen atoms, which react with triplet oxygen to give ozone (Eq. 5.9). The ozone is decomposed photochemically to ^1D oxygen atoms, which in turn react with water vapor to produce hydroxyl radicals (Eq. 5.10). One may picture this combination of steps as a cyclical machine that converts one mole of rather unreactive triplet oxygen into *four* moles of very reactive hydroxyl radicals (Eq. 5.11).

$$N_2 + O_2 \xrightarrow{1000^\circ C} 2\,\cdot NO \xrightarrow{O_3 \text{ or } RO_2\cdot} 2\,\cdot NO_2 \xrightarrow{UV} 2\,\cdot NO + 2\cdot O\,(^3P) \qquad (5.8)$$

$$2\cdot O\,(^3P) + 2O_2 \longrightarrow 2O_3 \xrightarrow{UV} 2^1O_2 + 2\cdot O\,(^1D) \qquad (5.9)$$

$$2\cdot O\,(^1D) + 2H_2O \longrightarrow 4\cdot OH \qquad (5.10)$$

$$O_2 + 2H_2O \longrightarrow 4\cdot OH \qquad (5.11)$$

Leftover O_3 and $^1O_2$ join the list of bothersome air pollutants.

Although not investigated in the atmosphere until the 1970s, hydroxyl radicals soon were recognized as key players in atmospheric oxidations. In addition to the pathway shown, reaction of a peroxyl radical with nitric oxide and the photolysis of nitrous acid also are major sources. Hydroxyl reacts rapidly with many kinds of gaseous organic pollutants, including benzene, PAH, amines, dichloropropene, and methyl bromide (Finlayson-Pitts and Pitts, 1986).

Most hydroxyl is formed photochemically, meaning only during daylight hours, but many of its roles are assumed at night by nitrate radical, $\cdot O{-}NO_2$, formed principally from ozone and nitrogen dioxide (Eq. 5.12). The atmospheric reactions of nitrate include hydrogen abstraction from aldehydes and hydrocarbons (Eqs. 5.13 and 5.14)

$$O_3 + NO_2 \longrightarrow O_2 + \cdot ONO_2 \tag{5.12}$$

$$\cdot NO_3 + RCH{=}O \longrightarrow HNO_3 + R\dot{C}{=}O \tag{5.13}$$

$$\cdot NO_3 + RH \longrightarrow HNO_3 + R\cdot \tag{5.14}$$

$$\cdot NO_3 \xrightarrow{600\,nm} \cdot NO + O_2 \tag{5.15}$$

to form nitric acid, and sunlight eventually decomposes it back to nitric oxide (Eq. 5.15). The total of NO, NO_2, and NO_3 is termed "NO_x."

Conversion of NO to NO_2 requires an oxidant more reactive than triplet oxygen, and peroxyl ($ROO\cdot$) or hydroperoxyl ($HOO\cdot$) are the usual reactants. Oxidation of the ubiquitous methane (Section 2.3.1) leads sequentially to permethoxyl, hydroperoxyl, and NO_2 via a series of short-lived intermediates:

$$CH_4 + \cdot OH \longrightarrow \cdot CH_3 + H_2O \tag{5.16}$$

$$\cdot CH_3 + O_2 \longrightarrow \underset{\text{Permethoxyl}}{CH_3OO\cdot} \xrightarrow{NO} CH_3O\cdot + \mathbf{NO_2} \tag{5.17}$$

$$CH_3O\cdot + O_2 \longrightarrow H_2C{=}O + \underset{\text{Hydroperoxyl}}{HOO\cdot} \tag{5.18}$$

$$HOO\cdot + NO \longrightarrow HO\cdot + \mathbf{NO_2} \tag{5.19}$$

Hydroxyl also reacts with inorganic pollutants such as SO_2 and NO_2 to produce the strong mineral acids of "acid rain." Acid rain was recognized even in the 1700s and was described scientifically by Robert A. Smith in an 1872 book. At present, rain across the eastern United States often has a pH near 4, and a pH of 2 is not unheard of. Acid fog with a pH as low as 1.7 has been reported from coastal areas near Los Angeles and is of particular concern because the droplets are of a size to be respirable. At least half of the natural forests in Germany have been damaged by acid rain, and fish and other aquatic life are affected in our own country.

Sulfuric acid provides a persistent atmospheric haze, of ammonium sulfate particles in the Northern Hemisphere, formed with ammonia derived from synthetic fertilizers, and of the acid itself in the Southern Hemisphere. Near the ocean, particles are mostly sodium chloride thrown into the air by bursting bubbles, but around high concentrations of NO_x, the tiny salt crystals react with NO_2 to generate nitrosyl chloride and sodium nitrate (Finlayson-Pitts, 1983) (Equation 5.20).

$$NaCl + 2NO_2 \longrightarrow NOCl + NaNO_3 \tag{5.20}$$

5.3.3. Photochemical Smog

The atmospheric reagents react with airborne organic compounds, both natural and manmade (Chapter 2). The blue haze often seen between distant mountains, giving the Blue Ridge Mountains their name, is "natural smog" that results from atmospheric oxidation of terpene hydrocarbons volatilizing from trees. Human activities likewise release huge amounts of hydrocarbons into the atmosphere, especially from petroleum. The oxidation of petroleum hydrocarbons, sulfur compounds, and other substances takes place in internal combustion engines, releasing not only large proportions of carbon monoxide, sulfur oxides, and unburned fuel but providing temperatures high enough to generate NO from nitrogen and oxygen (Eq. 5.8) to add to that from natural processes.

Oxidant generation on a typical, sunny, summer day in Pasadena, California, is diagrammed in Fig. 5.5 (McRae and Seinfeld, 1983). Background levels of atmospheric NO_x, present before dawn, increase sharply as the morning rush hour spews more into the atmosphere. As sun-light intensity increases, NO_2 is photolyzed (Eq. 5.8), and the resulting oxygen atoms generate ozone (Reaction 5.9), which peaks near midday. The O_3 then declines as it reacts with other atmospheric pollutants and is converted to oxygen atoms and hydroxyl in sunlight (Reaction 5.10). A spurt of NO_2 appears again at evening rush hour but drops off overnight due to dark reactions with reagents such as $\cdot NO_3$; no ozone is formed. By sunset, an acrid pall of organic oxidation products, sulfur oxides, and airborne particles—"photochemical smog"—is swirling around the street lamps.

Appendix 5.2 shows some principal atmospheric reaction types and products. One especially pernicious class, the peroxyacyl nitrates $R\text{--}COO\text{--}O\text{--}NO_2$, is illus-

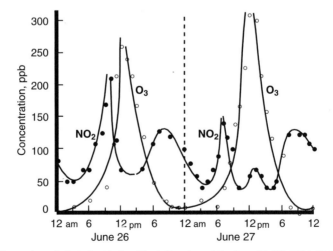

Figure 5.5. Generation of photochemical oxidant, Pasadena, CA, June 26–27, 1974. Data from McRae and Seinfeld, 1983.

trated by peroxyacetyl nitrate (PAN), whose formation from propane is summarized in Scheme 5.21. PAN is the smog constituent responsible for the familiar eye and throat irritation, while peroxybenzoyl nitrate generated from toluene vapor is killing Ponderosa pines around Los Angeles.

$$CH_3CH_2CH_3 \xrightarrow{\cdot R, O_2} CH_3COCH_3 \xrightarrow{UV} CH_3CO\cdot + \cdot CH_3 \qquad (5.21)$$

$$CH_3CO\cdot + O_2 \longrightarrow CH_3CO-O_2\cdot \xrightarrow{\cdot NO_2} CH_3COO-ONO_2$$
$$PAN$$

What finally becomes of the organic atmospheric pollutants? They are increasingly oxidized, eventually to inorganic form: Carbons turn into CO_2, hydrogens become water, sulfur forms sulfate, chlorines produce chloride, and so on. In other words, they are "mineralized." Is smog all bad? At the least, it demonstrates that the environmental self-cleaning mechanisms of photooxidation developed over the ages still operate, albeit they sometimes become overloaded.

5.3.4. Oxidations in Water

Surface waters in contact with the atmosphere dissolve oxygen, to the extent of 62.3 mg/L (1.95 mM) at 4°C and 39.3 mg/L (1.23 mM) at 25°C. This, oxygen too, enters into free radical reactions. The English chemist Michael Faraday showed in the 1850s that a closed jar containing tetrachloroethylene and water, placed on a sunny windowsill, soon became filled with a mass of white crystals. These proved to be trichloroacetic acid, formed by photooxidation and subsequent rearrangement of the tetrachloroethylene epoxide.

$$Cl_2C=CCl_2 \xrightarrow{[O]} Cl_2C\overset{O}{-}CCl_2 \xrightarrow{UV} Cl_3C-COCl \xrightarrow{H_2O} Cl_3COOH \qquad (5.22)$$

Removal of the 2,4-D sidechain (Eq. 5.23, R = 2,4-dichlorophenyl) illustrates another type of aqueous photooxidation. In sunlight, a phenoxyacetate anion loses CO_2 to generate a free radical that reacts with triplet O_2, and, via intermediates familiar from atmospheric oxidations, forms 2,4-dichlorophenol (Crosby and Tutass, 1966). The anionic charge becomes an hydrated electron, e_{aq}^-, a species to be discussed later.

$$ROCH_2COO - \xrightarrow[-e_{aq}^-, -CO_2]{UV} ROCH_2\cdot \xrightarrow{O_2} ROCH_2O_2\cdot \xrightarrow[-H_2O]{RH} ROCH=O \qquad (5.23)$$

$$ROCH=O \xrightarrow{-CO} R-OH$$

Hydroxyl radicals are now recognized as very common aquatic oxidants. They are generated photochemically by a variety of natural pathways (Table 5.2), including both the Fenton reaction and from cupryl ions (Table 5.2, reactions 7 and 8). Ferryl (FeO^{2+}) and cupryl (CuO^+) represent the hydroxyl radical anion, $\cdot O^-$, coordinated to the respective metals in a form similar to that existing in oxidizing enzymes in

TABLE 5.2
Generation of Hydroxyl Radicals in Water

1. $H_2O_2 \xrightarrow{<300 \text{ nm}} 2HO\bullet$

2. $NO_3^- \xrightarrow[H^+]{<350 \text{ nm}} NO_2^- + HO\bullet$

3. $NO_2^- \xrightarrow[H^+]{<350 \text{ nm}} NO + HO\bullet$

4. $O_3 + H_2O \xrightarrow{<315 \text{ nm}} O_2 + 2HO\bullet$

5. Humic acids $+ O_2 \xrightarrow{<350 \text{ nm}}$ Oxidized humic acids $+ HO\bullet$

6. $Cu^{2+} + H_2O \xrightarrow{UV} Cu^+ + H^+ + HO\bullet$

7. $Fe^{2+} + H_2O_2 \longrightarrow Fe^{3+} + OH^- + HO\bullet$ Fenton Reaction

8. $Cu^+ \xrightarrow{O_2} \underset{\text{Percupryl}}{CuO_2^+} \xrightarrow{2H} \underset{\text{Cupryl}}{CuO^+} \xrightarrow{H_2O} Cu(OH)^{++} \longrightarrow Cu^{++} + HO\bullet$

vivo (Section 6.2.3). They are broad-spectrum oxidizing agents that typically oxidize benzene to phenol by the route shown in reaction 2 of Appendix 5.2.

The principal source of $\bullet$OH in water probably is the photolysis of H_2O_2 and NO_3^-. Although the steady-state concentration of $\bullet$OH in sunlit water is only about 10^{-14} M, continuous generation and high reactivity (Table 5.3) cause it to dominate many aquatic oxidations. Its rate constants near, 10^{10}, are considered extremely high; even 10^7 is unusual and generally means that the reactions are controlled primarily by the rate of the diffusion that brings the reactants together. Although the concentration of hydrogen peroxide in field water reaches 7 μM, and the O–O bond dissociation energy is only 213 kJ/mole (51 kcal/mole), its UV absorption is low ($\epsilon = 1.2$ at 300 nm). UV absorption by nitrate is somewhat larger ($\epsilon = 7.1$ at 302 nm), and concentrations can exceed 500 μM in areas treated with nitrate fertilizers. Consequently, $\bullet$OH is prominent in agricultural environments (Mabury and Crosby, 1994) and an important factor in the degradation of many pesticides. The photolysis of dissolved organic matter (DOM), particularly humic acids, also contributes.

These reactions cause *indirect photodegradation,* where molecular species other than substrate absorb the UV radiation and generate the $\bullet$OH reactant. Although the herbicide thiobencarb does not appreciably absorb sunlight UV, it is oxidized when exposed to light in sterile field water, although not in distilled water. The degradation products are the same as those formed in dark reactions with hydroxyl (Fig. 5.6), including oxidation of –S– to –SO– and –SO$_2$–, oxidative dealkylation of N, and aromatic hydroxylation. The first represents addition of O, the others H abstraction.

TABLE 5.3
Typical Reaction Rates of
Hydroxyl Radicals[a]

Reagent	Rate[b]
Hydrogen sulfide	2×10^{10}
Tyrosine	2×10^{10}
Iodide ion	2×10^{10}
Benzene	4×10^{9}
Leucine	2×10^{9}
Glucose	2×10^{9}
Carbonate ion	4×10^{8}
Chloroform	1×10^{7}

[a] Dorfman and Adams, 1973.
[b] Rate constant, $M^{-1} s^{-1}$.

Certain inorganic semiconductors generate oxidants in sunlight. Perhaps the most important of these is the white paint pigment, titanium dioxide (Helz et al., 1994), which absorbs radiation near 380 nm, well into the sunlight UV region. The absorbed energy dislodges a valence electron from its normal level and promotes it to a higher-energy, conduction band (Eq. 5.25). This electron then migrates to the particle surface and reacts there with dissolved or adsorbed oxygen (Eq. 5.26), while the remaining

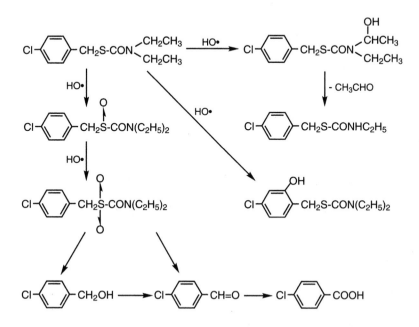

Figure 5.6. The indirect photooxidation of thiobencarb in sterile field water.

"hole" (h^+) reacts with water to generate •OH (Eq. 5.27). Titanium dioxide serves only as catalyst and is regenerated.

$$TiO_2 + UV \longrightarrow TiO_2 (h^+) + e^- \qquad (5.24)$$

$$e^- + O_2 \longrightarrow •O-O^- \qquad (5.25)$$

$$TiO_2 (h^+) + H_2O \longrightarrow TiO_2 + H^+ + •OH \qquad (5.26)$$

The high efficiency and rate of the overall process has led to a number of potentially valuable methods for the oxidative destruction of aqueous wastes and water pollutants (Special Topic 15).

Aqueous oxidants are found in nearly all natural waters. Triplet oxygen, singlet oxygen, hydrogen peroxide, and hydroxyl radicals have all been detected in raindrops, cloud droplets, and the sea, as well as inland waters, and metal-associated oxidants like percupryl ion also occur widely.

5.3.5. Oxidations in Soil

Most soils contain air or oxygenated water in the interstices between solid particles (Section 4.3). Normal soils are aerobic to a depth of a meter or more, but oxygen content diminishes continuously with depth due to microbial respiration—microorganisms are the main cause of oxidations in soil. A number of oxidative enzymes are present in soils (Crosby, 1976), originate from dead organisms, and probably are stabilized by adsorption to soil particles. Copper, manganese, iron, and other transition metal species also are present and can catalyze abiotic oxidations. Although sunlight does not penetrate more than a millimeter or so below the soil surface, the upward "wicking" of water and solutes presents chemicals at the soil surface for photooxidation (Miller et al., 1989).

5.4 | REDUCTION

5.4.1. Redox Potentials

Reduction technically means a gain of electrons, as when ferric ion (Fe^{3+}) changes to ferrous (Fe^{2+}). It is the reaction type most characteristic of the soil environment, and while reduction by the soil's microbial community is important (Section 6.2), many chemicals can be reduced readily whether the soil is sterile or not. Reductions generally take place in pore water under anaerobic conditions caused primarily by water saturation and microbial metabolism, The determinant of reducing power is the *redox potential.*

This potential represents a voltage that drives the reversible flow of electrons from reductant to oxidant. The "half-reactions" illustrated in Appendix 5.3 follow conventions that show electron flow (reduction) from left to right and the relative redox potential of hydrogen as 0.00 V. The reverse represents a corresponding oxidation, and each reduction must be coupled with an oxidation; there cannot be one without the other. The more negative the redox potential, the stronger the reducing agent.

Standard redox potentials (E°) are those existing in 1 M aqueous solutions at 25°C and 1 atmosphere pressure, although few of the large number that have been tabulated (*e.g.*, see Clark, 1960) are environmentally relevant. The difference in potential between any two half-reactions, $\Delta E°$, determines the equilibrium constant for the overall reaction as calculated by Eq. 5.27 where n is the number of electrons transferred and $\mathcal{F}$ the Faraday constant (96,490 coulombs per mole).

$$\log K_{eq} = \frac{2.303n\,\mathcal{F}\Delta E°}{RT} = \frac{n\Delta E°}{0.059} \qquad (5.27)$$

For example, the calculated $\Delta E°$ for the reduction of nitrobenzene to aniline by iron protoporphyrin IX would be 0.16 V (Appendix 5.3), making the equilibrium point far to the right ($K_{eq} = 1.86 \times 10^{16}$) and the reaction virtually complete.

E° values are measured under rigidly defined conditions, which may differ greatly from those existing in the environment. The Appendix includes a second column of redox potentials, $E°(W)$, recalculated for a natural water at pH 7, 1 mM in Cl^-, and 0.01 mM in Br^- (Schwarzenbach et al., 1993). Potentials actually measured electrochemically under field conditions are generally more useful, a typical value for the top 2.5 cm of a water-saturated sediment being about $+300$ mV. Schwarzenbach et al. (1993) provide a good discussion of redox reactions.

5.4.2. Reductions in Soil

Soil is a complex chemical mixture (Section 2.5.1). Typically, it has a mineral base of sand (SiO_2), silt, clay, water, and air, as well as an organic component largely responsible for the sorption of organic pollutants (Section 3.5). Microorganisms are important in natural reductions, but soil also contains a wide variety of chemical constituents (Crosby, 1976) which might serve as reducing agents (Table 5.5). Few natural reductants other than Fe^{2+} and H_2S have been specifically identified in the field, although others surely include complexed Fe^{2+} (*e.g.*, iron porphyrins), sulfhydryl compounds, and hydroquinones (Macalady et al., 1986).

Reducing sugars such as glucose also are logical candidates, since glucose reduces aromatic nitro compounds to amines *in vitro* and then couples the two to form azoxybenzenes. Such abiotic reductions must play a significant role in pollutant dissipation, as the degradation rates of many pesticides such as amiben, amitrole, atrazine, pronamid, and simazine are the same in either sterilized or unsterilized soil.

Just as oxidation often denotes oxygenation, the term *reduction* usually signifies hydrogenation. Aliphatic and aromatic organohalogen compounds are reductively dechlorinated (Vogel et al.,1987), sulfoxides form sulfides, and the nitro substituent of the insecticide methylparathion is converted to the corresponding amine with a half-

TABLE 5.4
Typical Reactions of Hydrated Electrons, e_{aq}^- [a]

Reaction		pH	Rate[b]
$e_{aq}^- + H_2O \longrightarrow H\bullet + OH^-$		8.4	16
$e_{aq}^- + H_3O^+ \longrightarrow H\bullet + H_2O$		4.3	2.1×10^{10}
$e_{aq}^- + H_2O_2 \longrightarrow HO\bullet + OH^-$		7.0	1.2×10^{10}
$e_{aq}^- + O_2 \longrightarrow \bullet O_2^-$		2.0	2.0×10^{10}
$e_{aq}^- + NO_3^- + H_2O \longrightarrow NO_2 + 2OH^-$		—	8.5×10^9
$e_{aq}^- + R_2S \longrightarrow RS^- + R\bullet$	$(CH_3)_2S$	6.0	2.0×10^7
	Methionine	6.0	3.5×10^7
$e_{aq}^- + RX \longrightarrow R\bullet + X^-$	1-Bromobutane	6.6	1.0×10^{10}
	Chlorobenzene	11.0	5.0×10^8

[a] Hart and Anbar, 1970.
[b] Rate constant, $M^{-1}\,s^{-1}$.

life of <60 s (Wolfe et al., 1986). Aromatic nitro compounds progress through the intermediate nitroso and hydroxylamino compounds on the way to amines (Eq. 5.28–5.30), although the mechanism of hydroxylamine reduction to amines still is tentative.

Nitro reduction involves the alternating transfer of a total of 6 electrons and 6 hydrogen ions, but nonetheless is among the most facile of the abiotic reductions.

$$R-N\overset{O}{\underset{+\,O^-}{\diagdown}} \xrightarrow{e^-} R-\overset{\bullet}{N}\overset{O^-}{\underset{+\,O^-}{\diagdown}} \xrightarrow{H^+} R-\overset{\bullet}{N}\overset{OH}{\underset{+\,O^-}{\diagdown}} \xrightarrow{e^-} R-N\overset{OH}{\underset{O^-}{\diagdown}} \xrightarrow{H^+} R-N{=}O + H_2O \qquad (5.28)$$

Nitro   Nitroso

$$R-N{=}O \xrightarrow{e^-} R-\overset{\bullet}{N}-O^- \xrightarrow{H^+} R-\overset{\bullet}{N}-OH \xrightarrow{e^-} R-\overset{-}{N}-OH \xrightarrow{H^+} R-NH-OH \qquad (5.29)$$

Hydroxylamine

$$R-NH-OH \xrightarrow{e^-} R-\overset{\bullet}{N}\overset{OH}{\underset{H}{\diagdown}} \xrightarrow[-H_2O]{H^+} R-\overset{\bullet}{N}\overset{}{\underset{H}{\diagdown}} \xrightarrow{e^-} R-\overset{-}{N}\overset{}{\underset{H}{\diagdown}} \xrightarrow{H^+} R-NH_2 \qquad (5.30)$$

Amine

5.4.3. Reductions in Water

Many natural waters contain reducing agents, that is, solutes that readily donate electrons, and reduction becomes increasingly prominent as the water loses its dissolved oxygen. For example, long after the use of DDD was halted at Clear Lake, California, substantial residues of this insecticide were still detectable. As it turned out, dissolved DDT from applications to nearby pear orchards was being reduced by natural iron porphyrins to form DDD (Miskus et al., 1965).

Among the most interesting of the natural aquatic reducing agents are hydrated electrons—free electrons trapped in a polymeric cage of water molecules. They have lifetimes on the order of microseconds, absorb visible light with a maximum at 715 nm, and are thought to be responsible for the blue color of deep ice, where they are trapped in solid water under pressure. One important source is the photolysis of hydroxide ion in an alkaline water such as seawater;

$$OH^- \xrightarrow{\ UV\ } HO\bullet + e_{aq}^- \qquad (5.31)$$

The solar UV energy at >325 nm that is absorbed by seawater is estimated to generate a steady 3×10^{13} e^-/kg/s, so the sea-surface concentration of hydrated electrons in bright sunlight can reach 10^{-15} M (Swallow, 1969). The radioactive decay of natural ^{40}K and the impact of cosmic rays also contribute. As the most powerful reducing agent possible, hydrated electrons react rapidly with many common environmental substrates (Hart and Anbar, 1970), including dissolved oxygen and nitrate ion (Table 5.4). It is unclear whether photoreductions in water, such as that of nitrofen (Fig. 5.7), are due to hydrated electrons, H abstraction by radicals, or other reductants.

TABLE 5.5
Some Natural Constituents of Soils[a]

Chemical Class	Authenicated Examples
Amines	Ammonia, ethanolamine, glucosamine
Amino acids	Arginine, histidine, lysine, methionine
Carbohydrates	Glucose, uronic acids, cellulose
Carboxylic acids	Formic acid, benzoic acid, glutamic acid
Hydrolases	Amylase, glucosidase, lipase, trypsin
Oxidases	Catalase, catechol oxidase, peroxidase
Phenols	Resorcinol, vanillic acid, p-hydroxybenzoic acid
Phosphate esters	Inositol hexaphosphate, nucleotides
Purines and pyrimidines	Cytosine, guanine

[a] Crosby, 1976.

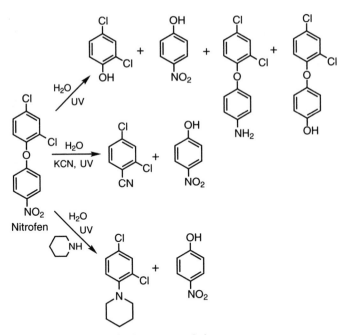

Figure 5.7. Photodegradation of nitrofen in aqueous solution.

5.5 | HYDROLYSIS

5.5.1. Water

Imagine that you are gazing down through perfectly clear lake or ocean water. You can see sand ripples and shells, so light obviously is reaching the bottom. You cannot see all the myriad chemical reagents and reactions that are present, and what you are looking through is itself a layer of reactive chemical, hydrogen oxide. Water is a dynamic reaction medium with *dielectric* properties such that both ionic and covalent reactions can occur in it.

Natural water is not just H_2O. It also contains dissolved organic matter (DOM), ionized salts, oxygen, and varying amounts of H_3O^+ and OH^-. It absorbs atmospheric carbon dioxide and converts it into ionizable carbonic acid:

$$H_2O + CO_2 \underset{\text{Slow}}{\rightleftharpoons} H_2CO_3 \underset{k_1 = 4.3\times10^{-7}}{\rightleftharpoons} H^+ + HCO_3^- \underset{k_2 = 5.6\times10^{-11}}{\rightleftharpoons} 2H^+ + CO_3^{2-} \quad (5.32)$$

Where carbonate precipitates as $CaCO_3$ or $MgCO_3$, CO_2 continues to dissolve, maintaining a buffered equilibrium between water and atmosphere only slightly affected by small changes in pH. Most inland waters become acidic at night as H_2CO_3 dissolves, but then go to as high as pH 9 at midday as algae utilize the dissolved CO_2

in photosynthesis. The carbonic and humic acids causes many ponds and lakes to be acidic, sometimes pH 3, a situation aggravated by acid rain (Section 5.3.2). On the other hand, the average pH of clean seawater is 8.2, and some well water reaches pH 10.

5.5.2. Hydrolysis in water

Water dissolves virtually any organic compound, and its reactions, principally hydrolysis, can be catalyzed by either acid or base. The rates of ester hydrolysis depend on pH (Table 5.6), with alkaline hydrolysis usually much more rapid than that under acidic conditions. The negative hydroxide ion attacks the polarized carbonyl of the ester in an S_N2 reaction (Section 5.1) to displace alkoxide ion:

$$RCH_2-\overset{\overset{O}{\|}}{C}-OR' \longleftrightarrow RCH_2-\overset{\overset{O^-}{|}}{\underset{\underset{OH^-}{+}}{C}}-OR' \longrightarrow RCH_2-\overset{\overset{O^-}{|}}{\underset{\underset{OH}{|}}{C}}+ {}^{-OR'} \longrightarrow RCH_2-\overset{\overset{O}{\|}}{C}-OH \qquad (5.33)$$

Structural features that increase the carbonyl's positive charge (*e.g.*, $R = Cl$) speed reaction, while electron-donating substituents slow it; changes in R' have less effect.

Amides are hydrolyzed in the same way as esters, but much more slowly. Where the calculated $t_{1/2}$ for hydrolysis of ethyl acetate is 2 years at pH 7 (Mabey and Mill, 1978), that of *N*-methylacetamide is 20,000 years and that of ethyl *N,N*-dimethylcarbamate 50,000 years. Hydrolysis rate is pH sensitive, so the $t_{1/2}$ for ethyl acetate becomes only 20 days at pH 8.5 and 25°C (Schwarzenbach et al., 1993). Here, the OH^- acts as a **nucleophile,** an electron-rich species that seeks out and reacts with a positively charged center. Conversely, the electropositive atom is termed an **electrophile,** as it attracts electrons (-phile coming from the Greek word *philos,* meaning "friendly").

TABLE 5.6
Hydrolysis Rates of 2,4-D Esters[a]

	$t_{1/2}$ (h)	
Ester R	pH 9	pH 6
2-Octyl	37	36,000
2-Propyl	17	17,040
1-Butyl	5.2	5,280
1-Octyl	5.2	5,280
Methyl	1.1	1,056
2-Butoxyethyl	0.6	624

[a]Zepp et al., 1975.

$$Cl-\underset{\overset{|}{Cl}}{\underset{}{\bigcirc}}-OCH_2COO\text{-}R \xrightarrow{H_2O} Cl-\underset{\overset{|}{Cl}}{\underset{}{\bigcirc}}-OCH_2COOH + R\text{-}OH$$

Alkyl halides such as ethylene dibromide (EDB) and 1,3-dichloropropene (D-D, Telone®) undergo rapid alkaline hydrolysis to an alcohol, while the hydrolysis of most aromatic halides is extremely slow. However, absorption of UV radiation by the aromatic ring causes a dramatic increase in reactivity. For example, the herbicide 2,4-D (2,4-dichlorophenoxyacetic acid) absorbs UV energy (Fig. 5.2), and exposure of an aqueous solution of 2,4-D to sunlight results in a rapid replacement of first one and then both ring chlorines by hydroxide ions from water. Similar irradiation of nitrofen in water causes displacement of nitrophenate ion, like chloride a good leaving group, even at neutral pH (Fig. 5.7); no reaction occurs in the dark (Nakagawa and Crosby, 1974). These are *photonucleophilic substitutions,* where a nucleophile such as hydroxide, amine, or cyanide rapidly attacks a light-activated aromatic ring but not its normal ground state (Crosby et al., 1972).

$$Cl-\langle\ \rangle-OR \xrightarrow{UV} Cl-\langle\ +\rangle-OR \xrightarrow{OH^-} HO-\langle\ \rangle-OR + Cl^- \quad (5.34)$$

Like carboxylic esters, phosphate esters undergo alkaline hydrolysis, the hydroxide ion attacking an electropositive P atom (Section 15.8.1). The hydrolysis rate of phosphorothionates, $(RO)_2P(S)–OR'$, a class of insecticides (Section 15.4), is slowed due to the lower electronegativity of sulfur but accelerated by coordination of the S atom with cupric ions, which increases the positivity of the phosphorus.

Hydrolysis under acidic or neutral conditions uses a different mechanism, where the electronegative oxygen of water is attracted to the positive carbonyl followed by proton transfer and loss of alkoxide. This is the most common of 8 known mechanisms of acid hydrolysis of esters (March, 1992), but rates are so low as to have little environmental significance (Section 5). Mabey and Mill (1978) provide a critical review of environmental hydrolytic reactions.

5.5.3. Hydrolysis in Soil

Hydrolysis reactions are common in moist soils. Although the soil organic fraction participates, hydrolyses also are catalyzed by clay minerals, as the acid strength of a clay surface can rival that of sulfuric acid. The hydrolytic replacement of Cl by OH in chlorotriazine herbicides, and the cleavage of organophosphate insecticides, are catalyzed by clay minerals (Wolfe et al., 1989). The presence of free esterases in soil probably contributes, too.

Although UV radiation penetrates less than 1 mm into soil, it energizes chemical transformations on the surface (Miller et al., 1989). However, the top few millimeters of soil are different from the rest. Solid, solution, and vapor phases are close together and subject to intense solar radiation, temperatures exceeding 55°C, and continuous introduction of contaminants from atmospheric fallout or intentional applications. Consequently, some degradation reactions at the soil surface form products other than those found in aqueous solution. For example, photolysis of the herbicide bentazon produces a nitroso derivative not observed otherwise.

5.6.1. Alkylation

Alkaline hydrolysis extends to alkyl halides (Section 5.5.2). Here, it can be viewed as an ***electrophilic*** reaction in which a positive charge generated on an alkyl halide reacts with nucleophilic hydroxide ion to "alkylate" it (Equation 5.35). Any electron-rich atom or group can be alkylated, and among common nucleophiles, the order of reactivity is roughly $RS^->CN^->I^->OH^->RNH_2>Br^->Cl^->H_2O$ (Edwards and Pearson, 1962).

$$H-\underset{\underset{Br}{|}}{\overset{\overset{H}{|}}{C}}-\underset{\underset{H}{|}}{\overset{Br}{C}}-H \longrightarrow H-\underset{\underset{Br}{|}}{\overset{\overset{H}{|}}{C}}-\underset{\underset{H}{|}}{\overset{OH}{C}}-H + Br^- \qquad (5.35)$$

$$^-:OH$$

Actually, any aliphatic molecule that can develop a positive center can be considered an alkylating agent, including sulfate and phosphate esters, certain reactive aromatic halides, and epoxides. Ethylene oxide was used extensively at one time for sterilizing packaged fruit but was found to react with chloride ions in the food to generate toxic 2–chloroethanol. Electrophiles have become very important in toxicology and will be discussed in detail later. More on alkylating agents is to be found in Section 15.2.

5.6.2. β-Elimination

Elimination reactions somewhat resemble hydrolysis and, as in the case of EDB, may occur at the same time. They require an hydrogen on the atom adjacent ("β") to one bearing a leaving group, such as halogen, which is removable by base:

$$-\underset{\underset{H}{|}}{\overset{|}{C}}-\overset{X}{\underset{|}{C}}- \longrightarrow BH + \overset{\backslash}{\underset{/}{C}}=\overset{/}{\underset{\backslash}{C}} + X^- \qquad (5.36)$$

$$:B^-$$

A frequently encountered example is the dehydrohalogenation of DDT and other chlorinated hydrocarbons. DDE is formed from DDT in soil and water under even mild alkaline conditions (Fig. 14.1), helping to account for its prevalence in the environment. As eliminations are *trans,* the stereochemistry of cyclic organohalogen compounds is important (Section 14.3.1), γ-hexachlorocyclohexane (lindane) being readily dehydrochlorinated in the environment while the β-isomer is not. Similar eliminations generate a nitrile from an oxime ester such as the insecticide, aldicarb ($-CH=N-OR$ rather than $-CHCHX$), or methyl isothiocyanate from sodium *N*-methyldithiocarbamate (metam-sodium) which contains an $-NH-C(S)-SH$.

5.6.3. Chlorination

Elemental chlorine is a common biocide used for disinfection of swimming pools and public water supplies, algae control in industrial cooling towers, and control of aquatic animals and algae in power plant water intakes (Special Topic 13). As with the hypochlorite it generates in water, chlorine reacts rapidly with phenols to generate foul-tasting chlorophenols. The reaction of chlorine with natural humic acids in water to give chloroform is now widely recognized (Section 2.4.2) and involves halogenation of aliphatic carbonyl components followed by hydrolysis of the resulting trichloromethyl compounds (R = humic acid):

$$RCOCH_3 \xrightarrow{3Cl_2} RCOCCl_3 \xrightarrow{OH^-} RCOOH + CHCl_3 \qquad (5.37)$$

Chlorination of ammonia and amines produces volatile and very toxic N-chloramines. The parent compound, $ClNH_2$, has caused numerous household poisonings after someone mixes bleach (sodium hypochlorite) with cleaning ammonia in the toilet bowl, puts down the lid, and then receives a toxic dose of chloramine when the cover is lifted again. The chlorination of aqueous solutions of certain amino acids, such as aspartic acid, methionine, or tryptophan, produces the corresponding N,N-dichloramine, which decomposes to mutagenic and carcinogenic dichloroacetonitrile. While yields from pure amino acids are <10% (Trehy and Bieber, 1981), chlorination of algae-containing water for disinfection leads to much larger amounts (Oliver, 1983).

5.7 | SUMMARY OF ABIOTIC REACTIONS

Oxidation, Reduction, Hydrolysis, Elimination

Although we have considered each of these key reaction types separately, they operate simultaneously and continuously in our environment. A dilute aqueous solution of pentachlorophenol in outdoor sunlight soon forms degradation products representing reduction (tri- and tetrachlorophenols), hydrolysis (dihydric phenols), and oxidation (quinones and acids). Within a few days, the carbons have been converted to CO_2 and the chlorines to chloride ions (Wong and Crosby, 1981)—the end products of mineralization.

Despite the hundreds of reactions one is supposed to learn during a course in organic chemistry, only a few types actually are common in the environment (Appendix 5.4). Oxidations, reductions, hydrolyses, photonucleophilic substitutions, and a few others predominate. Their rates are highly dependent on both chemical structure and environmental conditions, but reductions often proceed the most easily, followed perhaps by base-catalyzed eliminations, then hydrolysis and oxidation. Photochemical reactions can be rapid where the UV radiation is intense. The principal abiotic transformations are roughly prioritized in Appendix 5.4, and specific practical examples are described in later chapters. Note that these are all organic reactions; abiotic inorganic reactions are considered in Chapter 11.

5.8 │ REFERENCES

Baker, K. S., and R. C. Smith. 1982. Spectral irradiance penetration in natural waters, in *The Role of Solar Ultraviolet Radiation in Marine Ecosystems* (J. Calkins, ed.), Plenum Press, New York, NY, pp. 233–46.

Calvert, J. G., and J. N. Pitts, Jr. 1966. *Photochemistry,* John Wiley & Sons, New York

Clark, W. M. 1960. *Oxidation–reduction Potentials of Organic Systems,* Williams and Wilkins, Baltimore, MD.

Crosby, D. G. 1976. Nonbiological degradation of herbicides in the soil, in *Herbicides: Physiology, Biochemistry, Ecology* (L. J. Audus, ed), 2nd Ed., Academic Press, New York, NY., Vol. 2, pp. 65–97.

Crosby, D. G. 1983. Atmospheric reactions of pesticides. In *Pesticide Chemistry: Human Welfare and the Environment* (J. Miyamoto and P. C. Kearney, eds.), Pergamon Press, Oxford, U.K., pp. 327–32.

Crosby, D. G., K. W. Moilanen, M. Nakagawa, and A. S. Wong. 1972. Photonucleophilic reactions of pesticides. in *Environmental Toxicology of Pesticides* (F. Matsumura, G. M. Boush, and T. Misato, eds.), Academic Press, New York, NY.

Crosby, D. G., and H. O. Tutass. 1966 photodecomposition of 2,4-dichlorophenoxyacetic acid. *J. Agr. Food Chem.* **14**: 596–99

Dorfman, L. M., and G. E. Adams. 1973. *Reactivity of the Hydroxyl Radical in Aqueous Solutions,* National Bureau of Standards, Washington, DC, Publ. NSRDS-NBS-46.

Dulin, D., and T. Mill. 1982. Development and evaluation of sunlight actinometers. *Environ. Sci. Technol.* **16**: 815–20.

Edwards, J. O., and R. G. Pearson. 1962. The factors determining nucleophilic reactivities. *J. Amer. Chem. Soc.* **84**: 16–24.

Finlayson-Pitts, B. J. 1983. Reaction of NO_2 with NaCl and atmospheric implications of NOCl formation. *Nature* **306**: 676.

Finlayson-Pitts, B. J., and J. N. Pitts, Jr. 1986. *Atmospheric Chemistry: Fundamentals and Experimental Techniques,* Wiley-Interscience, New York, NY.

Franklin, J. 1993. The atmospheric degradation and impact of 1,1,1,2-tetrafluoroethane (Hydrofluorocarbon 134a). *Chemosphere* **27**: 1565–601.

Hart, E. J., and M. Anbar. 1970. *The Hydrated Electron,* Wiley-Interscience, New York, NY.

Helz, G. R., R. G. Zepp, and D. G. Crosby. 1994. *Aquatic and Surface Photochemistry,* Lewis Publishers, Boca Raton, FL.

Hiberty, P. C. 1983. The diradical character of 1,3-dipoles. *Israel J. Chem.* **23**: 10–20.

Lide, D. R. 1992. *Handbook of Chemistry and Physics, 1992–1993,* 73rd Edition, CRC Press, Boca Raton, FL, pp. 9–129 to 9–137.

Mabey, W., and T. Mill. 1978. Critical review of hydrolysis of organic compounds in water under environmental conditions. *J. Phys. Ref. Data* **7**: 383–415.

Mabury, S. A., and D. G. Crosby. 1994. The relationship of hydroxyl reactivity to pesticide persistence, in *Aquatic and Surface Photochemistry* (G.R. Helz, R.G. Zepp, and D. G. Crosby, eds.). Lewis Publishers, Boca Raton, FL, pp. 149–61.

Macalady, D. L., P. Tratnyek, and T.J. ф. Grundl. 1986. Abiotic reduction reactions of anthropogenic organic chemicals in anaerobic systems: A critical review. *J. Contam. Hydrology* **1**: 1–28.

March, J. 1992. *Advanced Organic Chemistry: Reactions, Mechanisms, and Structure,* 4th ed., McGraw-Hill, New York, NY.

McRae, G. J., and J. H. Seinfeld. 1983. Development of a second generation mathematical model for urban air pollution. II. Evaluation of model performance. *Atmos. Environ.* **17**: 501–22.

Miller, G. C., V. R. Hebert, and W. W. Miller. 1989. Effect of sunlight on organic contaminants at the atmosphere–soil surface. in *Reactions and Movement of Organic Chemicals in Soils* (B. L. Sawhney and K. Brown, eds.), Soil Science Society of America, Madison, WI, pp. 99–110.

Miskus, R. P., D. P. Blair, and J. E. Casida. 1965. Conversion of DDT to DDD by bovine rumen fluid, lake water, and reduced porphyrins. *J. Ag. Food Chem.* **13:** 481–483.

Montzka, S. A., J. H. Butler, R. C. Meyers, T. M. Thompson, T. H. Swanson, A. D. Clarke, L. T. Lock, and J. W. Elkins. 1996. Decline in the tropospheric abundance of halogen from halocarbons: Implications for stratospheric ozone depletion. *Science* **272:** 1318–22.

Nakagawa, M., and D. G. Crosby. 1974. Photodecomposition of nitrofen. *J. Agr. Food Chem.* **22:** 849–53.

Oliver, B. G. 1983. Dihaloacetonitriles in drinkng water: Algae and fulvic as precursors. *Environ. Sci. Technol.* **17:** 80–83.

Schwarzenbach, R. P., P. M. Gschwend and D. M. Imboden. 1993. *Environmental Organic Chemistry,* John Wiley & Sons, New York, NY.

Stumm, W., and J. J. Morgan. 1981. *Aquatic Chemistry,* Wiley-Interscience, New York, NY.

Swallow, A. J. 1969. Hydrated electrons in seawater. *Nature* **222:** 369–70.

Trehy, M. L., and T. L. Bieber. 1981. Detection, identification, and quantitative analysis of dihaloacetonitriles in chlorinated natural water, in *Advances in the Identification and Analysis of Organic Pollutants in Water* (L. H. Keith, ed.), Ann Arbor Science Publishers, Ann Arbor, MI, 1981, pp. 941–75.

Vogel, T. M., C. S. Criddle, and P. L. McCarty. 1987. Transformations of halogenated aliphatic compounds. *Environ. Sci. Technol.* **21:** 722–36.

Wolfe, N. L., B. E. Kitchens, D. L. Macalady, and T. J. Grundl. 1986. Physical and chemical factors that influence the anaerobic degradation of methyl parathion in sediment systems. *Environ. Toxicol. Chem.* **5:** 1019–26.

Wolfe, N. L., M. E. -S. Metwally, and A. E. Moftah. 1989. Hydrolytic transformations of organic chemicals in the environment, in *Reactions and Movement of Organic Chemicals in Soils* (B. L. Sawhney and K. Brown, eds.), Soil Science Society of America, Madison, WI, pp. 229–42.

Wong, A. S., and D. G. Crosby. 1981. Photodecomposition of pentachlorophenol in water. *J. Agr. Food Chem.* **29:** 125–30.

Zepp, R. G., N. L. Wolfe, J. A. Gordon, and G. L. Baughman. 1975. Dynamics of 2,4-D esters in surface waters. *Environ. Sci. Technol.* **9:** 1144–50.

Special Topic 5. Free Radicals and the Ozone Layer

Any child who has played with a toy electric train is familiar with the acrid, musty odor of ozone. The blue gas, O_3, is generated by reaction of molecular oxygen with oxygen atoms generated in an electric discharge, so its odor also is recognizable near power-generating equipment and lightning storms. It is a toxic component of photochemical smog (Section 5.3), generated in the troposphere by a series of reactions involving nitrogen oxides, oxygen atoms, and molecular oxygen (Equations 5.8–5.9).

Ozone can exist in many resonance forms:

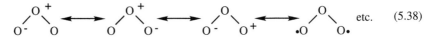

$$\text{(5.38)}$$

allowing it to serve as an electrophile, nucleophile, or free radical (Hiberty, 1983). Many other atmospheric pollutants are reactive free radicals (triplet molecular oxygen being a notable exception, Section 5.3.1), and ozone reacts with nitric oxide ($\cdot N = O$), nitrogen dioxide ($\cdot NO_2$), atomic oxygen ($\cdot O \cdot$), chlorine or bromine atoms to release molecular oxygen:

$$\cdot O - O - O \cdot + \cdot X \longrightarrow \cdot OX + \cdot O - O \cdot \tag{5.39}$$

A thin layer of **stratospheric ozone** surrounds the earth at an altitude of 30–40 km (18–24 miles) above the surface. On the outer edge of the stratosphere, the intense ultraviolet radiation of the sun dissociates oxygen into atoms which then combine with more O_2 to form O_3 (Eq. 5.9). This ozone diffuses downward until, somewhere between 20 and 30 km (12–18 miles), its generation is balanced by reactions with tropospheric gases. This "ozone layer," if condensed to a liquid, would be only 3 mm thick, but ozone's UV absorption is so great that this thin barrier effectively blocks the most energetic solar radiation, below 295 nm, that would literally "fry" all life on the planet.

The natural tropospheric constituents include halides such as methyl chloride and bromide (Section 2.3) and the rather unreactive nitrous oxide, N_2O. A small proportion of these gases becomes photolyzed at the ozone layer, and the resulting halogen atoms or NO react according to Eq. 5.39 to generate ClO, BrO, and ClONO at the expense of O_3. This is a cyclic process (Scheme 5.40), so a small amount of reagent can destroy a lot of ozone.

In recent years, the production and use of chlorofluorocarbons (CFCs) for refrigerants, aerosol propellants, and foamed plastics have grown rapidly, as is the case with many other halogenated organics. The Freons®, which are chlorofluorocarbons such as 1,2-dichloro-1,1,2,2-tetrafluoroethane, are especially resistant to tropospheric breakdown and readily diffuse into the stratosphere where they dissociate photochemically to provide chlorine atoms.

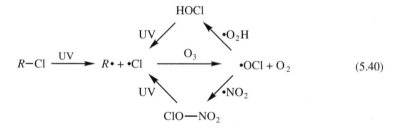

$$\tag{5.40}$$

Humans have been able to tilt the age-old ozone equilibrium, with the result that its protective screen is thinner. Initially, the thinning was observed only where chlorinated chemicals concentrate at the poles (Section 4.6), but later measurements revealed as much as 10% ozone loss even near the equator.

The industrialized nations now have agreed to reduce their use of the worst CFC offenders and eventually to cease it altogether. An oxidizable hydrofluorocarbon, 1,1,1,2-tetrafluoroethane (HFC 134a), is being substituted (Eq. 5.41), but its stable breakdown product, trifluoroacetic acid (TFA), is falling to earth in rain (Franklin, 1993). The law of conservation of mass should long ago have told us that nothing disappears.

$$CF_3CH_2F \xrightarrow{HO\cdot} CF_3CHF\cdot \xrightarrow{O_2} CF_3CHO_2\cdot \xrightarrow[O_2]{NO} CF_3CO-F \xrightarrow{H_2O} CF_3COOH \tag{5.41}$$

HFC 134a   TFA

As of early 1996, the good news is that atmospheric concentrations of several of the most prevalent organochlorine compounds such as methylchloroform and carbon tetrachloride have started to decline (Montzka et al., 1996). The bad news is that the levels of newer HFC replacements are rising sharply. This should tell us that the battle for a stabilized ozone layer is not over yet.

APPENDIX 5.1.
Bond Dissociation Energies

Bond	D (kcal/mole[a]	D (kJ/mole)	λ (nm)[b]
C_6H_5–H	111	464	258
CH_3O–H	104	437	275
$CH_3CH_2CH_2$–H	100	420	286
$(CH_3)_2CH$–H	96	401	298
$(CH_3)_3C$–H	93	390	307
CH_3–CH_3	90	376	318
CH_3–SCH_3	77	323	371
CCl_3–Cl	73	306	392
O=N–O	73	306	392
CH_3–$As(CH_3)_2$	67	280	427
CH_3–$HgCH_3$	61	255	469
Cl–Cl	59	243	485
CH_3CH_2–$Pb(C_2H_5)_3$	55	230	520
HO–OH	51	213	561
HO–NO	49	206	583
CH_3O–OCH_3	38	157	752

[a] Lide, 1992, D = dissociation energy.
[b] Radiant energy corresponding to D.

APPENDIX 5.2.
Some Major Atmospheric Reactions[a]

A. Alkanes

1. $RCH_2CH_3 \xrightarrow{HO\bullet} RCH_2CH_2\bullet \xrightarrow{O_2} RCH_2CH_2O_2\bullet \xrightarrow{RH} RCH_2CH_2OOH \xrightarrow{-H_2O} RCH_2CH=O$

B. Aromatics

2.

C. Olefins

3. $RCH=CH_3 \xrightarrow{O_3}$ $\xrightarrow{H_2O} RCH=O + O=CHCH_3$

4. $RCH=CH_3 \xrightarrow{R'OO\bullet} RCH-CH_3$

D. Organohalogens

5. $CH_3Br \xrightarrow{HO\bullet} \bullet CH_2Br \xrightarrow{O_2} \bullet OOCH_2Br \xrightarrow{NO} \bullet OCH_2Br \xrightarrow{O_2} O=CHBr$ (Troposphere)

6. $CH_3Br \xrightarrow[-\bullet CH_3]{UV} \bullet Br \xrightarrow{O_3} \bullet BrO \xrightarrow{NO_2} BrONO_2 \xrightarrow{UV} \bullet Br$ (Stratosphere)

E. Inorganics:

7. $H_2S \xrightarrow{HO\bullet} SO_2 \xrightarrow{HO\bullet} [HOSO_2] \xrightarrow{O_2} SO_3 \xrightarrow{H_2O} H_2SO_4$

8. $NO_2 + HO\bullet \longrightarrow HNO_3$

[a] Finlayson-Pitts and Pitts, 1986.

APPENDIX 5.3.
Some Standard Redox Potentials of Environmental Interest[a]

Half-Reactions[b]	$E°$ V[c]	$E°$(W), V[d]
$RS\ SR + 2H^+ + 2e^- \rightleftharpoons 2RSH$	-0.360	-0.390
$Hg_2Cl_2 + 2e- \rightleftharpoons 2Hg + 2Cl^-$	-0.268	–
$2H^+ + 2e^- \rightleftharpoons H_{2(g)}$	0.000	–
$S_{(s)} + 2H^+ + 2e- \rightleftharpoons H_2S$	-0.410	-0.240
Fe^{3+} (PP)[f] $+ e^- \rightleftharpoons Fe^{2+}$ (PP)	+0.170	+0.230
$Cu^{2+} + 2e^- \rightleftharpoons Cu$	+0.337	–
$Fe^{2+} + 2e^- \rightleftharpoons Fe$	+0.440	–
$C_6H_5NO_2 + 6H^+ + 6e^- \rightleftharpoons C_6H_5NH_2 + 2H_2O$	+0.680	+0.420
$CHCl_3 + H^+ + e^- \rightleftharpoons CH_2Cl_2 + Cl^-$	–	+0.560
$NO_3^- + 2H^+ + 2e^- \rightleftharpoons NO_2^- + H_2O$	+0.830	+0.420
$NO_3^- + 10H^+ + 8e^- \rightleftharpoons NH_4^+ + 3H_2O$	+0.880	+0.360
$Cl_2 + 2e^- \rightleftharpoons 2Cl^-$	+1.360	–
$O_3 + 2H^- + 2e^- \rightleftharpoons O_2 + H_2O$	+2.070	–

[a] Data from Clark, 1960; Stumm and Morgan, 1981; and Schwarzenbach et al., 1993.
[b] 1 M aqueous solution, 25°C, pH 7, gases at 1 atm.
[c] Standard reduction potential, H electrode.
[d] $Cl^- = 10^{-3}$ M, $Br^- = 10^{-5}$ M.
[e] Cystine, $R = -CH_2CH(NH_2)COOH$.
[f] PP = protoporphyrin IX complex, PH 9.

APPENDIX 5.4.
Summary of Abiotic Transformations

A. Reduction

    1. Nitro $-NO_2 \longrightarrow -NH_2$

    2. Aliphatic halide $C-Cl \longrightarrow C-H$

    3. Carbonyl $C=O \longrightarrow CH_2OH$

B. Elimination $CH-Cl \longrightarrow C=C$

C. Hydrolysis

    1. Epoxide

$$\overset{\displaystyle O}{C-C} \longrightarrow \overset{\displaystyle OH \quad OH}{\underset{}{C-C}}$$

    2. Ester (carboxylic) $-COOR \longrightarrow -COOH + ROH$

    3. Ester (phosphate) $P(O,S)\text{-}OR \longrightarrow P(O,S)\text{-}OH + ROH$

    4. Amide $CONHR \longrightarrow COOH + H_2NR$

D. Oxidation

    1. N-Oxidation $R_3N \longrightarrow R_3NO$

    2. Sulfoxidation $-S- \longrightarrow SO \longrightarrow OSO$

    3. Desulfuration $PS \longrightarrow PO$

    4. N-Dealkylation $NCH_3 \longrightarrow NCH_2OH \longrightarrow NH + HCHO$

    5. O-Dealkylation $-OCH_3 \longrightarrow -OCH_2OH \longrightarrow -OH + HCHO$

    6. Epoxidation $C=C \longrightarrow \overset{\displaystyle O}{C-C}$

    7. Aliphatic C-hydroxylation $C-H \longrightarrow C-OH \longrightarrow C=O$

    8. Aromatic hydroxylation

$$\bigcirc \longrightarrow \bigcirc\text{-OH}$$

E. Rearrangement $\overset{\displaystyle O}{C-C} \longrightarrow C-C=O$

F. Photonucleophilic Substitution

    1. Nitro hydrolysis $\bigcirc\text{-}NO_2 \longrightarrow \bigcirc\text{-OH}$

    2. Halo hydrolysis $\bigcirc\text{-}X \longrightarrow \bigcirc\text{-OH}$

Biotransformation processes are as common as life itself. All kinds of living organisms, from bacteria to humans, are able to convert organic chemicals—and many inorganic ones—to other substances. The term **biodegradation** is widely applied to this process, and especially to microbial transformations, but it suggests a breakdown into smaller fragments which is not always the case. The term **metabolism** also is used frequently, but as it normally refers to the transformations of natural substrates necessary for life, the more cumbersome **xenobiotic metabolism** must be specified. **Biotransformation** seems to be the most generally satisfactory term.

Xenobiotic metabolism is divided into **primary (Phase I)** and **secondary (Phase II) metabolism** (Section 7.3). Primary metabolism refers to biotransformations that alter basic chemical structure, as in the stepwise oxidation of the methyl group of toluene to benzyl alcohol, benzaldehyde, and benzoic acid. Phase II metabolism, also called **conjugation,** involves modification of existing reactive functional groups, such as conversion of the benzoic acid to benzoylglycine. Benzoylglycine (hippuric acid), the first recognized **metabolite,** was isolated about 1840 by German chemist Wilhelm Keller from his own urine after he had swallowed several grams of benzoic acid. An up-to-date discussion of biochemical mechanisms in primary and secondary metabolism is provided by Josephy (1997).

Our world would be very different without biotransformation. Just as the vapor-phase oxidation of volatilized chemicals cleans the atmosphere, and photodegradation destroys water pollutants, biotransformations are the primary self-cleaning mechanism for terrestrial environments. Biotransformation allows living organisms to make environmental poisons less toxic, more readily excrete unwanted chemicals, and recycle nutrients. The process must have developed quite early in evolution, certainly before the appearance of the first heterotrophs requiring protection from natural toxic constituents of their food. Organisms that could not detoxify natural toxicants soon must have made way for those who could.

Like abiotic environmental reactions, biotransformations occur continuously and simultaneously. A sample of urine from a person exposed to traces of the insecticide parathion contains the Phase I metabolites aminoparathion from nitro reduction, para-oxon from *S*-oxidation (Eq. 6.2), and *p*-nitrophenol from ester hydrolysis. Phase II products are the *N*-acetyl derivative of the amine (Eq. 6.1.) and glucuronide and sulfate conjugates of the phenol (Eq. 6.3), all formed enzymatically (Table 6.1).

As in abiotic reactions, nitro reductions often are accomplished by hydride transfer, but now the donor is FAD or NADPH rather than a thiol or glucose. Biochemical oxidations (Eq. 6.2) generally utilize the familiar oxidants, perferryl (FeO^{3+}) or percupryl (CuO_2^{2+}) ions, and hydrolysis is catalyzed by a protein rather than a mineral base or acid. Synthesis (conjugation) involves high-energy reagents just as in the laboratory: amine acetylation by acetyl Coenzyme A, the biochemical equivalent of the chemist's acetyl chloride; methylation with the methylcarbonium ion generator, *S*-adenosylmethionine (SAM); and condensations of sugars or sulfate (*R* in Eq. 6.3) by displacement of a pyrophosphate leaving group by a nucleophile.

TABLE 6.1
Major Biotransformation Reactions

Type	Reaction[a]	Typical enzyme	Cofactor	Reactant
Oxidation	$C-C \rightarrow C-C$ (with epoxide O)	Aldrin epoxidase	NADPH	**FeO_3^+**
Reduction	$-NO_2 \longrightarrow -NH_2$	Nitro reductase	NADPH	**H** ... N-
Dehydrochlorination	$C-C \rightarrow C=C$ (H Cl)	DDT dehydrochlorinase	GSH	**HS-CH$_2$CH**
Hydrolysis	$-COOR \longrightarrow -COOH$	Lipase	Serine	**HO-CH$_2$ CH**
Acylation	$-NH_2 \longrightarrow -NHCOR$	Arylamine acyl-transferase	Acyl CoA	$-CH_2$**S-COR**
Methylation	$-OH \longrightarrow -OCH_3$	*O*-Methyltransferase	SAM[b]	**CH$_3$**$\overset{+}{S}$**-CH$_2$-** (CH$_3$)
Glycosidation	$R\text{-}OH \longrightarrow R\text{-}OX$	UDP transglucuronylase	UDPGA	**X-O-PPU**
Sulfation	$R\text{-}OH \longrightarrow R\text{-}OSO_3^-$	Sulfotransferase	PAPS[c]	**O-S-OPA** (O=S=O)

[a] GSH = glutathione, P = phosphate, X = sugar, U = uridine.
[b] SAM = S-adenosylmethionine.
[c] PAPS = 3'-phosphoadenosine-5'-phosphosulfate, A = adenosyl.

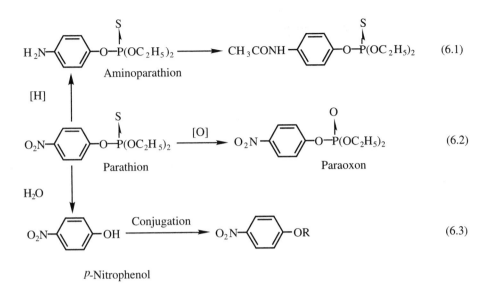

The principal reactions and reagents are summarized in Table 6.1. Expressing their common origin, the same basic biochemical processes of metabolism are found in microorganisms, flowering plants, and animals, although each exhibits its own individuality.

Metabolic reactions are primarily enzymatic, the cofactor (Table 6.1) being responsible for basic chemistry and the protein part for specificity. Excellent reviews of the enzymology of both primary and secondary metabolism are presented by Jakoby (1980) for animals and Hatzios and Penner (1982) for plants. However, a number of common-sense restrictions apply (Alexander, 1965). In 1966, microbiologist Martin Alexander "revealed" the Ten Commandments of Biodegradation (Alexander, 1966), which included the following (paraphrased):

    I. 𝕿𝖍𝖔𝖚 𝖘𝖍𝖆𝖑𝖙 𝖓𝖔𝖙 𝖉𝖊𝖌𝖗𝖆𝖉𝖊 𝖙𝖍𝖆𝖙 𝖋𝖔𝖗 𝖜𝖍𝖎𝖈𝖍 𝖙𝖍𝖔𝖚 𝖍𝖆𝖘𝖙 𝖓𝖔 𝖆𝖕𝖕𝖊𝖙𝖎𝖙𝖊 (there must be an enzymatic means of degradation available);

    II. 𝕿𝖍𝖔𝖚 𝖘𝖍𝖆𝖑𝖙 𝖓𝖔𝖙 𝖉𝖊𝖌𝖗𝖆𝖉𝖊 𝖙𝖍𝖆𝖙 𝖜𝖍𝖎𝖈𝖍 𝖙𝖍𝖔𝖚 𝖈𝖆𝖓𝖘𝖙 𝖓𝖔𝖙 𝖙𝖆𝖐𝖊 𝖎𝖓𝖙𝖔 𝖙𝖍𝖞 𝖇𝖊𝖑𝖑𝖞 (the substance must be absorbed or solubilized);

    III. 𝕿𝖍𝖔𝖚 𝖘𝖍𝖆𝖑𝖙 𝖓𝖔𝖙 𝖉𝖊𝖌𝖗𝖆𝖉𝖊 𝖙𝖍𝖆𝖙 𝖜𝖍𝖎𝖈𝖍 𝖜𝖎𝖑𝖑 𝖉𝖔 𝖙𝖍𝖊𝖊 𝖎𝖗𝖗𝖊𝖕𝖆𝖗𝖆𝖇𝖑𝖊 𝖍𝖆𝖗𝖒 (concentrations and toxicities must not be too great);

    IV. 𝕿𝖍𝖔𝖚 𝖘𝖍𝖆𝖑𝖙 𝖓𝖔𝖙 𝖉𝖊𝖌𝖗𝖆𝖉𝖊 𝖙𝖍𝖆𝖙 𝖜𝖍𝖎𝖈𝖍 𝖎𝖘 𝖊𝖓𝖙𝖎𝖗𝖊𝖑𝖞 𝖎𝖓𝖆𝖈𝖈𝖊𝖘𝖘𝖎𝖇𝖑𝖊 𝖙𝖔 𝖙𝖍𝖊𝖊 (the substance must not be strongly adsorbed to a surface or intercalated into other molecules);

    V. 𝕿𝖍𝖔𝖚 𝖘𝖍𝖆𝖑𝖙 𝖓𝖔𝖙 𝖉𝖊𝖌𝖗𝖆𝖉𝖊 𝖙𝖍𝖆𝖙 𝖜𝖍𝖎𝖈𝖍 𝖔𝖈𝖈𝖚𝖗𝖘 𝖎𝖓 𝖕𝖑𝖆𝖈𝖊𝖘 𝖉𝖊𝖋𝖎𝖈𝖎𝖊𝖓𝖙 𝖎𝖓 𝖆 𝖓𝖊𝖈𝖊𝖘𝖘𝖆𝖗𝖞 𝖗𝖊𝖆𝖌𝖊𝖓𝖙 (for example, if the reaction requires oxygen but the environment is anaerobic).

Although intended for microbial degradations, these commandments serve animals and higher plants equally well. However, as microbial biotransformations are both ubiquitous and representative of many in higher species also, we will start with them.

6.2 | TRANSFORMATIONS BY MICROORGANISMS

6.2.1. Microorganisms

Microorganisms are the principal biotic force in the transformation of chemicals. They include bacteria, actinomycetes, protozoans, fungi, yeasts, and algae (although the latter behave biochemically more like higher plants). Bacteria are the most numerous and biochemically active of all. They form single cells or colonies and are found everywhere—in soil and water, on plants, and in all kinds of animals. A single bacterial cell weighs only about 1 pg (10^{-12} g), but as one m^2 of grassland soil may contain as many as 10^{15} of these tiny degradation machines, their biomass can be significant. Species include the common water polluter, *Escherichia coli,* and powerful oxidizers of the genus *Pseudomonas.* They also include moldlike Actinomycetes such as *Streptomyces.*

Probably the most familiar fungi are the Basidomycetes, whose fruiting bodies we recognize as mushrooms. However, the metabolically most active belong to the class Phycomycetes (molds). Like most bacteria, the fungi do not photosynthesize, and, unlike higher plants, they produce no leaves, flowers, or seeds. In general, their cells form the long white filaments (hyphae) often seen spreading over rotting logs. A single gram of soil may contain 20 million fungal cells and 5 m of hyphae.

Most microorganisms are aerobic. Aerobes utilize oxygen to terminate their electron transport process, as do we (Eq. 6.4), and they are the oxidizers. Anaerobes, many of which cannot tolerate any oxygen, utilize sulfur, sulfate, nitrate, or bicarbonate instead (Eq. 6.5–6.8), and they are the reducers.

$$O_2 + 4H^+ + 4e^- \longrightarrow 2H_2O \qquad\qquad E_0 +1.22 \text{ V} \qquad (6.4)$$

$$NO_3^- + 10H^+ + 8e^- \longrightarrow NH_4^+ + 3H_2O \qquad\qquad E_0 +0.88 \text{ V} \qquad (6.5)$$

$$SO_4^{2-} + 10H^+ + 7e^- \longrightarrow H_2S + 4H_2O \qquad\qquad E_0 +0.25 \text{ V} \qquad (6.6)$$

$$S + 2H^+ + 2e^- \longrightarrow H_2S \qquad\qquad E_0 +0.17 \text{ V} \qquad (6.7)$$

$$HCO_3^- + 9H^+ + 8e^- \longrightarrow CH_4 + 3H_2O \qquad\qquad E_0 +0.17 \text{ V} \qquad (6.8)$$

As one descends through a water-saturated soil, the substrates show ever-decreasing oxidation potentials, the microbial populations change, and the metabolic products include, in order, ammonia, hydrogen sulfide, and methane. Aerobes generally predominate in terrestrial environments and anaerobes in aquatic ones; microbial oxidation is the most important route biotransformations near the surface, while reduction predominates in submerged sediments and animal intestines. Let us consider each of these processes.

6.2.2. Oxidation

Natural underwater oil seeps are common in coastal environments, yet the beaches are not covered with oil. Why? Lower hydrocarbons volatilize into the atmosphere, but microbial oxidation removes most of the rest (Special Topic 6). Bio-oxidation is a form of burning, but one that takes place at ambient temperature and is enzyme-catalyzed. Like the atmospheric oxidations described in Section 5.3, the reaction normally starts at either carbon 1 or 2 of an aliphatic chain and in each case results in a carboxylic acid (Fig. 6.1).

Each acid then forms a thiol ester with Coenzyme A, which undergoes subsequent **β-oxidation** that cleaves off a two-carbon unit in the form of acetyl CoA, a basic energy source for the organism. Each new long-chain ester is again subjected to β-oxidation until only a final acetate unit remains. Oxidation at carbon 2 results in similar acids having one C atom fewer than in the previous case, and formic acid or CO_2 is the final product. The initial oxidations are catalyzed in bacteria by an iron–sulfur hydroxylase, but β-oxidation does not require oxygen.

β-Oxidation cannot operate past a side chain. In the early days of commercial

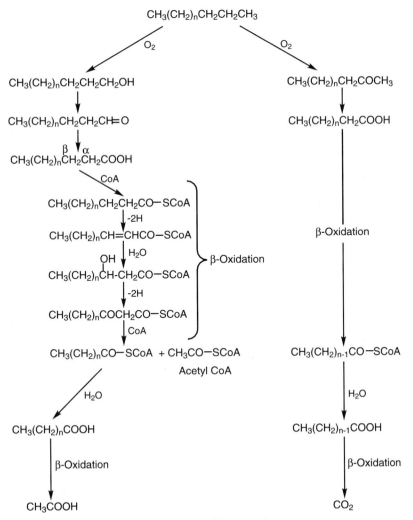

Figure 6.1. Microbial oxidation of alkanes. CoA = Coenzyme A.

detergents, branched alkyl benzenesulfonates were used widely because of their low cost. However, they proved to be virtually undegradable by bacteria, and the sight of thick foam on a river was not uncommon. Conversion to *n*-alkyl detergents solved the problem. Benzene rings are oxidized by bacteria by addition of atmospheric dioxygen across a double bond, with eventual formation of catechol (Eq. 6.9). Other aromatic compounds, such as benzoic acid, also are biotransformed to catechol by the same route. The initial reaction is catalyzed by two iron–sulfur proteins, and the necessary electrons are supplied by NADH via a flavoprotein reductase (Gibson and Subramanian, 1984). Even polycyclic aromatic hydrocarbons such as naphthalene, anthracene, and phenanthrene undergo oxidation to catechol (Fig. 6.2).

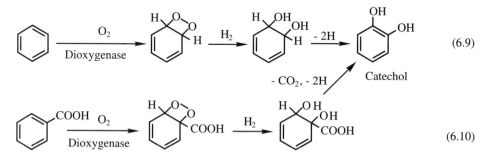

$$(6.9)$$

$$(6.10)$$

Substituted benzenes, C_6H_5R, are oxidized to R-substituted catechols at rates proportional to the electronegativity of the ring (oxidation is an electrophilic reaction). For example, the electron-rich rings of phenol ($R = OH$), aniline ($R = NH_2$), and anisole ($R = OCH_3$) are relatively nonpersistent in soil or water, while those possessing electron-withdrawing substituents such as sulfonate ($R = SO_3H$) or nitro ($R = NO_2$) are degraded much more slowly (Section 16.3). Benzoic acid, with its strongly electron-withdrawing carboxyl group, is the exception (Eq. 6.10). Even highly chlorinated compounds such as pentachlorophenol and polychlorinated biphenyls (PCBs) form the corresponding catechols (Reineke, 1984). *Meta*-substituted rings invariably are degraded more slowly than are the *o* and *p* isomers.

The catechol ring subsequently is cleaved by oxidation along either of two routes (Fig. 6.3). By the *ortho* pathway, oxidation between the hydroxyls results in an unsaturated diacid, one carboxyl of which adds across a double bond to form a lactone (cyclic ester) followed by isomerization, hydrolysis, and combination of the reactive β-ketoacid with Coenzyme A to generate acetyl CoA. The *meta* pathway involves oxidation adjacent to an hydroxyl, Michael addition of water, hydrolysis that releases the aldehyde function as formate, and reverse aldol cleavage of the remaining fragment to give acetaldehyde and pyruvate. This overall sequence of bacterial oxidations has great significance for modern-day humans, because ubiquitous environmental pollutants such as PAHs, PCBs, and those containing sulfur or nitrogen will be converted eventually to inorganic form as CO_2, Cl^-, H_2O, SO_4^{2-}, and N_2 by the process of *mineralization.*

An oil spill or natural seep provides enough hydrocarbon for microorganisms to utilize it for nutrition. However, even low levels of many pollutants are degraded rapidly and efficiently. In this case, the organism lives off other nutrient sources, and degradation of the xenobiotic is just incidental to normal metabolism. This phenomenon, called *cometabolism,* has been observed widely in microorganisms, plants, and animals and is carried out by the enzymes used for normal metabolism rather than by any special systems (Bollag and Liu, 1990).

For example, soil residues of the chlorinated herbicide 2,4-D persist for only a few days because of microbial degradation by the usual routes (Fig. 6.4). An application of 1 lb/acre of 2,4-D provides only about 100 mg of chemical per square meter of soil surface, or less than 0.0001% of the soil organic matter to a depth of 2 m. Once the side chain has been oxidized away, bacteria convert the resulting chlorophenol to catechols and eventually to Cl^-, CO_2, and succinate. However, fungi hydroxylate the ring by an initial epoxidation (Scheme 6.11) to form 5-hydroxy-2,4-D. An unexpected product, 4-hydroxy-2,5-D, results from epoxide rearrangement, the "NIH shift" (Faulkner and Woodcock, 1964), as illustrated by *p*-chlorophenoxyacetic acid (R = –OCH_2COOH):

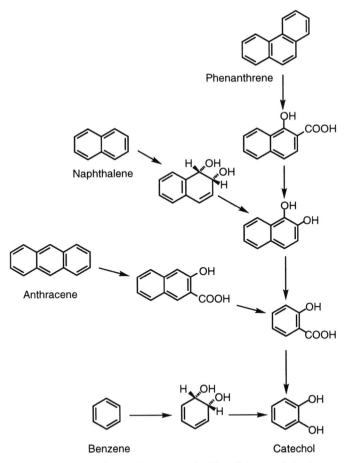

Figure 6.2. Microbial oxidation of aromatic compounds. Not all intermediate steps are shown. Ring oxidation by dioxygenases form *cis*-diols.

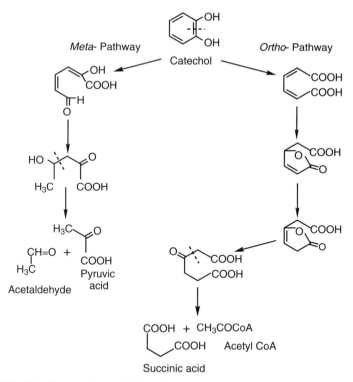

Figure 6.3.   Microbial cleavage of catechol by *ortho* and *meta* pathways.

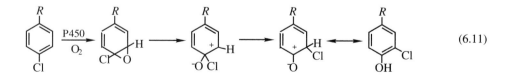

(6.11)

6.2.3. Cytochrome P450

The bacterial oxidations shown in Eqs. 6.9 and 6.10 are catalyzed by dioxy-genases, but other hydroxylations, epoxidations, and dealkylations rely primarily on mono-oxygenases. These enzymes are also called "mixed-function oxidases" (MFO), referring to the splitting of atmospheric dioxygen that donates one O atom to sub-strate and the other to water (Eq. 6.12). The most important mono-oxygenase system is *cytochrome P450,* an iron-containing enzyme whose designation arises from the UV absorption maximum at 450 nm in its reduced and CO-complexed form. The Fe is part of a heme molecule (Fig. 15.5), but with a cysteine-S replacing the imidazole N as ligand.

$$R-H + O-O + H^+ + NADPH \longrightarrow R-OH + H_2O + NADP^+$$ (6.12)

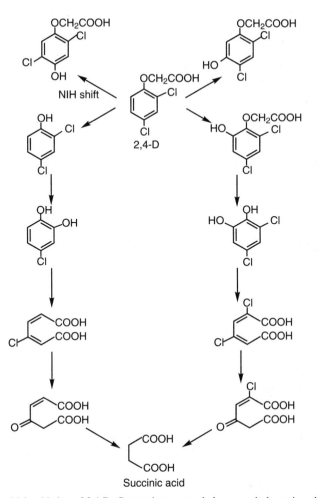

Figure 6.4.   Microbial oxidation of 2,4-D. Conversion to catechols occurs in bacteria, other ring hydroxylations in fungi.

P450 is coupled with NADPH-cytochrome P450 reductase as electron source and an iron–sulfur protein called a *redoxin*. Among the biochemical oxidations catalyzed by this system are aliphatic and aromatic C hydroxylations, epoxidation, dealkylation of N, S, and O compounds, deamination, sulfoxidation, and desulfuration of phosphorothionates to phosphates (Appendix 6.1). The heme coenzyme carries out the oxidation, and the protein apoenzyme determines specificity.

The chemical mechanism of P450 oxidations is shown in Scheme 6.13, the electrons supplied by NADPH. In the first step, ferric iron of the heme combines loosely with substrate RH, is reduced to the ferrous form (Step 2), which combines with triplet oxygen (Step 3), and the product is reduced again in Step 4. A bound oxygen radical is generated in Step 5, combines with substrate, and returns ferric iron (Step 7) to pick up another molecule of substrate.

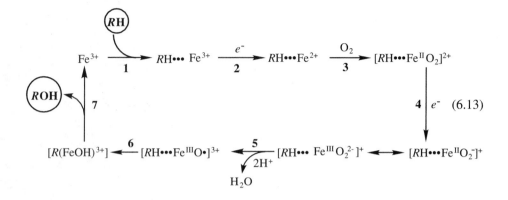

The key step 6, the rate-determining reaction of $(FeO)^{3+}$ with substrate, is still not entirely understood, but there is evidence for a radical mechanism (Eq. 6.14) for hydroxylation of an alkane (Guengerich and Macdonald, 1990).

$$-\overset{|}{\underset{|}{C}}-H + (FeO\bullet)^{3+} \longrightarrow \vdash \overset{|}{\underset{|}{C}}\bullet + (FeOH)^{3+}] \longrightarrow -\overset{|}{\underset{|}{C}}-OH + Fe^{3+} \quad (6.14)$$

Actually, cytochrome P450 is not a single entity but a broad group of isozymes (Section 6.3.1), dozens or even hundreds of them in bacteria alone (Sligar and Murray, 1986). They vary in their protein portion and substrate specificity but utilize identical oxidizing systems. The three-dimensional crystal structure of a bacterial P450 from *Pseudomonas putida,* determined by x-ray crystallography (Poulos, 1986), clearly shows the porphyrin ligand (Section 15.9) and the hydrophobic "pocket" into which the substrate fits while being oxidized.

Aromatic ring hydroxylations generally involve initial epoxidation of a ring double bond followed by either the 1,2-shift mentioned before or hydrolysis of the epoxide and subsequent dehydration of the dihydrodiol. This is illustrated by the conversion of naphthalene to naphthol and other products characteristic of epoxide reactions (Fig. 6.5). The epoxide, originally isolated and characterized by Jerina and Daly (1974) from mammalian cells, has been observed as a naphthalene metabolite in over 60 species of fungi as well as yeasts and algae, indicating the wide occurrence of mono-oxygenases. (Recall that bacteria utilize dioxygenases and produce catechols.) Polycyclic aromatic hydrocarbons such as benzo[a]pyrene are converted by fungi to the same carcinogenic diol-epoxides as are formed in mammals (Section 13.4).

Enzyme *induction* is an important feature of both prokaryotic and eukaryotic xenobiotic metabolism. Many enzymes, such as cytochrome P450, can be induced, meaning that their concentration can be made to increase. For example, the biotransformation of 2,4-D starts immediately in bacteria that earlier were grown on medium containing the herbicide, while those grown previously on glucose alone wait for 15 h before starting. Once prepared to do battle, such biotransformation processes can be mobilized again quickly at a later time.

The enzymes responsible for the bacterial transformations of alkanes, 2,4-D, naphthalene, and many other chemicals often are associated with *plasmids,* gene-

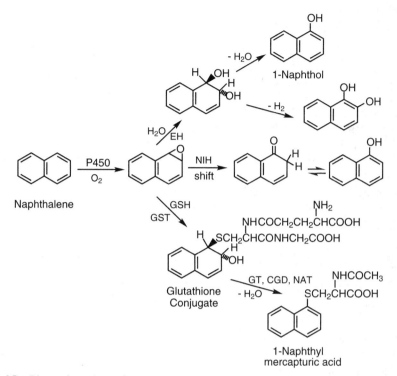

Figure 6.5. Biotransformations of naphthalene. Key enzymes include cytochrome P450, epoxide hydrolase (EH), glutathione *S*-transferase (GST), γ-glutamyl transferase (GT), cysteinylglycine dipeptidase (CGD), and *N*-acetyl transferase (AT).

carrying strands of DNA dispersed in the cytoplasm and transmitted between conspecific individuals and even between bacterial species during conjugation. For example, the gene for oxidation of naphthalene, encoded on the NAH plasmid of *Pseudomonas putida,* can be transferred to other Pseudomonads otherwise incapable of the reaction and so confer this metabolic ability; a Pseudomonad from which the plasmid has been removed no longer can carry out the epoxidation. In this way, biodegradation ability is distributed quickly and widely within a microbial community.

Not all biological oxidations involve oxidases. As in β-oxidation, the same products may be obtained by ***dehydrogenation,*** even in the absence of molecular oxygen. Although an alcohol such as ethanol can be oxidized to the corresponding aldehyde by P450, it more often is dehydrogenated by alcohol dehydrogenase and then further "oxidized" to carboxylic acid by aldehyde dehydrogenase (Eq. 6.15). This is the process by which wine is converted to vinegar and by which, when the acetaldehyde accumulates, the pleasure of wine is converted to a hangover.

$$CH_3CH_2{-}OH \xrightarrow[\text{Dehydrogenase}]{-2H} CH_3{-}CH{=}O \xrightarrow[\text{Dehydrogenase}]{-H, +OH} CH_3{-}COOH \qquad (6.15)$$

6.2.4. Reduction

Many biological reductions, such as conversion of aromatic nitro compounds to primary amines, are also mediated by cytochrome P450. Here, the substrate rather than oxygen accepts the electrons from the NADPH–redoxin system. Other examples include the reduction of azo compounds to hydrazines and amines, and reductive dehalogenations such as conversion of DDT to DDD (Fig. 14.2). As this process competes with the normal P450 oxygenations, it is strongly inhibited by molecular oxygen, and reductions are most prevalent in the anaerobes found in soil, submerged sediments, sewage treatment plants, and the gut of animals. Fungi possess P450 and are effective reducers.

Reduction of nitro compounds to primary amines does not occur all at once. It takes place in discrete steps, as electrons are transferred from NADH or other reducing agent:

$$R-NO_2 \xrightarrow{2\,[H]} R-N{=}O \xrightarrow{2\,[H]} R-NHOH \xrightarrow{2\,[H]} R-NH_2 \qquad (6.16)$$

$$\underset{\text{Nitro}}{} \quad \underset{\text{Nitroso}}{\phantom{R-N{=}O}} \quad \underset{\text{Hydroxylamine}}{} \quad \underset{\text{Amine}}{}$$

The intermediates often can be isolated. Although bacteria do not utilize cytochrome P450 for such reductions, they have similar iron–sulfur enzymes and are very powerful reducers.

A practical example of the importance of reduction as a degradative process concerns the use of pesticides on rice. Rice is arguably the world's most important crop, and insecticides, herbicides, and fungicides are used heavily on flooded rice fields. The rate at which they are inactivated in soil is dependent on whether conditions are aerobic or anaerobic (Table 6.2). Those such as DDT and trifluralin, which contain reactive chlorines or nitro groups, are relatively stable in aerobic soils but are

TABLE 6.2
Pesticide Persistence in Aerobic and Anaerobic Soils[a]

Pesticide	Type	Principal Degradation	Halflife (days) Aerobic	Anaerobic
DDT	Chlorinated	Reduction	ND[b] (28 days)	10–45
BHC	Chlorinated	Reduction	ND (28 days)	20
Diazinon	Phosphate	Hydrolysis	35	15
Carbofuran	Carbamate	Hydrolysis	95	20
PCP[c]	Chlorophenol	Reduction	50	30
Thiobencarb	Thiolcarbamate	Oxidation	10–26	30–60
2,4-D	Phenoxy acid	Oxidation	9	28
Trifluralin	Dinitroaniline	Reduction	150	3–10

[a] Sethunathan and Siddaramappa, 1978.
[b] ND = no degradation detected within the period indicated.
[c] Pentachlorophenol.

rapidly reduced under anaerobic conditions; those such as readily oxidized 2,4-D are degraded best in aerobic soils.

6.2.5. Hydrolysis

Most microorganisms are very effective at the hydrolysis of esters, amides, epoxides, and acetals (Table 6.3). The enzymes are high-molecular-weight proteins that, like the acetylcholinesterases of Section 9.4.1, rely on a specific configuration of amino acids at their active site to accelerate the reaction of water and substrate. The sequence of these amino acids in hydrolases generally is gly–glu/asp–ser–gly/ala, and the steric relationship between the serine OH, histidine –NH, and the COOH of glutamic or aspartic acid is crucial.

An important example of an amidase is penicillinase (β-lactamase), the hydrolytic enzyme that deactivates the antibiotic penicillin (Section 9.6.1). Penicillinase generated by *Staphylococcus* in response to its exposure to penicillin, or via a plasmid, causes that particular strain to become penicillin resistant. In many hospitals, nursing staff carry the resistant organisms in their nasal passages from frequent low-level contact with penicillin, and wide use of the antibiotic in agriculture may lead to similar generations of resistant bacteria.

6.2.6. Other Transformations

Conjugation reactions, that is, synthesis, are less common in microorganisms than in animals and plants. Microorganisms are adept at acetylation of amines (Eq. 6.1) and methylation of phenols (Eq. 6.3, R = CH_3), but there is little evidence of the glycosidation or peptide conjugation so common among more advanced organisms. Glutathione conjugation has been observed in a few protozoans, fungi, and algae, but not in bacteria. If the ostensible purpose of glycosidation and mercapturation is to make lipophilic xenobiotics more water soluble and so more readily excreted, the microbes—only a cell wall away from the outside world—might have little need for it.

TABLE 6.3
Examples of Biochemical Hydrolysis

Compound Type	Enzyme	Typical Substrate
Carboxylic ester	Carboxyesterase	Malathion
	Cholinesterase	Acetylcholine
	Lipase	Glyceryl stearate
Phosphate ester	Phosphatase	Paraoxon
Amide	Aryl acylamidase	Propanil[a]
	Penicillinase	Penicillin
Epoxide	Epoxide hydrolase	Naphthalene epoxide
Acetal (diether)	Glucosidase	p-Nitrophenyl glucoside
	Glucuronidase	Phenyl glucuronide

[a] 3,4-Dichloropropionanilide.

Rather than to confer water solubility, the purpose of methylation is probably to inactivate nucleophiles. Every biologist since Lister can attest to the bactericidal properties of phenols, and organisms must early have evolved ways of permanently disabling natural phenols, including those of their own making. This reaction requires a source of methylcarbonium ions, usually the sulfonium salt *S*-adenosylmethionine (Table 6.1) analogous to the methyl iodide or methyl sulfate employed by organic chemists. However, the unique ability of many microorganisms to methylate Hg, As, Sn, and Se (Special Topic 11) requires CH_3^- rather than CH_3^+, and this is provided by Vitamin B_{12} (methylcobalamine):

$$CH_3:CoX_5 + Hg^{2+} + H_2O \longrightarrow H_2O:CoX_5 + CH_3:Hg^+ \qquad (6.17)$$

Methylcobalamine   Methylmercuricion

Most of the above reactions take place simultaneously. Microbial degradation of parathion in soil, outlined in Eqs. 6.1–6.3, quickly provides all the products shown, as well as others. Not all microbial species biotransform equally well, so the end result represents a community effort where some members prepare the way for others (see Special Topic 15). Of course, the process does not end with these reactions but continues to ever-smaller fragments, until mineralization is complete.

There are circumstances under which microbial transformations fail. ***Recalcitrant molecules*** structurally resistant to reaction include chlorinated hydrocarbons such as DDE, dieldrin, and hexachlorobenzene (Chapter 14). They are unreactive due to size, the electron-withdrawing capacity of multiple chlorines, or adverse stereochemistry. Some, such as chlorinated dibenzo-*p*-dioxins and dibenzofurans (Section 14.4) resist for all these reasons. An interesting example is the polyfluorinated surfactants (Structure 6.18), which simply provide no point for enzymatic attack. Other examples include phthalate plasticizers, nonylphenol from surfactants and contraceptives, and most solid polymers.

$$CF_3CF_2CF_2CF_2CF_2CF_2CF_2CF_2CF_2CF_2CF_2CF_2CF_2CF_2CF_2CF_2CF_2COOH \qquad (6.18)$$

Another problem relates to availability. At some level in soil or water, the concentration of even a degradable chemical may become so low that the organism's ability to gather, concentrate, and provide for its transformation becomes ineffective and the concentration-based reaction rates negligible (Alexander, 1994). The ***biodegradation threshold*** in water is about 2 μg/L for 2,4-D and 0.07 μg/L for di(2-ethylhexyl) phthalate, while that for naphthalene in soil is 0.5 μg/kg. As analysis at such levels is now routine, the implications for setting and enforcing regulatory standards seem clear: There will generally be some ng/L level of concentration below which a chemical will not be removed from water or soil at an appreciable rate by physical, chemical, or biological forces. To paraphrase another of Dr. Alexander's "Commandments":

𝕿𝔥𝔬𝔲 𝔰𝔥𝔞𝔩𝔱 𝔫𝔬𝔱 𝔡𝔢𝔤𝔯𝔞𝔡𝔢 𝔱𝔥𝔞𝔱 𝔴𝔥𝔦𝔠𝔥 𝔱𝔥𝔬𝔲 𝔠𝔞𝔫𝔰𝔱 𝔫𝔬𝔱 𝔣𝔦𝔫𝔡.

6.3 | TRANSFORMATIONS IN ANIMALS AND HIGHER PLANTS

6.3.1. Primary (Phase I) Metabolism

Plants and animals reflect the microbial transformations in their oxidation, reduction, hydrolysis, and conjugation. Although the specifics may differ, most basic mechanisms remain the same (Appendix 6.1). While transformations in prokaryotes occur in the cytosol, many eukaryotic reactions take place on the lipoprotein membranes of the subcellular structures called endoplasmic reticulum (ER). As most substances transformed in the ER are lipophilic, an enzyme system such as MFO and its substrate are brought into close proximity on the membrane. However, microorganisms still are responsible for some of the reductions and hydrolyses observed in higher organisms. For example, laboratory rats fed antibiotics or otherwise freed of enteric bacteria seem unable to carry out nitro reduction.

Genetic evidence suggests the existence of cytochrome P450 as far back as 3 billion years, not long after the origins of life on earth. Perhaps the most important of all xenobiotic-transforming systems, P450 is found throughout the animal kingdom, from sponges to killer whales and people (Stegeman and Hahn, 1994) as well as in plants (and of course microorganisms). It is very similar to that in bacteria, except that the iron-sulfur protein for electron transfer is cytochrome b_5 in animals and plants.

In all that time, many genetic variations of P450 have evolved. These now have been systematized and each designated as a CYP, followed by the gene family as an Arabic numeral, the subfamily as a capital Roman letter, and the gene itself as another Arabic numeral. For example, the most common P450 group in animals is CYP1A, which mammals express in two forms: CYP1A1 catalyzes C-oxidations such as epoxidation of benzo[a]pyrene, and CYP1A2 oxidizes primarily N or S and also dealkylates. There are 36 gene families, and a single animal species may contain as many as 100 specific P450 genes, but not all families are found in all species.

The significance of these various isoforms lies in their differing levels, specificities, and susceptibility to induction. CYP1A1 catalyzes the oxidation of PAH and planar PCBs, while CYP1A2 oxidizes acetanilide and estradiol but not the 1A1 substrates. However, PAH, including those in tobacco smoke and charbroiled meat, as well as PCBs, induce both. Therefore, those animals with a low level of CYP1A1 are the more likely to be affected by exposure to a PAH, and will simultaneously suffer an increased loss of the estradiol necessary for reproduction. This kind of knowledge should go far in explaining the observed susceptibilities of the various species toward xenobiotics.

While P450 is induced in bacteria by substrate, it can be induced in eukaryotes by other substances including PAH and HAH (*H*alogenated *A*romatic *H*ydrocarbons). In animals including mammals, fish, amphibians, and reptiles, CYP1A level is regulated by a factor called the A*h* receptor (Whitlock, 1990), or AHR (Hankinson, 1995); an example of this mechanism is shown in Fig. 9.5 for the inducer 2,3,7,8-tetrachlorodibenzo-*p*-dioxin (TCDD). Linked to A*h* receptor, TCDD is moved into the cell nucleus, where it stimulates messenger RNA to replicate CYP1A1 (and do other things). The oxidative action of P450 isoforms can be inhibited by certain xeno-biotics (Halpert, 1995), a prime example being the methylenedioxyphenyl synergist, piperonyl butoxide, commonly incorporated into insecticide formulations to slow oxidative detoxication in the intended victim.

Although most cells are capable of xenobiotic metabolism, this activity is most pronounced in the liver or hepatopancreas of animals, as illustrated by the selective hydrolysis of malathion by hepatic carboxyesterases. This chemical was introduced into the United States in 1950 as a broad-spectrum insecticide of low toxicity to mammals and birds. Its acute oral LD_{50} is 1375 mg/kg in male rats but 8.4 mg/kg in the American cockroach, *Periplaneta americana*. Malathion still is used extensively on home gardens and in other applications where a wide margin of safety is required (Section 15.2). The reason for malathion's spectacular species selectivity is hydroly-sis. Both mammals and insects easily activate the insecticide via oxidation (Scheme 6.19), but mammals rapidly hydrolyze the resulting malaoxon to a nontoxic carbox-ylic acid via a carboxyesterase, while insects employ the much slower phosphatase and so build up a lethal dose of oxon (Krueger and O'Brien, 1959).

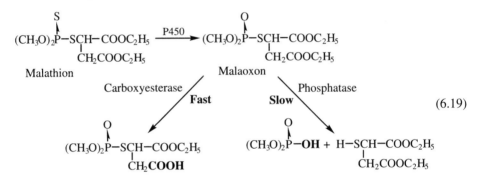

(6.19)

Cleavage of carboxylic esters also may be accomplished by oxidative dealkylation rather than hydrolysis, as CYP3A4 can hydroxylate the α-carbon of the alcohol to a hemiacetal ester, $RCOOCH(OH)R'$, readily converted to aldehyde $R'CH=O$ and the acid.

Dehydrochlorination can be important in animals. This glutathione-dependent re-action results in the detoxication of DDT and is the principal cause of resistance to the insecticide. The resulting DDE, now the predominant and stable DDT relative in the environment, has been a major cause of eggshell thinning in birds. The insecticide lindane likewise provides nontoxic γ-pentachlorocyclohexene (γ-PCCH), which is then metabolized by further dehydrohalogenation, oxidation, and conjugation (Fig. 6.6). Figure 6.6 also illustrates the complexity of most xenobiotic transformations, the steps occurring concurrently via distinctive enzymes often located in separate parts of the cell.

Although many types of eukaryotic cells can carry out reductions, such as con-version of nitro and azo to amino, both hepatic microsomes and cytosol are especially proficient. The nitro- and azo-reductases are Fe-containing enzymes perhaps identical to cytochrome P450 isoforms, although many reductions are so facile that they can

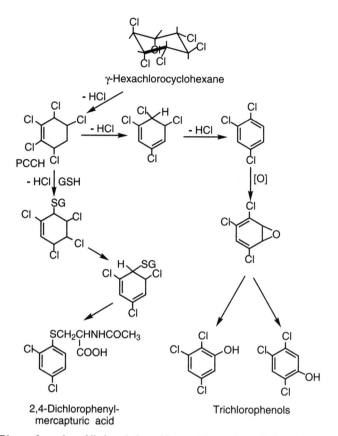

Figure 6.6. Biotransformation of lindane (γ-hexachlorocyclohexane) to substituted benzenes. The chlorophenols undergo further oxidation and conjugation.

be brought about nonenzymatically by a circulating H donor such as reduced glutathione. Parasitic microflora are responsible for much of the reduction observed in animals (Section 6.2.4).

6.3.2. Secondary (Phase II) Metabolism

While microorganisms excel at primary metabolism, secondary metabolism (conjugation) is the specialty of animals and plants. This may reflect the need of multicellular organisms to eliminate or sequester reactive metabolites such as phenols, amines, and carboxylic acids in nontoxic and water-soluble form. A common means is glycosylation, the derivatization of –OH, –NH, –SH, and COOH groups with glucuronic acid (in most animal species) to give **glucuronides** or with glucose (insects and higher plants) to give **glucosides.**

For example, 4-nitro-3-trifluoromethylphenol (TFM, Eq. 6.20) is highly toxic to the sea lamprey *(Petromyzon marinus)*, a parasitic eel that preys upon commercial fish in the Great Lakes.

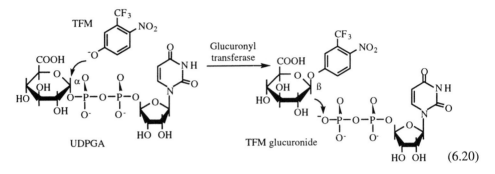

$$(6.20)$$

The chemical is virtually nontoxic to other fish species, which detoxify it by forming the glucuronide while the sea lamprey cannot. In this and related glycosidations, the phenolate anion or other electronegative center displaces the high-energy pyrophosphate of uridine-5'-diphospho-α-D-glucuronic acid (UDPGA), with inversion of configuration, to give the phenyl β-D-glucuronopyranoside (Eq. 6.20). Similarly, the high-energy 3'-phosphoadenosine-5'-phosphosulfate produces *sulfate esters* that often accompany glycosides in animals.

Combination of reactive electrophiles with the –SH group of the natural tripeptide, glutathione, also is an important means of detoxication. Alkyl halides, halobenzenes, nitroaromatics, and epoxides are frequent substrates, and the reaction is catalyzed by glutathione *S*-transferase (GST). Workers exposed to methyl bromide excrete *S*-methylglutathione or *S*-methylcysteine in their urine:

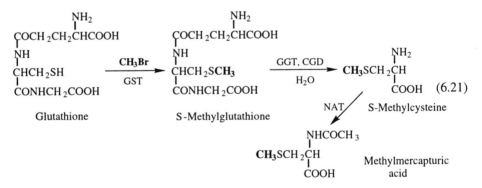

$$(6.21)$$

which affords an estimate of their degree of exposure. The glutathione conjugate often is hydrolyzed in turn by γ-glutamyl transferase (GT) and cysteinylglycine dipeptidase (CGD) to the *S*-substituted cysteine, which subsequently is acetylated to a *mercapturic acid* via *N*-acetyl transferase (NAT), as shown in Fig. 6.5. Degradation often proceeds further (Eq. 6.22), first via cysteine β-lyase to form the thiol and then, upon methylation, the corresponding thioether. Oxidation (Reaction 3) can lead to a methyl sulfone, the terminal metabolite from even such stable substances as HCB and DDT (Section 14.2).

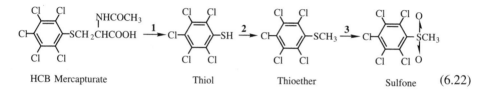

| HCB Mercapturate | Thiol | Thioether | Sulfone | (6.22) |

Mammals are especially adept at mercapturation, and urinary thioether levels have been used to detect toxic stress in workers. Chemical workers showed higher levels than did office workers, and rubber and tire workers exposed to a wide variety of reactive chemicals had the highest thioether levels of all. Glutathione conjugation is a major route for the detoxication and elimination of some of the most dangerous xenobiotics, including aflatoxins, vinyl chloride, alkyl bromides, and benzo[*a*]pyrene. However, the end result is not always detoxication. The widely used fumigant ethylene dibromide (EDB) reacts rapidly with GSH (Eq. 6.23), but the initial conjugate can cyclize to form a DNA-reactive alkylating cation (Anders and Dekant, 1994).

$$BrCH_2CH_2Br + G\!-\!S^- \xrightarrow{GST} \begin{matrix} CH_2Br \\ | \\ CH_2\!-\!SG \end{matrix} \xrightarrow{-Br^-} \begin{matrix} CH_2 \\ / \ \backslash \\ CH_2S^+ \!\!\diagdown_G \end{matrix} \xrightarrow{DNA} \begin{matrix} CH_2\!-\!DNA \\ | \\ CH_2S \diagdown_G \end{matrix} \quad (6.23)$$

Another common type of conjugate forms with amino acids or peptides. Carboxylic acids react with CoA to form a thioester, which subsequently reacts with an amino acid to form the amide:

$$RCOOH \xrightarrow[\text{Synthetase}]{CoA-SH} RCO-SCoA \xrightarrow[\text{NAT}]{R'NH_2} RCONHR' + CoA-SH \quad (6.24)$$

Conjugation with glycine is common in animals, and glutamic or aspartic acid forms the usual conjugate in plants, but there is a good deal of species specificity (Section 6.4). Some animals form conjugates with unusual amino acids such as taurine (2-aminoethanesulfonic acid) in fish, and birds and reptiles acylate either or both of the two amino groups of ornithine (2,5-diaminopentanoic acid).

6.4 | COMPARATIVE METABOLISM

Organisms are not all equally capable of biotransformation. Although the ability broadly follows phylogenetic lines, specificity varies with class, species, and to some extent, individuals. Distinct and sometimes marked biochemical differences exist even between human identical twins. Biotransformation rates are closely associated with that of intermediary metabolism (see Boxenbaum, 1984), itself proportional to the 0.75 power of body weight in animals as diverse as mice, humans, and elephants (Fig. 10.5).

Table 6.4 provides a broad outline of some species capabilities. A few mamma-

TABLE 6.4
Comparative Metabolism of Xenobiotics[a]

Reaction	Bacteria	Plants	Molluscs	Arthropods	Fish	Amphibians	Reptiles	Birds	Mammals
Ring hydroxylation	++	+	−	+	+	++	+	++	++
Nitro reduction	++	+	++	±	+	−	+	++	++[b]
Ester hydrolysis	++	++	+	+	+	+	+	++	++
Amine acetylation	+	+	++	++	++	+[c]	−	++[d]	++[e]
Glucosylation	±	++	+	+	−	−	+	−	−
Glucuronylation	−	−	+	+	++	+	++	++	++[f]
Sulfation	−	−	+	+	−	+	++	++	++
Peptide conjugation	−	Asp	Gly	Gly	Gly	Gly	Orn	Orn	Gly
Glutathione conjugation	−	++	++	++	±	+	++	++	++[g]
O-Methylation	++	+	+	+	+	+	+	+	+

[a] Strong ++, moderate +, weak ±, insignificant −. See Parke (1968), Hatzios and Penner (1982).
[b] Sometimes microbial.
[c] Frogs.
[d] Chickens.
[e] Dogs.
[f] Cats.
[g] Guinea pigs.

lian and bacterial species have received the bulk of attention, generally because of their relation to human health, and certain insect species have attracted interest because of their resistance to pesticides. By contrast, biotransformations in molluscs have received little attention, and cnidarians have been almost completely ignored. Most biotransformation information from animals comes from the presence of the associated enzymes, which are isolated and studied with relative ease, but investigation of the metabolites actually produced or excreted has not kept pace.

Aerobic species obviously are capable of oxidation, and some bacteria excel. However, rates normally decline as the phylogenetic tree of animals is descended, insects excepted, and fish and molluscs usually lag behind mammals in metabolic activity. Reduction, too, is a general reaction, although the results often are due more to gut microflora than to the animals themselves. Ruminants and rodents harbor very large populations of microbial reducers, while humans possess relatively few. The little work on amphibians indicates that they may be poor reducers. Apparently, everyone can hydrolyze, and for their size, bacteria again are very good at it. Not surprisingly, present-day species must have evolved because their ancestors could detoxify a wide variety of natural poisons.

When it comes to secondary metabolism, there seems to be much more species variation. For example, dogs, chickens, and certain people are poor at the acetylation of amines, so that chocolate often is toxic to dogs due to its relatively high amine content. The consequences for some humans also can be serious: Over 60% of people in Britain and Southern India, and more than half of both black and white Americans, are "slow acetylators," in whom amine-containing drugs and pollutants cause an increased incidence of peripheral neuropathy, lupus, or bladder cancer.

Plants and insects tend to form glucosides of phenols and glucose esters of carboxylic acids, while most animals make glucuronides. Humans excrete about 23% of a dose of phenol as glucuronide and 71% as sulfate ester, while rats excrete 25% and 68%, and guinea pigs 78% and 17%, respectively (Hodgson and Levi, 1987). This suggests that rats provide a satisfactory surrogate for humans in this case, but guinea pigs would not. Cats seem incapable of glucuronidation, which suggests they should not be fed phenols such as the laxative phenolphthalein. O-Methylation is not prominent among animals, perhaps because they are so intent upon solubilizing xenobiotics, and at least some of the observed methylation may again be due to intestinal microorganisms.

The widespread conjugation of carboxyl groups with amino acids or peptides, such as Herr Keller's hippuric acid, shows considerable variation among species (Table 6.4). Plants most often conjugate carboxyls with aspartic and glutamic acids, while many animals use glycine and humans also utilize glutamine. Birds and reptiles both employ the terminal amino group of the uncommon amino acid ornithine [$H_2NCH_2CH_2CH_2CH(NH_2)COOH$], or sometimes both amines, and such metabolic similarity is seen often enough to support the purportedly close evolutionary relationship between the two phyla. A few animal species, including fish, substitute the unusual 2-aminoethanesulfonic acid (taurine), and conjugation of acids with glutamine, arginine, and serine is not uncommon.

The reaction of electrophiles with glutathione also is very general, although the resulting conjugates usually are converted to the substituted cysteines and ***mercapturic acids*** found in the urine. Molluscs, which are not all that effective at other biotransformations (Table 6.4), readily form mercapturates, while chickens do not. Higher plants frequently employ glutathione to detoxify herbicides, the water-soluble conjugates being stored in the cells' vacuoles.

The foregoing examples illustrate how a particular type of biotransformation varies between species, but what does a single species do with a series of related compounds? Over a period of years, Bray and coworkers (1957) detailed exactly how rabbits metabolize each of the 19 possible chlorinated nitrobenzenes (Table 6.5). Although the proportions of each type of product differed, nitro reduction, amine acylation, and oxidation to and conjugation of phenols were always observed, and mercapturation was frequent. That is, all of the major biotransformation routes were represented, as is the usual outcome.

Knowledge about comparative metabolism is surprisingly incomplete, considering its significance for the development of more selective drugs and pesticides, ways to combat resistance in insects and other pests, and understanding how to protect from pollution the thousands of small but important links in the global food web. A single square meter of meadow contains hundreds of animal species alone, including spiders, tardigrades, nematodes, earthworms, terrestrial molluscs, and others whose biotransformation potential remains completely unknown. Detoxication and activation by biotransformation are the key to the toxicity of chemicals in *any* species.

We now recognize that the basic biotransformation processes are roughly similar between prokaryotes (bacteria) and the Eukarya (animals and plants). However, determination of the complete genome of a deep-water microorganism, *Methanococcus jannaschii,* places it among the Archaea, a third form of life (Bult et al., 1996). While 44% of its genes also are found in the other "kingdoms," the function of the other 56% remains unknown, and only two of the 1738 genes represent recognized detoxication pathways. Are archaeons capable of general detoxication? If not, how have they survived? What can they tell us about toxicology at the beginnings of life and possibly even in extraterrestrial forms? Our knowledge of biotransformation may be at an even earlier stage than we suppose.

TABLE 6.5
Metabolism of Chloronitrobenzenes in Rabbits[a]

Cl Position	Mercapturate (%)	Free Amine (%)	Acylamine (%)	O-Glucuronide (%)	Sulfate Ester (%)	Total
4-	7	9	4	19	21	60
3,4-	45	17	5	13	12	92
3,4,6- (2,4,5-)	30	21	8	40	4	103
2,3,4,6-	37	11	20	33	0	101
2,3,4,5,6,-	37	10	21	*b*	0	>68

[a]Bray et al., 1957.
[b]Present but analysis unreliable.

6.5 | SUMMARY OF BIOTRANSFORMATIONS

Even a brief examination of biotransformations reveals that the products of Phase I metabolism are the same as those from corresponding abiotic reactions: *oxidation, reduction, hydrolysis, elimination,* etc. (Appendix 6.1). For example, biotransformation of parathion by any of a wide variety of organisms results in the same initial products as are formed by parathion photolysis in water (Eqs. 6.1–6.3). Indeed, biotransformation reactions are basically the same as their abiotic counterparts (Table 6.1), but the reagents generally are cofactors associated with enzymes. As in the aquatic environment, oxidations are carried out primarily by reactive Fe–O complexes, reductions by hydride transfer, hydrolyses by an oxyanion (the serine O^-), and conjugation by displacement of a good leaving group (pyrophosphate) by the nucleophile. Again, the close relation between environmental toxicology and chemistry is apparent.

6.6 | REFERENCES

Alexander, M. 1965. Biodegradation: Problems of molecular recalcitrance and microbial fallibility. *Adv. Appl. Microbiol.* **7:** 35–80.

Alexander, M. 1966. Biodegradation of pesticides, in *Pesticides and Their Effects on Soils and Water* (M. E. Bloodworth, ed.), Soil Science Society of America, Madison, WI, pp. 78–84.

Alexander, M. 1994. *Biodegradation and Bioremediation,* Academic Press, New York.

Anders, M. W., and W. Dekant. 1994. *Conjugation-Dependent Carcinogenicity and Toxicity of Foreign Compounds,* Academic Press, New York, NY. Advances in Pharmacology, Vol. 25.

Bollag, J.-M., and S.-Y. Liu. 1990. Biological transformation processes of pesticides, in *Pesticides in the Soil Environment: Processes, Impacts, and Modeling,* Soil Sci. Soc. of America, Madison, WI, pp.169–211.

Boxenbaum, H. 1984. Interspecies pharmacokinetic scaling and the evolutionary-comparative paradigm. *Drug Metab. Rev.* **15:** 1071–121.

Bray, H. G., S. P. James, and W. V. Thorpe. 1957. Metabolism of 2,3-, 2,6-, and 3,5-dichloronitrobenzene and the formation of a mercapturic acid from 2,3,4,5-tetrachloronitrobenzene in the rabbit. *Biochem. J.* **67:** 607–16.

Bult, C. J., and 39 coauthors. 1996. Complete genome sequence of the methanogenic archeon, *Methanococcus jannaschii. Science* **273:** 1058–73.

Faulkner, J. K., and D. Woodcock. 1964. The metabolism of 2,4-dichlorophenoxyacetic acid (2,4-D) by *Aspergillus niger* van Tiegh. *Nature* **203:** 865.

Floodgate, G. D. 1973. A threnody concerning the biodegradation of oil in natural waters, in *The Microbial Degradation of Oil Pollutants* (D. G. Ahearn and S. P. Meyers, eds.), Louisiana State University, Baton Rouge, LA, pp. 17–24.

Gibson, D. T., and V. Subramanian. 1984. Microbial degradation of aromatic hydrocarbons, in

Microbial Degradation of Organic Compounds (D. T. Gibson, ed.), Marcel Dekker, New York, NY, pp. 181–252.

Guengerich, F. P., and T. L. Macdonald. 1990. Mechanisms of cytochrome P450 catalysis. *FASEB J.* **4**: 2453–59.

Halpert, J. R. 1995. Structural basis of selective cytochrome P450 inhibition. *Ann. Rev. Pharmacol. Toxicol.* **35**: 29–53.

Hankinson, O. 1995. The aryl hydrocarbon receptor complex. *Ann. Rev. Pharmacol. Toxicol.* **35**: 307–40.

Hatzios, K. K., and D. Penner. 1982. *Metabolism of Herbicides in Higher Plants,* Burgess Publ. Co., Minneapolis, MN.

Hodgson, E., and P. E. Levi. 1987. *A Textbook of Modern Toxicology,* Elsevier, New York.

Jerina, D. M., and J. W. Daly. 1974. Arene oxides: A new aspect of drug metabolism. *Science* **185**: 573–82.

Jakoby, W. B. 1980. *Enzymatic Basis of Detoxication,* Academic Press, New York, NY, Vols. I and II.

Jordan, R. E., and J. R. Payne. 1980. *Fate and Weathering of Petroleum Spills in the Marine Environment,* Ann Arbor Science, Ann Arbor, MI.

Josephy, P. D. 1997. *Molecular Toxicology,* Oxford University Press, New York, NY.

Krueger, H. R., and R. D. O'Brien. 1959. Relationship between metabolism and differential toxicity of nalathion in insects and mice. *J. Econ. Entom.* **52**: 1063–67.

National Research Council. 1985. *Oil in the Sea,* National Academy Press, Washington, DC.

Parke, D. V. 1968. *The Biochemistry of Foreign Compounds,* Pergamon Press, Oxford, UK.

Poulos, T. L. 1986. The crystal structure of cytochrome P-450$_{cam}$, in *Cytochrome P-450: Structure, Mechanism, and Biochemistry* (P. R. Ortiz de Montellano, ed.), Plenum Press, New York, NY, pp. 505–23.

Reineke, W. 1984. Microbial degradation of halogenated aromatic compounds, in *Microbial Degradation of Organic Compounds* (D. T. Gibson, ed.), Marcel Dekker, New York, NY, pp. 319–60.

Sethunathan, N., and R. Siddaramappa. 1978. Microbial degradation of pesticides in rice soils, in *Soils and Rice,* The International Rice Research Institute, Los Banos, Philippines, pp. 479–97.

Sligar, S. G., and R. I. Murray. 1986. Cytochrome P450$_{cam}$ and other bacterial P450 enzymes, in *Cytochrome P-450: Structure, Mechanism, and Biochemistry* (P. R. Ortiz de Montellano, ed.), Plenum Press, New York, NY, pp. 429–503.

Stegeman, J. J., and M. E. Hahn. 1994. Biochemistry and molecular biology of monooxygenases: Current perspectives on forms, functions, and regulation of Cytochrome P450 in aquatic species, in *Aquatic Toxicology* (D. C. Malins and G. K. Ostrander, eds.), Lewis Publishers, Boca Raton, FL, pp. 87–206.

Wheeler, R. B. 1978. The fate of petroleum in the marine environment. Exxon Production Research Co., Special Report.

Whitlock, J. P. 1990. Genetic and molecular aspects of 2,3,7,8-tetrachlorodibenzo-*p*-dioxin action. *Ann. Rev. Pharmacol. Toxicol.* **30**: 251–77.

Special Topic 6. Anatomy of an Oil Spill

Crude oil is found almost everywhere on earth. The 1993 world oil production was 2.84×10^9 gal/day or 1.04×10^{12} gal/year, principally from the Arab nations, South America, Indonesia, Siberia, and Alaska. As these producers are far distant from the major consumers— Western Europe, Japan, and the United States—most oil must be transported *via* water, although it is stored and used primarily on land.

Oil enters the environment by a variety of routes. In the sea, an estimated 8% seeps naturally from the earth, 9% is recycled from the atmosphere, almost 40% comes from waste and runoff (mostly from cities), and 45% comes from its transportation for a total input of 3.2 billion kg annually (National Research Council, 1985). On land, most pollution by crude oil results from production and refining, as relatively little crude is transported there, although petroleum products such as diesel fuel are shipped extensively by truck and rail (Section 4.5). Major spills occur frequently: the 1970 IXTOC I well blowout in the Gulf of Mexico (176 million gals), the 1978 grounding of the Amoco Cadiz off the French coast (66 million gals), and the 130 million gallons released into the Persian Gulf during the 1991 Iraq war make the notorious 1989 Exxon Valdez spill of 10 million gallons in Alaska seem small by comparison.

Oil spilled on land is largely absorbed and virtually immobilized by soil, but that spilled at sea quickly spreads (Fig. 6.7). The thin surface film that results is crucial for any subsequent dissipation. Components of $<C_{10}$ volatilize into the atmosphere within a few hours, and the more water-soluble phenols, nitrogenous constituents, and polar hydrocarbons dissolve into the water column (Jordan and Payne, 1980). The remainder forms oil-in-water and water-in-oil (mousse) emulsions or sinks as tar balls and particles. However, if the spill is large and close to land, much of heavy semiliquid residue may wash ashore.

Besides physical dispersion, spilled oil is degraded by sunlight, air, and living organisms. In both soil and water, the principal microbial actors are *Pseudomonas* species, although fungi such as *Cladosporium* and even algae also are important. All together, over 200 species of microorganisms are known to metabolize petroleum (Jordan and Payne, 1980). The oxidative biodegradation of both aliphatic and aromatic hydrocarbons, discussed in Section 6.2, produces alcohols, phenols, aldehydes, ketones, acids, and esters. Although the hydrocarbons of crude oil seem to be degraded simultaneously, *n*-alkanes and monocyclic aromatics react the most rapidly, branched alkanes, cycloalkanes, and polycyclic aromatics are slow to be metabolized, and high-molecular-weight hydrocarbons such as asphaltenes and polycycloalkanes (*e.g.*, steranes) remain virtually unchanged.

Except for the universal activity of *Pseudomonas*, petroleum biodegradation varies greatly with microbial species. It generally is restricted to only a limited range of structural types for any given species, slows as molecular weight increases, and is limited by the supply of nutrients and oxygen. However, the oil is confronted by a mixed microbial population that consequently is able to perform cooperatively a wide range of degradation reactions (Special Topic

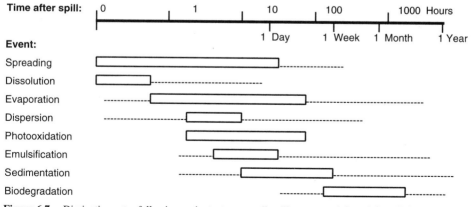

Figure 6.7. Dissipation rates following an instantaneous oil spill on water. Adapted from Wheeler, 1978.

15). Microbial metabolism is essential to the eventual fate of oil; steam cleaning of oily shores may be good public relations, but it is about the worst thing that can be done to remediate a spill.

Aquatic and soil invertebrates also are capable of absorbing and degrading oil, although information about their activities is sparse (National Research Council, 1985). Uptake can be from food or directly from the water, and the precipitation of oil in the form of fecal pellets is well established. Vertebrates such as fish readily absorb and metabolize oil, but their influence is limited by both a small total biomass and a well-developed avoidance response.

In contrast to spills, most other oil pollution arises continuously from many dispersed sources such as aerial fallout, seeps, and waste discharges from roads, parking lots, and boats. Most biodegradative enzymes are inducible, and organisms living in proximity to constant discharges have become very efficient at, as well as genetically selected for, the degradation of oil. While the animals and plants use cometabolism, many of the microorganisms are adapted to rely on oil for their nutrition.

What is the ultimate fate of environmental oil? Most of the continuous input is oxidized chemically and biologically to small organic fragments and CO_2, although some microorganisms actually biosynthesize oil-like mixtures. However, as the light components evaporate, dissolve, and are metabolized, the remainder becomes heavier, less reactive, more recalcitrant. The solid or semisolid, high-asphaltene tarballs, for example, are almost immune to photodegradation, and they are so dense and contain so little oxygen that microorganisms cannot deal with them. These dark blobs are often seen on beach rocks and can be very old. Like the tree resins that turned into amber, they may be fossils in the making.

Having said all this, a word of caution may be appropriate. "The term microbial degradation of oil in the marine environment means the degradation of a complex and variable mixture of hundreds of substrates by unknown populations of microorganisms in an erratically changing medium [where] other processes such as evaporation, solubilization, photooxidation and other abiological mechanisms operate at the same time. The task of untangling the skein of events is formidable, and bold generalizations are ill advised" (Floodgate, 1973).

APPENDIX 6.1.
Summary of Major Biotransformations

A. Reduction

   1. Nitro $-NO_2 \longrightarrow -NH_2$

   2. Aliphatic halide $C-Cl \longrightarrow C-H$

   3. Carbonyl $C=O \longrightarrow CH_2OH$

B. Elimination $CH-Cl \longrightarrow C=C$

C. Hydrolysis

   1. Epoxide

$$\overset{O}{\overset{\diagup\!\!\backslash}{C-C}} \longrightarrow \overset{OH\ OH}{\underset{|\quad|}{C-C}}$$

   2. Ester (carboxylic) $-COOR \longrightarrow -COOH + ROH$

   3. Ester (phosphate) $P(O,S)-OR \longrightarrow P(O,S)-OH + ROH$

   4. Amide $CONHR \longrightarrow -COOH + H_2NR$

D. Oxidation

   1. N-Oxidation $R_3N \longrightarrow R_3NO$

   2. Sulfoxidation $-S- \longrightarrow -SO- \longrightarrow -SO_2$

   3. Desulfuration $PS \longrightarrow PO$

   4. N-Dealkylation $NCH_3 \longrightarrow NCH_2OH \longrightarrow NH + HCHO$

   5. O-Dealkylation $-OCH_3 \longrightarrow -OCH_2OH \longrightarrow -OH + HCHO$

   6. Epoxidation

$$C=C \longrightarrow \overset{O}{\overset{\diagup\!\!\backslash}{C-C}}$$

   7. Aliphatic C-hydroxylation $C-H \longrightarrow C-OH \longrightarrow C=O$

   8. Aromatic hydroxylation

E. Conjugation

   1. Glycosylation $-OH \longrightarrow -O\text{-Sugar}$

   2. Sulfation $-OH \longrightarrow -O\text{-}SO_3K$

   3. Methylation $-OH \longrightarrow -OCH_3$

   4. Acylation $-NH_2 \longrightarrow -NH\text{-}COCH_3$

   5. Amidation $-COOH \longrightarrow -CO\text{-}NHRCHCOOH$

   6. Mercapturation

$$C-Cl \longrightarrow C-\overset{\overset{\displaystyle NHCOCH_3}{|}}{SCH_2CHCOOH}$$

$$\overset{O}{\overset{\diagup\!\!\backslash}{C-C}} \longrightarrow C-C\text{-}\overset{\overset{\displaystyle NHCOCH_3}{|}}{SCH_2CHCOOH}$$

7.1 | THE INTOXICATION PROCESS

The dictionary defines *intoxication* as "an abnormal state that is essentially a poisoning." Most people automatically think "alcohol" when they hear the word, but to toxicologists, it means poisoning by any chemical. Contrary to the mystery stories, poisoning is never instantaneous but involves a series of overlapping steps. This intoxication process (Fig. 7.1) requires exposure to the chemical, absorption of it into the organism, distribution among organs and cells, short- or long-term storage, and eventual reaction with some biochemical target to produce an adverse response. Along the way, the substance may be removed mechanically, chemically altered and excreted, or converted to a more reactive and toxic form ("activated"). Intoxication has been called a "biochemical lesion," and it is characteristic of all life forms.

Intoxication first requires *exposure* to the chemical—no exposure, no toxicity. The exposure route makes a big difference; that is, whether the chemical is received via the mouth (*oral* exposure), skin (*dermal* exposure), the respiratory system (*inhalation* exposure), or by injection (*parenteral* exposure) (see Section 8.2). This is largely because each route has its own style of *absorption.*

7.2 | ABSORPTION AND DISPOSITION

7.2.1. Absorption

The first and necessary step toward intoxication is absorption of a chemical into the organism. Gross mechanical or chemical damage, such as that from sulfuric acid on skin, is not normally considered intoxication. In an animal, absorption can

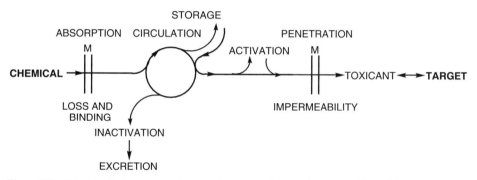

Figure 7.1. The intoxication process, from environment to biochemical target. M = Cell membrane.

take place through the skin, eyes, respiratory system, or gastrointestinal tract—any body surface. Plants absorb through leaves and roots, while microorganisms do so directly through the cell wall. In each case, the chemical must cross a lipid membrane, so it is not surprising that fatty (lipophilic) substances penetrate more readily than do hydrophilic ("polar") ones.

Cell membranes are remarkable structures. More alike than different among plants, animals, and microorganisms, they are composed primarily of a thin (often <1 nm) bilayer of long-chain phospholipids studded with and penetrated by bands of protein (Fig. 7.2). They behave like two-dimensional fluids, readily permitting diffusion of lipophilic substances through the fatty part while largely excluding those that are hydrophilic, although certain ions such as Ca^{2+} and some other polar solutes can enter via the protein *channels* (Section 9.2). Thus plasma membranes are said to be *semipermeable.* For example, swimming in seawater does not result in salt penetration through the skin, while suntan lotion is rapidly absorbed.

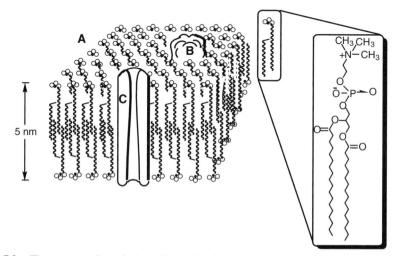

Figure 7.2. The structure of a typical membrane, showing the phospholipid bilayer (A), a globular transmembrane protein (B), and a protein "channel" (C).

Membranes are based on phospholipids (Structure 7.1), such as phosphatidylcholine (lecithin, $R^1 = R^2 = C_{16-18}$, $R^3 = CH_3$) and -ethanolamine (cephalin, $R^3 =$ H), which are the C_{16}–C_{18} fatty acid esters of phosphoglycerates. Note that the phospholipid structure is lipophilic at one end and polar (charged) at the other, leading to the formation of layers called *micelles* (Fig. 7.2) in which the polar side faces the aqueous phase and the hydrophobic side turns inward.

$$R^1CO-OCH_2CHCH_2O-\overset{\overset{O^-}{|}}{\underset{\underset{O}{|}}{P}}-OCH_2CH_2N(R^3)_3^+ \qquad (7.1)$$
$$R^2CO-O$$

Membranes can also contain triglycerides (fats), other fatty acid esters, and sterols. Phospholipids predominate in mammalian membranes, where a C_{20} tetra-unsaturated (arachidonic) fatty acid may esterify the center hydroxyl of the glycerol, and the sterol is cholesterol. Membranes of higher plants, called the *plasmolemma,* are similar, but those of fungi contain phosphatidylmyoinositol and ergosterol. Bacterial membranes are more complex, composed of diphosphotriglycerides, phosphatidylglycerol, and phosphatidylinositol, and they contain no sterol. Finean et al. (1978) provide more information.

Albert (1985) distinguishes four types of membranes. Type 1, about 5 nm thick, is by far the most common. Nonpolar and moderately polar substances penetrate them mostly by simple diffusion, and few molecules with more than three hydrophilic groups (*e.g.*, glycerine) or molecular weights over 150 can cross the barrier. The rarer Type 2 membranes allow facilitated transport of essential nutrients via protein carrier molecules, and those of Type 3 take up solutes such as Na^+ and K^+ against a concentration gradient into mammalian cells, inorganic nutrients into plant roots, and inorganic ions, sugars, and amino acids into bacteria. The last, Type 4, is characterized by nanometer-sized "pores" that admit solutes on the basis of size. Most toxicants enter cells through Type 1 membranes.

Entry is controlled by distribution (partition) of a solute between lipid and water as measured by the *partition coefficient* (Section 3.3).

$$\text{Partition coefficient} = K_p = C_{lipid}/C_{water}$$

K_p (or K_{ow}) is a measure of lipophilicity, and except in instances of assisted transport, a higher K_p means easier diffusion across membranes. A nonpolar substance such as toluene ($K_p = 590$) has a much greater affinity for lipid (glyceryl trioleate) than for water and will be absorbed rapidly into a cell, whereas ionic sodium chloride ($Kp = <1$) would not. Some representative K_p values between water and glyceryl trioleate are presented in Table 7.1. It is important to understand that K_p is *not* the ratio of the individual saturation solubilities in the two solvents.

If a substance is only partially ionized in aqueous solution, the equilibrium concentration of its un-ionized (and hence absorbable) form will determine the maximum proportion that can enter a cell at any given pH:

$$B: + H^+ \rightleftharpoons BH \tag{7.2}$$

$$\frac{[H^+][B:]}{BH} = K_a \tag{7.3}$$

$$pK_a = -\log K_a \tag{7.4}$$

$$\% \text{ HA ionized} = 100/[1 + \text{antilog}(pK_a - pH)] \tag{7.5A}$$

$$\% B: \text{ ionized} = 100/[1 + \text{antilog}(pH - pK_a)] \tag{7.5B}$$

Benzidine (pK_a 4.66) is highly carcinogenic, but, of course, it must first be absorbed into the blood stream. What proportion of an oral dose of benzidine will be absorbed from the stomach (pH 2)? Equation 7.5B tells us that 99.8% of any dose of the carcinogen will be in ionic form at pH 2, so only 0.2% will be available for absorption. However, 98.9% will be available at the average pH 6.6 of the intestine.

7.2.2. Disposition

Disposition (disposal) refers to what happens to the chemical after it has been absorbed. In animals, the next step normally is circulation in the blood or hemocoel that redistributes the substance away from the site of absorption toward the various organs. The movement in blood often is assisted by carrier proteins, which can loosely bind many types of chemicals. In higher plants, movement is termed *translocation* and takes place in aqueous solution upwards in the xylem tubes and downwards in the sieve tubes of the phloem. Microorganisms, their cells exposed directly to the environment, have little need for circulation.

Fat is a normal constituent of most cells, and especially of adipose tissue, and lipophilic substances tend to concentrate there. Chemicals generally enter and leave fat by simple dissolution and diffusion, so it is not surprising that lipophiles become concentrated and stored. Like absorption, storage in fat is related to the partition coefficient, and even rather polar substances probably enter the lipid pool briefly after

TABLE 7.1
Triolein–Water Partition Coefficients[a,b]

Chemical	log K_p	Chemical	log K_p
Aniline	0.91	1,2,4-Trichlorobenzene	4.12
Benzaldehyde	1.58	Hexachloroethane	4.21
Nitrobenzene	2.15	1,2,4,5-Tetrachlorobenzene	4.70
Benzene	2.25	2,4'-Dichlorobiphenyl	5.30
Anisole	2.31	2,4,4'-Trichlorobiphenyl	5.52
Chlorobenzene	2.97	2,5,2',5'-Tetrachlorobiphenyl	5.62
Ethylbenzene	3.27	4,4'-DDT	5.90
1,4-Dichlorobenzene	3.55	2,4,5,2',4',5'-Hexachlorobiphenyl	6.23

[a]Chiou, 1985.
[b]Triolein is glyceryl trioleate.

absorption. However, significant storage occurs only with those whose K_p exceeds 10,000 (Table 7.1). Such high values usually are presented in log form; that is, log K_p = 4.0 in this case.

Disposition also includes irreversible binding to tissue, xenobiotic metabolism (Chapter 6), and elimination. Some substances, especially metal derivatives, react chemically with cell components such as proteins and nucleic acids to bond so tightly that they are lost only by the normal turnover of the structural macromolecules.

7.2.3. Elimination

Absorbed chemicals are eliminated from most organisms automatically, rapidly, and continuously. They simply diffuse out of microbial cells into the surrounding medium, and even fish eliminate significant proportions in the same way via the gills. Terrestrial animals release them in exhaled breath, milk, or sweat, but mostly by *excretion* in urine, bile, and feces. Higher plants face a more difficult problem, in that while small proportions of some chemicals can be transpired away from leaf surfaces or diffused out of roots, plants do not have the other usual means of excretion. Instead, they concentrate toxicants into the water-filled vacuoles of leaf cells, usually as detoxified water-soluble metabolites, and at some point simply drop the leaves.

As circulation generally proceeds more rapidly than either absorption or elimination, one might expect that most chemicals would have a good chance at reaching a biochemical target or receptor and producing a toxic response. Receptors will be discussed in Section 9.2. However, biotransformation generally interferes and, at low exposures, may sidetrack the intoxication process entirely.

7.3 | FACTORS GOVERNING INTOXICATION

7.3.1. Biotransformation (Metabolism)

Metabolism is the balance-wheel of intoxication. We refer here to xenobiotic metabolism, as differentiated from the intermediary metabolism that converts food into energy and living structure. In most instances, the toxic effects of a chemical are inversely proportional to its metabolic detoxication and subsequent elimination. Simply the more efficient the permanent removal of the chemical from the intoxication process, the less of it will be available at the receptor. One may think of the intoxication process as a horizontal water pipe with a pump (the absorbing membrane) at one end and a flame (the receptor) at the other. If water (the toxic chemical) is pumped in, it will eventually extinguish the flame. However, metabolism may be viewed as a side tube leading to the drain; if this tube is large enough and not blocked, the water will be diverted out of the system and little or none can reach the flame.

Xenobiotic metabolism involves both direct chemical modification of toxicant structure by Phase I reactions, such as oxidation, and Phase II synthesis (conjugation) which combines the original or altered molecule with natural sugars, amino acids, etc. (Chapter 6). For example, a nitrobenzene is reduced to the corresponding amine, which is then acetylated to acetanilide (Eq. 6.1). These are largely enzymatic transformations, and most kinds of living organisms are capable of carrying out at least some of them (Section 6.4). Metabolism has been thought to confer increased water solubility that will afford easier elimination of the xenobiotic in urine or movement into plant cell vacuoles; for example, the aqueous solubility of nitrobenzene is increased by conversion to aniline. However, the aniline's solubility actually is *decreased* by acetylation, and so conjugation may serve instead to mask the reactivity of nucleophiles such as amine, hydroxyl, and carboxyl groups to prevent further reactions with receptors. Indeed, with a few notable exceptions, conjugates are substantially less toxic than their parents.

On the other hand, metabolic transformation can convert some classes of chemicals into more toxic products. A good example is the oxidative activation of insecticidal thiophosphate esters such as parathion into their oxygen analogs (oxons) (Section 10.2); another is the oxidation of relatively inert polycyclic aromatic hydrocarbons such as benzo[*a*]pyrene to epoxides that can react with DNA and initiate tumors (Section 13.3.2). Fluoroacetate is activated to fluorocitrate, nitrobenzenes to phenylhydroxylamines, and arsenate to arsenite. The substance that finally reacts with the target is called the ***proximate toxicant.***

7.3.2. Bioavailability

Bioavailability is defined by environmental scientists as the degree to which a substance is free to move into or onto an organism (Hamelink et al., 1994), while the pharmcologist's definition is that it is the proportion of chemical that enters the systemic circulation following oral administration (absorption is ignored). It is easy to understand the difference. The physician who gives the patient a pill wants to know the concentration in the bloodstream immediately available for action, while the environmental toxicologist is concerned with the proportion of chemical available from the environment via all exposure routes. For a fish swimming in polluted water, for example, the concentration of chemical ready to enter the intoxication process may be affected by such factors as adsorption onto suspended particles, the hardness of the water, the ambient temperature, and any residues already in the food.

7.3.3. Individual and Species Differences

The various species, and even individuals, may react differently toward any particular chemical. Plants, animals, and microorganisms generally do not respond in the same way to any given chemical, due partly to exposure routes, partly to types of receptors, and partly to metabolic ability. For example, plants photosynthesize and animals do not, so photosynthesis inhibitors tend to be toxic to one and not the other. Such differences lead to ***selective toxicity.***

Selectivity forms the basis of effective pesticides and antibiotics, chemicals intended to affect one kind of organism without affecting others. Much of the interest, research activity, and controversy in environmental toxicology have centered on this important subject, and the notion that selectivity among species is not only possible but commonplace seems to have escaped the general public. While no section of this book specifically addresses this aspect, it is cited frequently and is implicit in much of Chapters 6, 8, and 9. For a thorough discussion, see Albert (1985).

7.3.4. Other Factors

Many other factors influence intoxication, including the age of the organism, its stage of development, its sex, and its nutritional status. Within limits, each individual must be thought of as toxicologically distinctive, as discussed in the next chapter.

7.4 | TOXIC EFFECTS

7.4.1. "Poisoning"

Have you ever been poisoned? If so, you know some of the common symptoms: blurred vision, sweating, headache, muscular weakness, nausea, vomiting. Severe cases may include convulsions, unconsciousness, and even death. Although probably not aware of it, you reached the end of the intoxication process, and the reaction of the proximate toxicant with biochemical targets was translated into physiological responses. This *acute* poisoning was the result of just a single exposure, while continued exposure to smaller doses could have resulted in *chronic* intoxication. We cannot ask other creatures how they feel, but many of their symptoms often are similar to our own. Every living thing is susceptible to poisoning.

There are thought to be two types of toxicity: (1) nonspecific (nonreactive), where there is no specific biochemical target, and (2) specific (reactive) in which a biochemical process is affected. As suggested by the name, *nonspecific toxicity* does not result from the making or breaking of chemical bonds and is related only to the amount of toxicant absorbed. Its effects, too, are nonspecific and appear as an increasing metabolic and sensory depression *(narcosis)*. Anesthetics such as halothane (1,1,1-trifluoro-2,2-dichlorobromoethane), diethyl ether, and chloroform provide familiar examples.

Of course, toxicity requires that there *must* be a physiological target of some sort. As no biochemical reaction seems to be involved, the toxicant's physical properties dominate. K_{ow} and bioconcentration control the effects, and whether in amoeba, nematodes, fish, or mammals, the onset of narcosis is associated with a constant *internal critical level* of toxicant in membrane lipids. As neurons of the central nervous system (CNS) are surrounded by a fatty sheath, one hypothesis says that the added cellular volume caused by uptake of a lipophilic substance increases cellular pressure and leads to a depression of nerve impulses (Albert, 1985). Many chemicals cause nonspecific toxicity, including common solvents, volatile components of petroleum, and drugs such as phenobarbital, and the narcosis is observed in a wide variety of organisms. The significance of this effect has only recently been recognized outside human medicine (Connell and Markwell, 1992).

However, contrary to popular belief, most toxic chemicals do have a *specific toxicity* that results from reaction with a biochemical target and ultimately generates the observed symptoms. Some of the most important of these are discussed in the next section.

7.4.2. The Nervous System

Many toxic chemicals affect the nervous system in some way. The vertebrate nervous system is made up of two major parts: the central nervous system (CNS) comprising the brain and spinal cord, and the peripheral nervous system (PNS) that includes everything else. A simplified organization chart, adapted from Albert (1985), is shown in Scheme 7.6. Advanced invertebrates such as insects and molluscs possess a rudimentary brain, and others obviously have some system of nerve coordination. Even tiny cnidarian coral polyps display, move, and retract their tentacles and swallow their prey in an organized manner.

$$\text{(7.6)}$$

I. CNS
II. PNS:
    A. sensory nerves (afferent)
    B. Motor nerves (efferent):
        1. Somatic (voluntary)
        2. Autonomic (involuntary):
            a. Sympathetic (excitatory)
            b. Parasympathetic (inhibitory)

The brain and spinal cord serve as headquarters. Sensory information is received via afferent (input) sensory nerves from all parts of the body and processed automatically according to preset rules. Orders are sent out via efferent (output) nerves that cause the designated body parts to take action. Some actions may be intentional and voluntary on the individual's part, but they are carried out against a background of continual, involuntary action that keeps the organism operating (*e.g.*, breathing, heart pumping, food digesting, etc.) and fine-tuned by excitatory and inhibitory signals. Headquarters (the CNS) coordinates and integrates it all.

Everything is accomplished through *nerve impulses,* self-propagating bands of negative charge that are generated in the nerve cells or *neurons* (Fig. 7.3). An electric potential difference of 50–100 mV exists across the neuron's plasma membrane due to concentration differences between Na^+ on the outside and K^+ on the inside. Momentary loss of this potential (depolarization) causes a transient increase in membrane permeability to sodium ions, which pass via protein "channels" (Figs. 7.2 and 9.1) into the neuron and reverse the potential. This is followed, in microseconds, by out-migration of potassium ions and restoration of the original potential. The adjacent area of membrane by then has become depolarized, and the process is repeated. Where an electric current would travel at millions of meters per second down a wire, nerve impulses are propagated at 1–100 meters per second. Neurons showing the most rapid impulses often are insulated by a thin fatty sheath of a substance called *myelin.*

When an impulse reaches the end of the neuron, it stimulates the release of a chemical ***neurotransmitter*** from the synaptic vesicles; the common vertebrate neurotransmitters include acetylcholine (ACh), aralkylamines (catecholamines), and γ-aminobutyric acid (GABA), as detailed in Section 9.4. The transmitter then diffuses across the 20–400-nm-wide, fluid-filled gap or synaptic cleft between cells and inter-acts with a receptor in the postsynaptic membrane of the next neuron to generate a new impulse or an action in a muscle cell (Fig. 7.3). The transmitter subsequently is returned or enzymatically deactivated. An example is ACh, which is deactivated by hydrolysis catalyzed by acetylcholinesterase, AChE:

$$(CH_3)_3\overset{+}{N}CH_2CH_2O^-COCH_3 + H_2O \xrightarrow{\text{AChE}} (CH_3)_3\overset{+}{N}CH_2CH_2OH + HOCOCH_3 \qquad (7.7)$$

Acetylcholine   Choline   Acetic acid

Figure 7.3 shows only a single connection between two neurons, but the extensive interconnections between dendrites and axonic branches actually form a complex net. Plants and microorganisms appear to lack any such type of nervous system, although they obviously respond to certain stimuli.

Neurotoxic poisons   Of the many substances known to depress the activity of the CNS, probably the most familiar is ***ethanol.*** The dictionary definition of the word *intoxication* also relates it to alcohol. Where only 2% of coroner's reports found

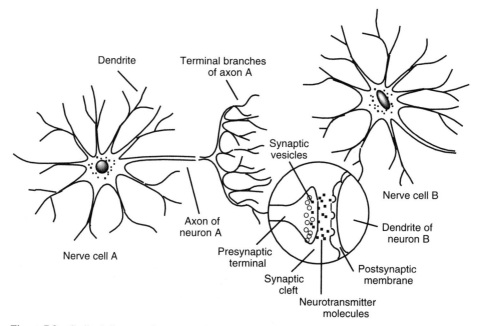

Figure 7.3. Stylized diagram of neurons, with an expanded view of the synapse that separates them.

methanol in the blood of poisoning victims, an astounding 40% of the blood samples contained ethanol (Sunshine, 1986). About 100 mg of ethanol per 100 mL of blood (0.1%), caused by drinking roughly 6 ounces of whiskey or 8 cans of beer by a 150-pound person, causes incoordination and blurred vision. At about 0.2%, there is visual impairment, staggering, and slurred speech; 0.3–0.5% causes stupor and convulsions; and over 0.5% results in coma and death (Rumack and Lovejoy, 1986). The difference between the anesthetic and fatal doses is dangerously small.

Although everyone recognizes the symptoms of alcohol intoxication, the action of this drug is poorly understood. The principal effect is depression of the CNS (anesthesia, reduced visual and sensory perception, increased pain threshold), the cause of which remains unknown. Chronic exposure to ethanol results in fat accumulation in the liver (cirrhosis) and embryotoxicity leading to mental deficiency in the newborn. The effects may be due in part to metabolites, as ethanol is converted to acetaldehyde and acetic acid (Sections 6.2.2 and 13.2.2). The undesirable effects of acetaldehyde are illustrated by the familiar hangover.

Stimulant and narcotic alkaloids, such as nicotine (Section 12.2), also act primarily on the CNS. A ***narcotic*** is a drug that relieves pain and numbs the senses (depresses the CNS), but the term also represents a broad legal classification illustrated by the alkaloid cocaine. Cocaine is called a narcotic even though it initially causes stimulation. It is extracted from *Erythroxylon coca,* a shrub whose leaves have been used for centuries by the mountain people of Peru and Ecuador to increase endurance, high-altitude body temperature, and a sense of well-being under harsh conditions. Poisoning symptoms are much like those of nicotine—restlessness, excitability, hallucinations, dilated pupils, numbness, and vomiting—but cocaine is less toxic. The usual dose by inhalation or injection is about 25 mg, although human adults can die from as little as 7 mg/kg (450 mg) taken orally or a total of 20 mg by injection. Cocaine produces a pronounced euphoria, but, as half of the substance disappears from the blood stream in less than an hour, the user may require a dose every 30–40 min to maintain the "high."

As an ester, oral cocaine is detoxified rapidly by hydrolysis in the stomach (Eq. 7.8), but chewing *Coca* leaves releases the drug slowly into the mouth and throat, and most absorption takes place there.

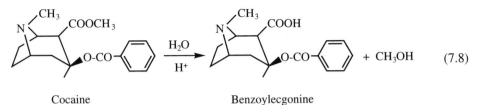

$$\text{Cocaine} \qquad\qquad\qquad\qquad \text{Benzoylecgonine} \qquad\qquad\qquad (7.8)$$

As with other narcotics, repeated use leads to increasing tolerance, so that the much larger doses eventually required lead to insomnia, weight loss, disinterest in food, and growing physical dependence (addiction). There may be hallucinations—a sensation that insects are crawling over the body—and toxic psychosis. Clearly, CNS poisoning is the cause.

A common form of neurotoxicity is caused by organophosphorus insecticides such as the diazinon (Spectracide®), chlorpyrifos (Dursban®), and malathion familiar to homeowners (Section 15.2). Even mild exposure to these compounds affects transmission of nerve impulses across the synapse (Section 9.3) and results in visual disturbances (contracted pupils), dizziness, difficult breathing, and muscle weakness. Greater exposure causes salivation, nausea, sweating, and convulsions. The symptoms start within an hour and reach their maximum in humans in 2–5 hours. Fortunately, atropine and pralidoxime (2-PAM) can be used as *antidotes* to block the effects.

Neurotoxic effects are not always immediately detectable. Several organophosphorus esters, most notably the plasticizer and gasoline additive tri-*o*-cresyl phosphate (TOCP), require days or weeks before poisoning is evident. The toxic effect is a slow degeneration of peripheral axons, termed *delayed axonal neuropathy,* resulting in increasing sensory disturbance, weakness, and eventual paralysis in the legs. The protective myelin nerve sheath may deteriorate. Many mammal and bird species are susceptible, especially humans for whom a toxic oral dose may be as little as 150 mg. Thousands of people were poisoned during Prohibition by drinking a liquor concocted from contaminated ginger extract, hence the name "ginger jake" paralysis (WHO, 1990). Thousands more have been poisoned by cooking oil diluted with this widely used industrial chemical. Intoxication is not due to the TOCP *per se,* but rather to a dioxaphorin metabolite (Casida et al., 1961, Eq. 7.8), whose biochemical mechanism of action remains unclear. The neurotoxicity of alkylmercury compounds, which behave similarly, is discussed in Section 11.5.

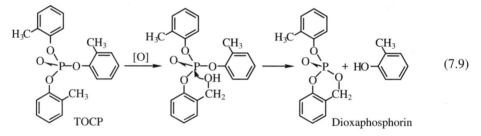

$$(7.9)$$

7.4.3. The Skin

Skin may be considered as a continuous and flexible envelope that contains and protects the body's organs and their components and serves as the barrier between them and the outside environment. It also helps to regulate internal temperature, provides sensory perception of the external world, and often is the first line of defense against toxicants. In vertebrates, its exterior is composed of hard, dead, and densely packed cells (the stratum corneum) that are continually abraded from the surface and replaced from the epidermis below (Fig. 7.4). An adult human's skin has an area of 1.5–2 m^2 and represents as much as 10% of the body weight, and it generally provides an individual's first contact with and reaction to most xenobiotics.

As with other exposure routes, the skin absorbs chemicals according to dose and K_p. The propensity for absorption varies according to location (scrotum >> forehead > scalp > back > forearm > palm) and to species (roughly, frog > rabbit > rat >

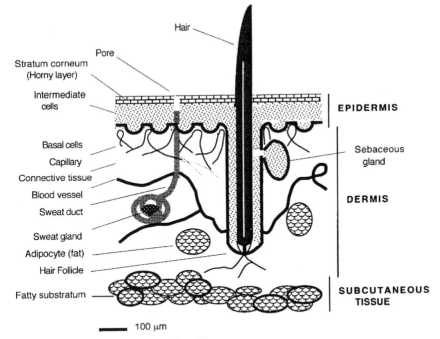

Hair

Pore

Stratum corneum (Horny layer)

Intermediate cells

EPIDERMIS

Basal cells

Capillary

Connective tissue

Blood vessel

Sweat duct

Sweat gland

Adipocyte (fat)

Hair Follicle

Fatty substratum

Sebaceous gland

DERMIS

SUBCUTANEOUS TISSUE

100 μm

Figure 7.4. Stylized diagram of mammalian skin.

pig > monkey > human). Absorption occurs largely by simple diffusion through Type 1 membranes of the epidermis or the walls of hair follicles, and the intercelluar spaces of the stratum corneum are filled with flat sheets of lipid. Obviously, liquids make better surface contact than do solids, but the skin usually is covered with a very thin waxy coating (sebum), produced by the sebaceous glands, which promotes absorption of lipophilic substances and repels polar ones. Numerous capillaries and blood vessels lie just below the surface (Fig. 7.4), as a scraped knee or elbow shows, and so absorbed chemicals quickly enter the blood stream if they do not undergo xenobiotic metabolism in the dermal cells.

Some chemicals such as nickel salts, benzocaine, epoxy resins, and formaldehyde cause *contact dermatitis,* in which the skin responds with reddening (erythema), swelling (edema), and blisters. Organic acids such as acetic acid and phenol cause painful burns because, being lipophilic, they are easily absorbed and then can work from the inside. Others cause acnelike eruptions, most notably the *chloracne* from exposure to halogenated aromatics such as the chlorodioxins (Section 14.4) and polychlorinated biphenyls. PCBs were responsible for "telephone dermatitis" on the faces of people who used old-style phones made of PCB-containing plastics.

Perhaps the most widespread toxic skin reaction is *allergic contact dermatitis* (ACD), typified by the rash that results from exposure to poison oak or poison ivy (Section 12.4). The toxic agent is absorbed, combines with carrier proteins on the Langerhans cells dispersed throughout the epidermis, and the resulting antigens are transported to regional lymph nodes. There they sensitize T-lymphocytes, which replicate and enter the circulating blood. Hapless individuals thus sensitized—about two-thirds of the U.S. population—can expect to suffer the characteristic erythema, edema, and blisters again and again upon even slight future exposures of any part of the body as the cells release inflammatory cytokines such as serotonin (Maurer, 1983). While partial and temporary immunity may be conferred by intentional treatment with the antigen, and immediate washing with soap or kerosene emulsion can reduce the spread of the toxicant, the only sure relief is avoidance.

Ultraviolet (UV) radiation penetrates the epidermis, about 20% at 300 nm and over 50% above 400 nm, and is absorbed by skin pigments such as melanin or the tryptophan and tyrosine of protein. If a xenobiotic is present and can absorb the radiation, it also may become chemically activated and capable of reacting with cell constituents. Subsequent reactions with protein can produce photoallergens, as caused by the sunscreen glyceryl p-aminobenzoate, and from there the process is similar to ACD to produce *phototoxicity.* In a famous example, phototoxic furocoumarins (methoxypsoralens) were administered orally to animals with the hope of finding a "suntan pill." UV irradiation did indeed produce tanning, but no way was found to halt the reaction before the skin turned to leather! Emmett (1986) gives a good account of the effects of chemicals on skin.

Insects and many other invertebrates do not possess skin as such, but rather a hardened exoskeleton or *cuticle.* In insects, this consists of layers of polysaccharide (chitin) and protein (arthropodin) covered by a 1–2 μm epicuticle composed largely of wax. Lipophilic chemicals can penetrate the cuticle, by diffusion and around hairs (setae), according to their K_p just as in vertebrates. Thus contact insecticides such as DDT need only to reach the insect exoskeleton to be effective. The surface of most other invertebrates consists of only one or a few layers of cells that are easily penetrated by chemicals. Plants do not have "skin," of course, but most are covered with a thin waxy cuticle, and chemicals can enter through this or through respiratory pores (stomata) on the leaves. Otherwise, the intoxication pathways are the same as in animals.

7.4.4. The Respiratory System

Terrestrial vertebrates breathe by means of lungs, fish and other aquatic animals have gills, and insects breathe through abdominal pores called *spiracles.* Respired oxygen is absorbed into the bloodstream and then delivered to the tissues to fuel oxidative metabolism. The mammalian respiratory system is composed of the trachea (main air tube), bronchial tubes of ever-decreasing size, and finally the minute alveoli from which air diffuses into the blood. The total alveolar area in an adult human is about 100 m^2, and most of the 11.7 L/min volume of air breathed by an adult male at rest (Appendix 10.1) is transferred to the blood while waste CO_2 moves out. The plasma membranes of the lung are exceptionally thin, so respiratory absorption of gaseous toxicants is very efficient; as the process is reversible, the lungs also serve as an effective route for elimination of chemicals via exhaled breath.

Many gases (chlorine, sulfur dioxide, ammonia) and vapors (toluene, methyl isocyanate) are acute pulmonary (lung) poisons, while others such as methyl bromide and trichloroethylene move from the alveoli into the bloodstream but act at other sites. A cloud of methyl isocyanate released during a 1984 industrial accident in Bhopal, India, killed more than 2000 people and injured as many as 100,000 more. Survivors suffered from blocked airflow, severe cough, chest pain, and pneumonitis. Another pesticide, the herbicide paraquat, is stored in and poisons the lungs no matter what part the body is exposed; it is often used for suicide, as its action is irreversible and death occurs within several days via massive pulmonary cell proliferation.

Carbon monoxide is a product of incomplete combustion in gas heaters, fireplaces, automobile exhaust, and tobacco smoke. It is responsible for over 2000 deaths each year in the United States, many of them suicides (Table 1.2). Respiratory exposure of nonsmokers to CO ordinarily is well below the 50 mg/m^3 necessary for the first signs of poisoning, but air above stalled traffic contains >50 mg/m^3, a closed car with a smoker in it affords >100, an unventilated garage provides at least 110, and a heavily traveled tunnel may contain >250 mg/m^3. In heater or automobile exhaust, CO can reach ~5% by volume, or about 60,000 mg/m^3. As an adult human male breathes about 20 m^3 of air each day, and CO is cumulative over short time periods, unventilated enclosures can become lethal.

Symptoms of CO poisoning begin with altered vision and mental responses at ~50 mg/m^3 and progress through headache, nausea, mental confusion, and unconsciousness. The skin takes on a bluish color characteristic of a lack of oxygen, as CO combines with hemoglobin in the blood to diminish its capacity to carry oxygen (Section 15.8). The toxic action is not on the lungs: many of the poisoning symptoms are typical of CNS damage, and death usually results from respiratory failure as the brain shuts down for lack of oxygen. CO accumulates during exposure, but clean air or oxygen reverses the toxicity; traffic police in downtown Tokyo are required to don facemasks and breathe pure oxygen at intervals throughout the day.

7.4.5. Blood and Internal Organs

Blood distributes oxygen and nutrients as well as toxic chemicals throughout the body, but it is cleansed by the liver and kidneys, which remove and degrade the chemicals and dispose of the wastes via bile and urine. The blood is forced through an animal's vascular system by a carefully controlled muscular pump, the heart. The simplest animals, such as sponges, cnidarians, and flatworms, apparently get along without this machinery, but it is clearly defined in earthworms, insects, and molluscs as well as in all vertebrates.

Some environmental chemicals exert a direct effect on the heart, for example, the steroid glycosides that make foxglove *(Digitalis)* and oleander *(Nerium)* so poisonous (Section 12.3). Digitalis is valuable in medicine for its ability to stimulate contractions in the heart muscle, but it may fatally disrupt heart rhythm in slightly larger doses. Natural sympathomimetic amines related to amphetamines occur in foods such as cheese and wine and can lead to fatally high blood pressure by constriction of blood vessels. As an extreme example, ergot alkaloids cause constriction that results in gangrene in the fingers and toes and was responsible for thousands of human deaths in medieval Europe.

The liver is porous, highly vascular, and penetrated by bile ducts which move fatty substances into the small intestine. It is responsible for much of the body's intermediary metabolism, such as fat digestion, and for most xenobiotic metabolism. It is especially active in oxidations via cytochrome P450, and is virtually a low-temperature incinerator. Substances harmful to the liver are said to be ***hepatotoxic,*** and as this organ is the principal site of metabolic detoxication, it can be affected adversely by many kinds of chemicals. In particular, chlorinated hydrocarbons such as carbon tetrachloride and trichloroethane produce liver necrosis and fat accumulation (cirrhosis).

The kidneys serve as selective filters, retaining valuable nutrients and moving dissolved wastes out of the body via the urine. Whether the flame cells of lower invertebrates such as rotifers and flatworms have this capacity is open to question, but it is seen in the hepatopancreas of molluscs and the nephridia of earthworms. Damage to these organs, or to the liver, threatens the animal's ability to deal with xenobiotics. Many substances, including mercury and cadmium compounds, damage the kidneys, and the poisoning symptoms always indicate a reduced ability to process wastes.

Toadstool poisoning provides an example of how specific some intoxication can be. Eating a single toadstool cap from a member of the common genus *Amanita,* such as *A. phalloides* ("destroying angel"), can be fatal. The poisoning symptoms commence after a period of 6–24 hours and include severe abdominal pain, nausea, bloody vomit and stools, inability to urinate, and jaundice, all clear signs of major damage to liver and kidneys. Death occurs in over 50% of cases, and recovery in survivors is slow and painful. See Section 12.5.

7.4.6. The Reproductive System

Reproduction is a basic requirement of life. Sexual reproduction begins far back in phylogeny, and its many stages invite toxic interference. In mammals, for example, male processes that could be affected include the formation of sperm, its viability, and its transport to the egg. Female processes include egg formation, ovulation, fertilization and implantation of the egg, and nourishment of the embryo and fetus. Despite the obvious importance, relatively little information has been available until recently on the effects of xenobiotics other than human contraceptives on reproduction in any species.

Extensive toxicity tests in chick embryos have been reported over the years but require injection of the chemical into a fertile egg, hardly a natural process. A more environmental example was the fatal thinning of the shells of eggs laid by some species of birds exposed to chlorinated hydrocarbons such as DDT (Faber and Hickey, 1973). About the same time, sterility was observed in men occupationally exposed to the nematicide 1,2-dibromochloropropane (DBCP), and evidence has accumulated since then that many other chemicals can produce reproductive dysfunction in both men and women (Appendix 7.1).

Higher plants reproduce by means of seeds, which imbibe water, develop a root system, and send out a shoot that makes its way upward toward the light (they **germinate**). Any of these processes can be altered or stopped by toxic chemicals, many used as ***preemergence herbicides,*** when applied to the soil. Some are natural products such as the 3-acetyl-6-methoxybenzaldehyde produced by the desert shrub, *Encelia*

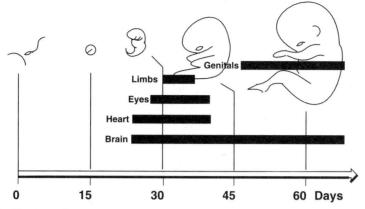

Figure 7.5.   The time schedule of normal human embryonic development.

farinosa, apparently to protect its water supply from competitors (Gray and Bonner, 1948).

Somewhat more is known about the ***teratogenic*** effects of chemicals, that is, a toxic action in an embryo or fetus that results in birth defects. The mammalian embryo is in contact with its mother through the placenta, which allows the transfer of nutrients but excludes many other chemicals; mother and baby have separate blood supplies. The embryo develops rapidly (Fig. 7.5), and toxicants that do get across the placental barrier can have specific effects on organogenesis, depending on the time at which exposure occurs. The best-known example of a teratogenic chemical is ethanol, whose consumption by a pregnant woman or animal results in mental retardation, reduced head size, and muscular incoordination in the newborn baby (fetal alcohol syndrome). Thalidomide was a sedative drug that produced spectacular teratogenic damage in humans, especially shocking in that the characteristic limb defects were not seen in rodents and so were unexpected.

A long-recognized example of teratogenicity in Western sheep is cyclopism, the result of the pregnant ewe eating wild skunk cabbage *(Veratrum californicum)* on the fourteenth day of gestation. The lambs are born with only a single eye, in the middle of the forehead, and malformed jaws. The toxic agent has been identified as an alkaloid called cyclopamine (Keeler, 1969). Another example is "crooked-calf disease," caused by the alkaloid anagyrine, in which a pregnant cow that has consumed wild lupines (*Lupinus* species) early in gestation bears calves with malformed forelegs. In 1980, a human baby in California's Trinity Mountains was born with badly deformed hands and forearms, and investigation revealed that the pregnant mother had been drinking milk from goats that fed on lupines rich in anagyrine (Kilgore et al., 1981). The incidence of teratogenic effects caused by environmental chemicals must be much greater than anyone realizes.

All aspects of reproduction are under hormonal control. There is increasing evidence that many environmental chemicals mimic estrogens at the estrogen receptor or inhibit estrogen or androgen action to produce reproductive dysfunctions (Colborn et al., 1993). For example, the eggshell thinning by DDT referred to above, is now thought to be due to estrogenic 2,4'-DDT, a major impurity in the commercial insecticide (Section 14.2). Although natural estrogens from plants have long been recognized, including formononetin and coumestrol in alfalfa and clover that cause reproductive disturbances in cattle, the estrogenic potential of persistent pollutants such as the insecticide kepone and hydroxylated metabolites of some PCB and PAH congeners has only recently been recognized (McLachlan, 1993). However, to put this threat into perspective, if the relative human reproductive hazard due to chlorinated hydrocarbons were set at 1, that due to natural plant flavonoids in food would be 4×10^7 and that from birth-control pills 6.7×10^9.

The effects of these "endocrine disrupters" are not restricted to mammals. Decreased fertility and hatching success, masculinization of females, and feminization of males have been observed in birds, fish, reptiles, and molluscs as well (Colborn et al., 1993). Effects are especially severe where the pollutants are bioconcentrated, for example, in PCB-contaminated Great Lakes fish and the animals that feed on them. As most endocrine disrupters are not acutely toxic, mutagenic, or carcinogenic, the hazard they present has been slow to be recognized. Although chlorinated hydrocarbons have been the favorite subjects of recent research, evidence is accumulating that sewage entering waterways contains urinary human estrogens, birth-control hormones, and the spermicide nonylphenol, all of which are potent endocrine disrupters.

7.4.7. Genes

Genes carry the information required for the life of every cell. They are precise sequences of nucleic acids (DNA), located at specific sites on the cell's chromosomes, and are largely responsible for regulating the biosynthesis of the enzymes that control all cellular processes. Changes in or damage to them can change the code and result in an altered cell *(mutation)* and may be caused by heat, radiation, or certain chemicals called **mutagens.** Mutations are not necessarily bad, and in fact are required for evolution. Also, genetic damage often can be repaired, and a seriously mutated cell seldom survives. Contrary to science fiction stories, the "symptoms" of mutation generally are not immediately noticable—no monsters.

The present flurry of interest in mutation and mutagens is due to the relation of mutagenicity to cancer. A tumor or cancer is the result of an unregulated proliferation of cells caused by changes in their genetic code, that is, by mutation. A chemical that causes the change is said to be **genotoxic.** The incidence of cancer in the United States has remained the same, or decreased, over many years, with the exception of lung cancer caused by smoking. About 80% of avoidable cancer deaths are estimated to be caused by environmental factors, primarily tobacco (30%), diet (35%), reproductive behavior (7%), and occupation (4%), while the popular causes such as pollution, food additives, pesticides, and UV radiation provide only about 5% (Doll and Peto, 1981).

"Cancer" is not just one disease. Leukemia affects white blood cells and their sources, sarcoma is cancer of connective tissue such as bone and cartilage, lymphoma affects the lymphatic system, and carcinoma—the most common form of cancer—is a disease of tissues such as skin that make up the interior and exterior of the body.

Tumors cause extensive swelling and disruption of tissue, but death is due to other causes such as infection or heart failure. Based on extensive observation, all animals appear to be subject to tumors, but few chemicals are known to cause tumors in plants.

Chemicals that cause tumors are called **carcinogens.** A few carcinogens act directly, but most reach the nucleus only in altered form, if at all (Special Topic 9). They can act in a variety of ways: **procarcinogens** like vinyl chloride require metabolic activation to become carcinogens; **promoters** like chlorinated hydrocarbons increase the potency of a primary carcinogen; and **cocarcinogens** such as TCDD enhance the overall carcinogenic process, for example, by increasing toxicological activation. The proximate carcinogen most often is an electrophile such as an epoxide, and it must have exactly the right balance of reactivity to allow it to avoid metabolic detoxication and still effectively attack DNA. This leads to considerable specificity in its site of action and selectivity with regard to species, sex, and organ.

An example is aflatoxin B_1, a carcinogenic lactone from the mold *Aspergillus flavus* (Section 12.6). Aflatoxins occur in oily foods such as old peanut butter that has been exposed to air, and they have been detected widely in shelled peanuts, including those from vending machines. Prolonged exposure to even a very small amount results in liver cancer in humans, rats, and monkeys. An epidemic of liver cancer in Nigeria, where peanuts are a staple and refrigeration scarce, was brought under control by regulation of peanut storage conditions to deter mold.

The International Agency for Research on Cancer (IARC) lists substances known to cause cancer in humans (Table 7.2). Many more compounds are suspect, based on animal tests, and still others undoubtedly remain to be discovered. Some are industrial chemicals, some drugs, and many are common natural products that occur in food (Ames, 1983). Humans and most other animals are exposed daily to chemicals that give positive results in laboratory tests for mutagenicity or carcinogenicity as well as the other types of toxicity discussed previously. It is a tribute to our selective absorption, biotransformation, and excretory abilities that so many of us survive.

TABLE 7.2
Some Recognized Human Carcinogens[a]

Carcinogen	Usual Exposure	Site
Aflatoxins	Food	Liver
Arsenic compounds	Agriculture, water	Liver
Asbestos	Buildings, dust	Lung
Benzene	Gasoline	Lung
Benzidine	Dyes	Bladder
Benzo[a]pyrene[b]	Food, oil, roads	Lung, stomach
Diethylstilbestrol	Food	Vagina
Dimethylnitrosamine[b]	Processed food, beer	Stomach
Ethylenethiourea[c]	Processed food	Thyroid gland
2-Naphthylamine	Dyes	Bladder
Phenacetin[b]	Headache remedies	Bladder
Vinyl chloride	Plastics	Liver

[a] IARC (1987).
[b] Probable human carcinogen.
[c] Listed by the U.S. National Toxicology Program but not by IARC.

7.5 | REFERENCES

Albert. A. 1985. *Selective Toxicity,* 7th Ed., Chapman and Hall, London, UK.

Ames, B. N. 1983. Dietary carcinogens and anticarcinogens. *Science* **221:** 1249–64.

Barlow, S. M., and F. M. Sullivan. 1982. *Reproductive Hazards of Industrial Chemicals,* Academic Press, London, U.K.

Casida, J. E., M. Eto, and R. L. Baron. 1961. Biological activity of a tri-*o*-tolyl phosphate metabolite. *Nature* **191:** 1396–97.

Chiou, C.T. 1985. Partition coefficients of organic compounds in lipid–water systems and correlations with fish bioconcentration factors. *Environ. Sci. Technol.* **19:** 57–62.

Colborn, T., F. S. vom Saal, and A. M. Soto. 1993. Developmental effects of endocrine-disrupting chemicals in wildlife and humans. *Environ. Health Perspect.* **101:** 378–84.

Connell, D. W., and R. Markwell. 1992. Mechanism and prediction of nonspecific toxicity to fish using bioconcentration characteristics. *Ecotoxicol. Environ. Safety* **24:** 247–65.

Doll, R., and R. Peto. 1981. *The Causes of Cancer,* Oxford University Press, New York, NY.

Emmett, E. A. 1986. Toxic responses of the skin, in *Casarett and Doull's Toxicology,* 3rd Ed. (C. D. Klaassen, M. O. Amdur, and J. Doull, eds.), Macmillan, New York, pp. 412–31.

Faber, R., and J. Hickey. 1973. Eggshell thinning, chlorinated hydrocarbons, and mercury in inland aquatic bird eggs. *Pestic. Monit. J.* **7:** 27–36.

Finean, J., R. Coleman, and R. Michell. 1978. *Membranes and Their Cellular Functions,* 2nd Ed., Oxford University Press, Oxford, U.K.

Gray, R. A., and J. Bonner. 1948. Structure determination and synthesis of a plant growth inhibitor, 3-acetyl-6-methoxybenzaldehyde, found in the leaves of *Encelia farinosa. J. Amer. Chem. Soc.* **70:**1249–53.

Hamelink, J. L., P. F. Landrum, H. L. Bergman, and W. H. Benson. 1994. *Bioavailability: Physical, Chemical, and Biological Interactions,* Lewis Publishers, Boca Raton, FL.

IARC. 1987. *Overall Evaluations of Carcinogenicity: An Updating of IARC Monographs Volumes 1 to 42.* International Agency for Research on Cancer, Lyon, France, Supplement 7.

Keeler, R. F. 1969. Toxic and teratogenic alkaloids of Western range plants. *J. Agr. Food Chem.* **17:** 473–92.

Kilgore, W. W., D. G. Crosby, A. L. Craigmill, and N. K. Poppen. 1981. Toxic plants as possible human teratogens. *Calif. Agric.* **35**(11/12): 6.

Maurer, T. 1983. *Contact and Photocontact Allergens,* Marcel Dekker, New York, NY.

McLachlan, J. A. 1993. Functional toxicology: A new approach to detect biologically active xenobiotics. *Environ. Health Perspect.* **101:** 386–87.

Nisbet, I. C., and N. J. Karch. 1983. *Chemical Hazards to Human Reproduction,* Noyes Data, Park Ridge, NJ.

Rumack, B. H., and F. H. Lovejoy. 1986. Clinical toxicology, in *Casarett and Doull's Toxicology,* 3rd Ed. (C. D. Klaassen, M. O. Amdur, and J. Doull, eds.), Macmillan, New York, NY, pp. 879–901.

Sunshine, I. 1986. Analytical toxicology, in *Casarett and Doull's Toxicology,* 3rd Ed. Macmillan, New York, NY, pp. 857–78.

WHO. 1990. *Tricresyl Phosphate,* World Health Organization, Geneva, Switzerland. Environmental Health Criteria 110.

Special Topic 7. Adaptation

When DDT was first introduced in the early 1940s, its control of disease-carrying mosquitoes and houseflies seemed almost miraculous. However, within only a few years, its effectiveness seemed to wane, and eventually even massive applications failed to provide the needed insect control. The same was true for penicillin and other "miracle drugs" used against pathogenic bacteria, and they became increasingly ineffective. It was said that the organisms had developed *resistance*. That is, they were genetically able to adapt to the presence of the toxicants.

Adaptation is not restricted to pests. Most smokers can recall how sick they felt after their first cigar, cigars they can now relish. They developed a *tolerance* for the nicotine. Resistance is a long-term and heritable ability to avoid intoxication and is generally a property of populations, while tolerance is defined as a more temporary and individual effect. Resistance to environmental toxicants is now observed in fish, amphibians, and mammals as well as in insects, microorganisms, and higher plants. The basis for tolerance and resistance is present in all living things.

The most common cause of tolerance is the ***induction*** of enzymes responsible for biodegradation. Exposure of the body to nicotine causes a rapid increase in the biosynthesis of the enzyme, cytochrome P450 (Section 6.2), which transforms it into less toxic products such as cotinine. The presence of penicillin stimulates bacteria to make the hydrolytic enzyme penicillinase, and insects respond to DDT by increasing the level of DDT dehydrochlorinase which converts the insecticide to nontoxic DDE. Chemicals that share the same mechanism of toxic action can stimulate ***cross-resistance,*** where an insect resistant to DDT also becomes resistant to the structurally-related methoxychlor but not to dieldrin (Section 9.4.1). Tolerance usually is temporary and declines within days or weeks after removal from exposure, and it is not heritable.

Continued exposure is different. As susceptible organisms are removed from the population, genetic variability and biochemical individuality ensure that the better-adapted individuals will pass resistance on to the next generation. Bacteria, insects, and some plants can go through many generations in a short time, and if a pressure such as a pesticide continues to operate, the selection process will repeat over and over. Where induction may double or triple an organism's adaptation level, genetic resistance eventually may increase it 1000-fold. Darwinian evolution is at work.

Biochemical detoxication is not the only route to adaptation. Exposure to cadmium or other inorganic toxicants induces the formation of a type of low-molecular-weight, high-sulfur proteins called *metallothioneines* (Section 11.6), which bind the metals and make them unavailable. Refractory chlorinated hydrocarbons are stored in lipid, and as long as they remain below toxic levels in the rest of the body, fat harmlessly accumulates concentrations that would be lethal if received all at once. Alternatively, an ingested chemical may be eliminated so rapidly that it never reaches a toxic level, the principal means by which the tobacco hornworm *(Protoparce sexta)* avoids nicotine intoxication.

Adaptation to intoxication appears inevitable. Over the centuries, it has permitted organisms to survive the natural toxic constituents in their food and to combat disease. Some actually benefit from food toxicants; larvae of the monarch butterfly *(Daneus plexippus)* concentrate toxic cardenolides from their food (milkweed) and themselves become toxic to predators. However, adaptation also presents serious problems for our modern chemical-oriented society. Continued use of pesticides results in ever-more-resistant competitors for food, health is at risk from uncontrollable microorganisms, and the effectiveness of medication is jeopardized by our own increasing biodegradative ability.

Every few years, a new pesticide such as the super-herbicide bensulfuron-methyl (Londax®) is introduced with the promise that *this* time, the target organism cannot resist. Just to make sure, resistance genes now can be spliced into crop plants to allow them to survive the larger applications of herbicides that force the demise of any recalcitrant weeds. However, evolution provides a different promise. Weeds already have adapted to Londax®, and the genetically altered crops are somehow sharing the new resistance genes with them.

APPENDIX 7.1.
Reproductive Dysfunction from Some Environmental Chemicals[a]

Chemical	Exposure[b]	Dysfunction Male	Dysfunction Female
Aniline	Industrial waste		+
Benzene	Gasoline		+
Carbon disulfide	Rubber	+	+
Chloroprene	Rubber	+	+
Dibromochloropropane	Drinking water	+	
Ethanol	Liquor, medicine		+
Ethylene dibromide	Food, gasoline	+	
Ethylene oxide	Hospitals	+	+
Formaldehyde	Indoor air	+	+
Glycol ethers	Solvent, nail polish	+	+
Hexane	Gasoline	+	+
Inorganic lead	Old paint	+	+
Organic lead	Gasoline	+	+
Methylmercury	Food		+
Nicotine	Tobacco		+
Phthalate esters	Plastics		+
Polychlorinated biphenyls	Recycled paper		+
Styrene	Plastics		+
Toluene	Household solvent		+
Vinyl chloride	Plastics	+	+

[a] Barlow and Sullivan, 1982; Nisbet and Karch, 1983.
[b] Other than occupational exposure.

8.1 | DOSE–RESPONSE RELATIONSHIPS

The intoxication process is mass-driven. That is, the greater the exposure, the more of a chemical will be absorbed per unit time and the higher will be the probability of its reaching the biochemical target. Thus toxic effects will be related to the degree of exposure, as recognized by Paracelsus hundreds of years ago (Section 1.6) and confirmed quantitatively in the early 1800s by Orfila. Whether in people, roses, or bacteria, "the dose makes the poison."

When the extent of any particular response is plotted against dose (exposure), a characteristic S-shaped curve is produced (Fig. 8.1A). At very low doses, biodegradation and/or elimination dominate the intoxication process, and no response is observed. The highest dose that still produces no response is the *no observable effect level* (NOEL), and the lowest dose to give the response is termed the *toxic threshold*. It is important to understand that finite NOELs exist for *any* chemical, as proven by the natural toxicants that make up our daily diet. With increasing doses, more of the test organisms respond, and eventually all do. In animals, the observed response often will be death, so a *lethal dose* (**LD**) or *lethal concentration* (**LC**) can be determined. Likewise, a nonlethal response may be quantified as an *effective dose* (**ED,**) or what pharmacologists sometimes call a *TD* (toxic dose).

Each individual organism is toxicologically distinctive because of variations in its absorption, metabolism, and response. Thus, for a given dose, some variation in response is to be expected, and a large enough test population provides responses that fit the bell-shaped curve of normal probability (Fig. 8.1B). *Hypersensitive* organisms respond at the low-dose end and *resistant* organisms at the other. If the test population were perfectly homogeneous, the dose–response line would be essentially vertical, but in real life, the steepness of the slope helps determine the actual degree of homogeneity of the affected population.

To account for this frequency distribution, a graphical system using probits

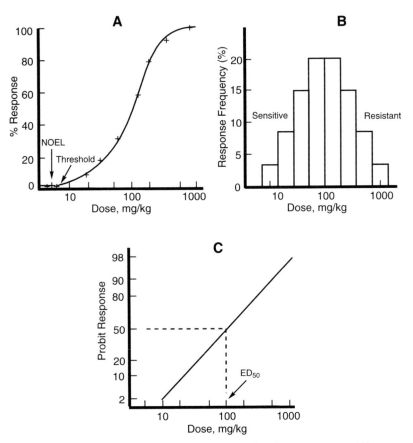

Figure 8.1. Three representations of the dose–response relation: Percent response (A), response frequency (B), and response in probability units (C). ED_{50} = 100 mg/kg.

(*prob*ability un*its*) has been adopted that reflects the standard deviations among the responses. When the y axis is thus corrected for probability and the x axis for cumulative (log) dose, the dose–response relationship becomes linear (Fig. 8.1C). The point of maximum probability (50% response) is the median, and the corresponding dose is called the median effective dose (**ED_{50}**). If that response is death, the ED_{50} becomes the median lethal dose, or **LD_{50}**. The LD_{50} of a chemical, then, is the dose that will kill a *statistical* 50% of a test population and is usually measured in milligrams of toxicant per kilogram of body weight (mg/kg). In the illustration, the calculated LD_{50} is 100 mg/kg, the toxic threshold is 6 mg/kg, and the highest NOEL is 5 mg/kg.

The LD_{50} has long been the most common measure of toxicity in animals, and extensive compilations of values are available. A few examples are listed in Table 8.1. Although the figures ordinarily are not precisely reproducible, an increase in the numbers of test animals and replicate experiments can reduce what otherwise may be a rather high variability. Variability in the rat oral LD_{50} values of Table 8.1 ranges from a usual ± 10% in the case of DDT to about ± 50% for carbaryl. While toxicity

TABLE 8.1
Acute Toxicities of Some Insecticides

Chemical	Rat[a] Oral LD$_{50}$ (mg/kg)	Rat[a] Dermal LD$_{50}$ (mg/kg)	Mallard Duck[b] Oral LD$_{50}$ (mg/kg)	Trout[c] 96 h LC$_{50}$ (μg/L)
Parathion	3.6 (3.2–4.0)[d]	6.8 (4.9–9.5)	1.9 (1.4–2.6)	1430 (962–2110)
Methyl parathion	24 (22–28)	67 (63–72)	10 (6.1–16.3)	3700 (3130–4380)
Dieldrin	46 (41–51)	60 (52–70)	381 (141–1030)	1.2 (0.9–1.7)
Diazinon	76 (66–87)	455 (379–546)	3.5 (2.4–5.3)	90 (—)
DDT	118 (106–131)	2510 (1931–3263)	>2240	8.7 (6.8–11.4)
Carbaryl	500 (307–815)	>4000	>2179	1950 (1450–2630)
Malathion	1000 (885–1130)	>4444	1485 (1020–2150)	200 (160–240)

[a] Gaines, 1960; each measurement represents 60–150 female rats.
[b] Tucker and Crabtree, 1970.
[c] Johnson and Finley, 1980.
[d] Range of 95% statistical confidence.

rank among a series of chemicals generally will remain roughly the same upon replication within a species, substantial differences exist between species due to different size, metabolic ability, and exposure conditions. Test species, chemical identity, and purity always are the most important factors in toxicity measurements and must be clearly specified.

8.2 | FACTORS AFFECTING QUANTITATIVE RESPONSES

The duration of exposure is a major factor in quantitative responses. The LD$_{50}$ values of Table 8.1 represent *acute toxicity,* that is, exposure to a single measured dose over a short period of time. Toxicity usually is greater (LD$_{50}$ is smaller) upon *subacute* (a few days) or *chronic* (longer-term) exposure, and so exposure time must always be specified. A duration of 24–96 h is usual for acute tests.

Physiological and hormonal differences often cause toxicity values to differ be-

TABLE 8.2
Toxicity of Pesticides in Male and Female Rats[a]

Chemical	Oral LD$_{50}$ (mg/kg)		Dermal LD$_{50}$ (mg/kg)	
	Male	Female	Male	Female
Parathion	13	3.6	21	6.8
Methyl parathion	14	24	67	67
Dieldrin	46	46	90	60
Diazinon	108	76	900	455
DDT	113	118	—	2510
Carbaryl	850	500	>4000	>4000
Malathion	1375	1000	>4444	>4444

[a] Gaines, 1960.

tween males and females of the same species, the female almost always being the more sensitive (Table 8.2). Absorption route also is very important: For a specified dose, *dermal* (skin) exposure usually results in lower toxicity than does **oral** (GI tract) exposure, and *respiratory* (lung or gill) exposure generally produces the highest toxicity. Dermal exposure is the most prevalent, certainly among humans, and respiratory exposure may be a close second, but oral dose is easiest to measure. In fish, the respiratory route provides about 90% of any exposure and dermal about 10%, while oral exposure often is negligible (see Table 3.3). The age or developmental stage makes a big difference, as the young are almost always the most sensitive. The various developmental stages of invertebrate animals often respond as though they were separate species.

Chemical purity directly affects the toxic outcome. For example, highly purified malathion insecticide shows negligible oral toxicity in rats (LD$_{50}$ 12,500 mg/kg), but the presence of only 1–2% of highly toxic impurities increases the acute toxicity 13-fold (see Table 15.1). The toxicity of a mixture of chemicals actually can be greater than the sum of the individual toxicities *(synergism)* if one of them inhibits the detoxication of another, or less than the sum if one interferes with the other's receptor *(antagonism)*. Synergism appears to be rarer, and antagonism more frequent, than might be expected.

8.3 | TOXICOKINETICS

Measurement of an LD$_{50}$ tells little about the *rates* at which uptake, distribution, biotransformation, and elimination take place. This is the realm of toxicokinetics. Much of our knowledge in this area comes from long experience with human medicine *(pharmacokinetics),* where such information is important in determining the efficacy and safety of specific doses of drugs. There has been relatively less application of toxicokinetics to environmental situations, where the emphasis most often has

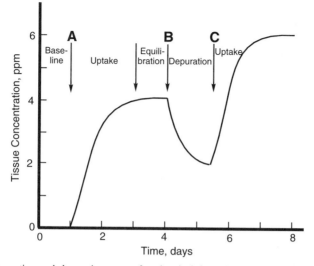

Figure 8.2. Absorption and depuration rates of a chemical from food or water. Exposure starts at A, depuration at B, and renewed exposure at C.

been on exposure. The two situations differ also in that medical applications usually involve single or sporadic doses, while environmental exposures often are low level and continuous.

The initial absorption of most un-ionized compounds into plants, animals, and microorganisms can be rapid. Figure 8.2 shows typical rates of uptake and loss *(depuration)* of a chemical from water or food. At some point, an equilibrium is reached as intake is balanced by metabolism and elimination (Section 3.3). If the organism is then removed from contact with the chemical, body content drops as net uptake is reversed, and upon reexposure at a lower concentration, a new equilibrium will be established. This equilibration occurs within a few hours in small aquatic species but may require many days or weeks in large terrestrial animals such as humans.

Once absorbed, the substance is distributed throughout the organism via normal circulation, more effectively in animals than in plants. The rates of its distribution into various organs and tissues, and the subsequent metabolic transformation and elimination, also can be estimated and provides a dynamic picture of persistence, fate, and concentration at toxicologically active sites. This will be discussed more fully in Section 10.3.

8.4 | TOXICITY MEASUREMENTS AND ENDPOINTS

8.4.1. Measurements

The purpose of most quantitative toxicity measurements is to estimate how much exposure to a particular toxicant an organism can tolerate (the NOEL) and how poisonous the chemical is relative to other substances (ED_{50}). One must decide at the outset exactly what response is to be measured and what endpoint will be used. Death is a crude but common toxicological endpoint, although it is not always easy to detect. For example, the small crustacean *Daphnia magna,* a frequent test animal, may appear dead after exposure to a toxic chemical only to return to life when moved to clean water. Growth rate often is an appropriate measure of toxicity in young plants and animals and is the usual endpoint in microbial cultures, but weight loss, respiration rate, and relative activity also are used. Toxicity sometimes is only inferred, one example being the measurement of blood cholinesterase depression to estimate exposure to organophosphate insecticides even when physiological symptoms of poisoning may be absent (Section 10.3).

8.4.2. Acute Toxicity

Probably the most common type of toxicity measurement for environmental chemicals is still acute lethality (LD_{50}, LC_{50}, etc.), and the white rat or mouse is the usual test organism. However, toxicity data now exist for many other species. In tests of pesticide toxicity to wildlife, Tucker and Crabtree (1970) measured LD_{50} values in 31 species (not all chemicals in all species), and extensive LC_{50} tables exist for some fish and aquatic invertebrates (Johnson and Finley, 1980). The number of animals required for a test depends on the desired level of statistical significance: With 50 individuals per dose level, even an 8% response is significant, but 4 out of a group of 10 must respond for significance, and *all* must respond if the total number is only three. Ten individuals is a common compromise.

Animals usually live in clean cages or other enclosures that hold 1–10 individuals, but the numbers of fish or other aquatic species may be greater, and they often are tested in large containers in flowing water. All are allowed food and water as they wish it *(ad libitum).* Rats and mice are highly inbred to make them as genetically similar as possible, and while this somewhat reduces variability in the results, it also removes those results ever further from what might have been seen in Nature. Acute toxic responses usually are measured within 96 h, although the treated animals may be held several weeks for observation. Needless to say, sick animals and dirty cages produce unsatisfactory results.

Introducing the chemical by mouth (oral, *per os*) can be by gavage, that is, direct introduction of a solution or suspension into the stomach by means of a tube down the throat, or via food or water. The dose also can be given by intraperitoneal (i.p.) injection into the body cavity, or injection into a vein (intravenous, i.v.) in larger animals, but these methods obviously circumvent normal absorption and maximize toxicity. For dermal administration, the hair is shaved from the back of a rat or rabbit, the test chemical is applied in a carrier such as corn oil, and the site is bandaged for a few hours to several days. Insects and other invertebrates can be dosed by external ("topical") application to the exoskeleton. Toxicity to the eyes (ocular toxicity) is measured by the controversial Draize Test in which a few drops of a solution of the chemical are introduced into one eye (usually in a rabbit), the other eye being used as control.

The respiratory route (inhalation) is the most difficult and expensive to use, as gaseous or highly volatile substances must be measured accurately and continuously into the airflow of a large chamber containing the animals. Adding known concentrations of solids or liquids as aerosols is even more difficult, especially as all the particles must be the right size for absorption (<1 μm in diameter), and skin absorption and other losses must be accurately accounted for.

An interesting twist in toxicity measurements is *phototoxicity.* Certain chemicals can appear relatively nontoxic under normal indoor lighting but exert a pronounced effect in sunlight, a phenomenon recognized for thousands of years. These include plant furocoumarins, the PAH fraction of petroleum, and drugs such as the tetracycline antibiotics, and they are effective whether applied to the skin or administered orally. In mammals, the response usually is a skin rash, but death is often the result in aquatic species (Section 7.4).

The longer *subacute* and *chronic toxicity* tests are conducted in the same way as the acute tests, although the purpose seldom is to measure an LD_{50}. Much more attention is paid to physiological and biochemical responses, especially changes in blood and organs, development of tumors, and reproductive effects. Because of the cost of maintaining the animals for months or even years, and the labor-intensive nature of the observations, such tests are both complex and very expensive.

An intriguing aspect of toxicological measurement arises when responses occur over long exposures to very low doses, as is often the case with carcinogens. The tumors develop so slowly, often over an entire lifetime, that the usual dose–probit plots cannot be used. Instead, the experimental data must be extrapolated into the unknown region that approaches zero dose and zero response. This is done with mathematical models that plot log dose against probability (tumor incidence); see Fig.10.2. The linear multistage model is probably that most widely used, while the FDA model, which connects the origin with the lowest point of experimental data, is the simplest (Gaylor, 1988). Errors can be very large, representing several orders of magnitude. There is more on this subject in Section 10.3, under Risk Characterization.

8.4.3. Life-cycle Tests

Standard life-cycle tests in fish or aquatic invertebrates require that animals be exposed to a chemical from embryo through a complete reproductive cycle, often over a period of more than 12 months (McKim, 1985). Common freshwater test species include the fathead minnow *(Pimephales promelas)*, bluegill *(Lepomis macrochirus)*, brook trout *(Salvelinus fontinales)*, and the invertebrate water flea *(Daphnia magna)*, amphipod *(Gammarus pseudolimnaeus)*, and snail *(Physa integra)*. A few marine species also may be included (Appendix 8.2). Tests often are run in complicated flowthrough systems under highly standardized conditions using automated serial dilutions, and like the corresponding mammalian tests, they are highly labor-intensive.

For this reason, attention has focused increasingly on shorter and more easily conducted tests employing only the early life stages (ELS). Typically, fish are exposed to the test chemical from egg fertilization through embryonic, larval, and juvenile stages, a process often requiring less than a month. Toxicity endpoints include hatching success, survival, growth rate, behavior, and deformities, and the goal is to determine an MATC (maximum acceptable toxicant concentration) for the chemical; the MATC is a concentration (range) between the NOEL and the toxic threshold. Although ELS tests do not provide data on adult reproductive factors such as spermatogenesis, early life stages prove to be the most sensitive period in the life of an aquatic animal, and ELS results correlate well with those from complete life-cycle tests (Table 8.3).

8.4.4. Other Bioassays

Quantitative measurement of toxicity is often referred to as "bioassay." Before the advent of modern methods of instrumental analysis, bioassays were used widely to detect and measure toxic chemicals in the environment. While any organism can be used, invertebrates probably have received the most attention, as they are relatively easy to rear and test in large, uniform numbers. Key species of insects, plants, and microorganisms long have been employed to screen chemicals for useful pesticide activity.

An interesting current example of the use of bioassays is *toxicity identification evaluation* (TIE). Wastewater samples, for example, are bioassayed for their toxicity to small groups of aquatic invertebrates such as *Daphnia magna* or *Ceriodaphnia dubia*. If toxicity is detected, it is followed as the sample goes through various separation procedures, such as thin-layer chromatography, until the responsible agent can be isolated and identified chemically (Mount, 1989).

8.4.5. Mutagenicity Tests

Although mutagenicity usually does not produce any immediate symptoms (Section 7.4), its association with carcinogenicity has generated great interest in practical short-term tests which might be used to screen for carcinogens. The most popu-

TABLE 8.3
MATC Values from Fish Life-Cycle and ELS Tests[a]

Chemical	Species	MATC-FLC[b]	MATC-ELS[c]
Aroclor 1254 (PCB)	Fathead minnow	1.8–4.6	1.8–4.6
Cadmium	Brook trout	3.4–7.3	3.4–7.3
Carbaryl	Fathead minnow	210–680	210–680
Copper	Brook trout	9.5–17.4	9.5–17.4
DDT	Fathead minnow	0.4–1.5	0.4–1.5
Lead	Brook trout	58.0–119.0	58.0–119.0
Lindane	Fathead minnow	9.1–23.5	No effect
Malathion	Flagfish	8.6–10.9	8.6–10.9
Trifluralin	Fathead minnow	2.0–5.1	5.1–8.2

[a] Adapted from McKim (1985).
[b] Full life-cycle test, $\mu g/L$.
[c] Early life stage test $\mu g/L$.

lar of these is the **Ames test** (Ames and McCann, 1981; Maron and Ames, 1983) based on the exposure of a mutant strain of *Salmonella typhimurium* bacteria to the test chemical. This strain is unable to biosynthesize histidine, a required amino acid, so it will not grow on a histidine-free agar plate. The idea is that a mutagen will cause a back-mutation (reversion) which allows the bacteria to make histidine again and so form readily counted colonies on the plate. The greater the number of colonies, the more potent the mutagen. As the proximate mutagen often is a metabolite of the original chemical, the medium usually is supplemented with a microsomal fraction isolated from rat liver to provide for biotransformation.

The benefits of such tests include the ease of preparation and interpretation, the large number of individual organisms (up to 10^8) which can be exposed on a single 10 cm plate, the number of generations provided by a 48 h test, and the correlation of test results with mammalian carcinogenicity in over 80% of more than 5000 compounds screened. The test is almost too easy, and one must realize that it is just a surrogate and that multiyear tests in mammals eventually must be performed on the most important candidates.

8.4.6. Phytotoxicity

Plants are much easier to work with than are animals. Toxic effects in intact plants are detected by spraying or brushing the chemical onto the leaves of the desired test species, often tomatoes *(Lycopersicon esculentum)* growing in soil, or by root uptake when incorporated into a nutrient solution in which the plant is growing. Relative growth rate, deformities, and loss of chlorophyll then are compared with those of untreated controls. Promotion or inhibition of growth can be measured conveniently and quantitatively by threading ~1 cm sections of the hollow leaf sheaths (coleoptiles) of wheat or oat onto the teeth of a comb, immersing in a solution of test substance in highly dilute nutrient medium overnight, and measuring any change in length. The length should more than double within a few hours if the medium contains a plant growth regulator such as 2,4-dichlorophenoxyacetic acid (2.4-D) in concentrations as low as 10^{-5} *M*.

Phytotoxic effects are readily measured by placing seeds of tomato, oat *(Avena sativa)*, or mung bean *(Phaseolus aureus)* on damp blotting paper in the dark and comparing percent germination, root elongation, and shoot growth in the presence and absence of test chemical. Growth inhibition also is used to screen chemicals such as antibiotics against cultures of fungi and bacteria.

8.4.7. Toxicity Tests for Pesticide Registration

Every pesticide must be registered with the U.S. Environmental Protection Agency before it can be applied in this country. The registration process requires that the Agency receive satisfactory data from a battery of specific toxicological (Appendix 8.1) and environmental tests (Appendix 8.2), although sometimes very limited experimental field applications may be allowed while long-term tests are still in progress. Not all the tests listed in the Tables necessarily will be required, as registrations are approved on a case-by-case basis. For example, if the pesticide is not expected to make significant contact with human skin, the 90-day dermal tests may be waived. Detailed guidelines have been established for conduct of the tests, and any deviations must be approved in advance by the EPA. The cost of such toxicity tests now exceeds $1,000,000 for each chemical.

8.4.8. Ethics and Alternatives

The American public increasingly demands safety, especially regarding chemicals in their environment. That safety requires tests in live animals, as required by federal and state laws, which has led to controversy and a concern over what rights these animals should have. If toxicity data would protect the test species in the future, the argument goes, perhaps a case could be made for the testing; after all, species differences in response sometimes require that only the one at risk can serve. However, if the results aid only humans or their pets, the motives become more questionable. Does not *every* creature have a right to control its own life and body?

On the other side of the argument, toxicologists and others point to the advances in knowledge, treatment, and safety that can be gained only through live *(in vivo)* animal tests. Nothing replaces the complex responses of an intact organism, and as toxicity testing in humans is both illegal and clearly unethical, the safety and welfare of our own species dictate that thorough but compassionate tests be made on other, presumably less-sentient, animals. Until recently, the sizable differences in metabolism, chemical fate, and time scale (Section 10.4.2) between humans and test animals were ignored.

Actually, the two sides often are not that far apart. Toxicologists long have recognized that accurate data come only from healthy, well-fed, normal animals. Both law and humanity require that test animals be treated with care and respect, and kindness often is included. Economics and oversight committees also reduce the number of animals to be used. Much of the public controversy seems to revolve around the use of warm, furry creatures, and increasing attention is being paid to other surrogates; few people get excited over killing houseflies or brine shrimp. Yet, if the premise of the sanctity of life is valid, one must question whether *any* species should be jeopardized. We humans have a poor record in this regard.

What are the alternatives? Despite a lot of research, there seem to be few. Tests with isolated enzyme systems, for example, ignore the rest of the intoxication process which largely controls toxic responses. Computer simulations often reveal only our lack of knowledge about the complex and variable activities within whole, living organisms. One practical approach might be to start reducing the numbers of animals required for each test. Considering the high variability inherent in most toxicity measurements and the relatively poor quantitative correlation between results in test animals and effects in humans, use of fewer individuals might produce results not seriously different from what we have now. Perhaps acute toxicity tests should attempt only rangefinding, and three points above the NOEL might well suffice for a dose–response curve.

Will toxicity testing in animals slow to a halt? Probably not. First, being big and dominant, we probably will continue to do as we please with smaller and less intelligent species. Second, the rearing and testing of laboratory animals has become an industry that would not willingly decline. Finally, government requirements specify extensive animal tests, as in the case of pesticide registration, and businesses must comply. Perhaps conservation efforts could start with these regulations.

8.5 | DATA FROM HUMANS

8.5.1. Toxicity Tests

Both ethics and law prohibit toxicity tests in humans. Unlike pharmaceuticals, most environmental chemicals have never been administered intentionally to people. There are a few exceptions: Some natural products, such as the heart stimulant digitoxin and many antibiotics, have valuable medicinal properties and so have received extensive human trials, and low doses of a few "safe" pesticides such as DDT have been given to volunteers. Still, considerable human toxicity data exist for many chemicals because of case reports from accidental poisonings and suicides, industrial hygiene records, and monitoring of field workers.

8.5.2. Mortality Statistics

Compilations of toxicity data sometimes list LD_{50} values for humans. These actually represent an **average** lethal dose rather than the more familiar statistical **median** lethal dose, and they reflect a range of ages, sizes, and exposures and so are not particularly precise or reliable. More often, an LD_{lo} is listed, the lowest reported lethal dose (*not* an LD_{10}), without regard for where the sensitivity of the victim would fit into that of the average human population. Lethality most often is extrapolated or inferred from rodent data, but as Table 8.4 shows, the extrapolation is notori-

TABLE 8.4
Comparison of Toxicity between Humans and Rats[a]

	Use	Human LD (mg/kg)[b]	Rat LD_{50} (mg/kg)[c]
Potassium cyanide	Metal plating	0.5–3.6	8.5
Nicotine sulfate	Insecticide	0.9	83
Parathion	Insecticide	1.7–2.1	13
Arsenic trioxide	Insecticide	1.8	20
Methomyl	Insecticide	12–15	17
Mercuric chloride	Disinfectant	15	37
DDT	Insecticide	16–120	118
Paraquat	Herbicide	17–43	100
Chlordane	Insecticide	28–56	335
Diazinon	Insecticide	90–400	250
Dieldrin	Insecticide	120	46

[a] Data adapted from Hayes (1982). Most rat data are from Gaines (1960).
[b] Adult males except for the lowest paraquat value (adult female).
[c] Adult males.

ously unreliable. Despite the frequency of human poisoning, accurate estimates of dose are seldom available. However, as a general rule, a lethal oral dose for a human is about an order of magnitude smaller than for the rat, and children may be killed at perhaps another order of magnitude below this. Surprisingly, there is almost no information on human intoxication in relation to size, gender, or race, although the census does classify poisoning fatalities roughly by age (USPHS, 1993).

Rat LD_{50} data have been extrapolated to provide probable lethal oral doses for humans (Table 8.5). This exercise may be useful if for nothing more than to illustrate how little of a really toxic chemical is required to do great damage. However, it is too general to consider age or gender, let alone individual sensitivities. In recent years, there has been growing awareness of the hypersensitivity and hyposensitivity of some people toward chemicals, that is, those toward the opposite ends of the response–frequency curve of Table 8.1B. Certainly, statistical measures such as the LD_{50} obscure toxic effects in any particular individual in a population, and "moderately toxic" may be little consolation for one who is at the low end of the curve!

8.5.3. Exposure Data

Data on levels that cause injury rather than death are more readily available and are generally more valuable than those on lethality. Oral exposure is relatively rare, while dermal exposure is commonplace, especially at work. At least 90% of skin exposure is on the hands and forearms, as the rest of the body normally is clothed, and skin irritation is the most frequent symptom. Skin sensitivity may be determined by a patch test, in which the test chemical is swabbed onto a measured area of arm or back, or a gauze patch soaked in the solution is taped on, and the results are examined after 1–4 days. The other arm, similarly treated but without chemical, serves as control. Comparisons in which a test area is exposed to UV radiation and an adjacent one kept covered identifies phototoxicity. A series of patches at different doses can provide a dose–response relation.

Extrapolations of dermal toxicity data from rats and rabbits to humans is unreliable, partly because of differences in absorption. The monkey has proved to be the best model (Table 8.6), but relatively few tests have been run on such expensive animals. However, a general conclusion is that human skin is the least permeable compared to most animal models (Section 7.4.3). Inhalation is also a frequent route

TABLE 8.5
Toxicity Ratings of Chemicals in Humans

Rating	Toxicity	Rat Oral LD_{50} (mg/kg)	Adult Human Oral LD	Example
1	Practically nontoxic	>15,000	More than a quart	Water
2	Slightly toxic	5000–15,000	A pint or more	Gin
3	Moderately toxic	500–5000	An ounce or more	Aspirin
4	Very toxic	50–500	A tsp to an ounce	Dieldrin
5	Extremely toxic	5–50	A few drops to a tsp	Parathion
6	Supertoxic	<5	A drop or two, a taste	Nicotine

TABLE 8.6
Permeability of Skin to Chemicals *in vivo*[a]

Chemical	% Applied Dose Absorbed[b]			
	Rabbit	Pig	Monkey	Human
DDT	46	43	2	10
Lindane	51	38	16	9
Parathion	98	15	30	10
Malathion	65	16	19	8
Nitrobenzene	—	—	4	2

[a] Bartek and LaBudde (1975); Feldman and Maibach (1974).
[b] Applied dose $4 \mu g/cm^2$.

of exposure, especially in the workplace, and the American Conference of Governmental Industrial Hygienists (ACGIH) has established an extensive list of threshold limit values (TLVs). A TLV represents the average maximum level of air contamination by a chemical that can be tolerated by workers upon continuous exposure for 8 hours each day, 40 hours each week (Appendix 8.3). A second TLV category is the short-term exposure limit (STEL), the maximum concentration tolerated for a 15-minute period, and another is a ceiling concentration (TLVC) to which people should not be exposed *under any circumstances.*

A quick TLV calculation can be very useful. Assume that you are removing grease from a skirt with trichloroethylene cleaning solvent, and knock the can over, spilling a half-pint (250 mL) of the heavy liquid (*d* 1.46) on the kitchen floor. If the volume of the room is 40 m^3 (8 × 10 × 20 feet), the concentration of TCE in the air could reach 9.1 g/m^3 (250 × 1.46/40). As the TLV of trichloroethylene is 0.27 g/m^3 (50 ppm) and the STEL is 0.54 g/m^3 (Appendix 8.3), it is obvious that remaining in the kitchen for even a short time could be very dangerous to your health.

Monitoring humans for environmental or occupational exposure provides especially valuable information. Skin exposure is measured by taping 100 cm^2 plastic-backed gauze pads to skin and clothing or wearing disposable coveralls, removal followed by extraction with solvent, and chemical analysis of the extracts. Inhalation is estimated via a small, battery-driven pump, fastened near the face, which draws air through a tube of adsorbent. These methods do not measure how much chemical actually enters the body, which must be determined by analysis of blood or urine.

Blood analysis offers many advantages: Xenobiotic levels remain relatively constant during the course of a day; blood levels reflect those at the target site; and the parent compound predominates. As little as 25 μL (a drop) of sample is required, which incidentally allows these tests to be used on birds and even smaller animals. Urine samples usually contain more metabolites than parent, and toxicant levels vary over even short time periods, but the ease of urine collection and the relatively large volumes available make it the most common source of samples. Often, blood and urine data can be related, and they both relate to the original internal dose or amount absorbed if the pharmacokinetics of the compound are known. The results can have far-reaching utility, as shown by the following account.

Twice in a decade, California agriculture was threatened by invasions of the

Mediterranean fruitfly *(Ceratitis capitata)* on fruit smuggled in on airliners from Hawaii. About 400 square miles of the Los Angeles basin and over 500 square miles of Northern California centered on the San Francisco International Airport were sprayed by helicopter with malathion bait, much to the consternation and eventual wrath of the sprayees. No human injuries were reported, but exposure levels were unknown, and an attempt was made to estimate the probable dose via a physiologically based pharmacokinetic (PBPK) model (Dong et al., 1994). Assumptions were made about tissue volume, blood perfusion rates in the various organs, metabolism (hydrolysis rates), skin permeability, and tissue–blood partition coefficients, viewing the body as a series of interconnected compartments that included skin, fat, liver, gastrointestinal tract, kidney and other highly vascularized organs, and muscle (Special Topic 10).

Through a series of equations, the pharmacokinetics in each compartment was estimated, and, working backward from urinary metabolite levels, the original exposures were calculated. For 70 kg adults and 14–34 kg children, the highest absorbed dose was suggested to total 1.3 and 0.4 mg, respectively, well below those required for symptoms. Malathion and its metabolites were detected in only a few subjects. As in many such retrospective investigations, data did not exist to allow the actual dose to be verified, but the calculated results were consistent with the spray application rate of 2 μg/cm^2 actually measured.

Indeed, malathion and malaoxon residues were detected in the environment after the California spraying, but they dissipated rapidly. In the end, the bulk of evidence favored the decision to spray. Despite critics' arguments, malathion has not been found to be carcinogenic, mutagenic, or teratogenic in extensive animal tests, and epidemiological data (Special Topic 8) suggest that no humans suffered permanent damage from the exposure. The maximum adult dose from dermal and respiratory exposure was estimated at 246 μg/kg/day, and eating unwashed home vegetables would have added at most another 80 μg/kg/day (Marty et al., 1994) for a total of something over 300 μg/kg/day. Harmless occupational exposures have reached 1000 μg/kg/*h* (Wolfe et al., 1967). The alternative to the spraying was the certain loss of much of the state's crop production, its principal industry, valued at $1.3 billion annually (Marx, 1981).

8.5.4. Epidemiology

A dictionary defines epidemiology as the branch of (medical) science that deals with the incidence, distribution, and control of disease in a population. It is the science of epidemics. A famous example concerns the great 1848 cholera outbreak in London, when physician John Snow was able to relate occurrences of the disease to the presence of sewage in drinking water. He introduced important sanitation precautions, although discovery of the microbial cause of the disease was still decades away. Epidemiology is now an important factor in the quantitative estimation of the human health impacts of environmental chemicals as well as protection of the public from them. Special Topic 8 provides more information and examples.

8.6 REFERENCES

ACGIH. 1994. *Threshold Limit Values for Chemical Substances and Physical Agents, and Biological Exposure Indices, 1994–1995,* American Conference of Governmental Industrial Hygienists, Cincinnati, OH.

Ames, B. N., and J. McCann. 1981. Validation of *Salmonella* tests: A reply to Rinkus and Legator. *Cancer Res.* 41: 4192–203.

Armstrong, R. W., E. R. Eichner, D. E. Klein, W. F. Barthel, J. V. Bennett, V. Jonsson, H. Bruce, and L. E. Loveless. 1969. Pentachlorophenol poisoning in a nursery for newborn infants. II. Epidemiologic and toxicologic studies. *J. Pedriat.* **75**: 317–25.

Bartek, M. J., and J. A. LaBudde. 1975. Percutaneous absorption *in vitro,* in *Animal Models in Dermatology* (H. I. Maibach, ed.), Churchill Livingston, New York, NY

Connell, D. W., and R. Markwell. 1992. Mechanism and prediction of nonspecific toxicity to fish using bioconcentration characteristics. *Ecotoxicol. Environ. Safety* 24: 247–65.

Dong, M. H., W. M. Draper, P. J. Papenek, J. H. Ross, K. A. Woloshin, and R. D. Stephens. 1994. Estimating malathion doses in California's medfly eradication campaign using a physiologically based pharmacokinetic model, in *Environmental Epidemiology* (W. M. Draper, ed.), American Chemical Society, Washington, DC. Advances in Chemistry Series 241, pp. 189–208.

Feinstein, A. R. 1988. Scientific standards in epidermiological studies of the menace of daily life. *Science* **242:** 1257–63.

Feldman, R. J., and H. I. Maibach. 1974. Percutaneous penetration of some pesticides and herbicides in man. *Toxicol. Appl. Pharmacol.* **28:** 126–32.

Gaines, T. B. 1960. The acute toxicity of pesticides to rats. *Toxicol. Appl. Pharmacol.* **2:** 88–99.

Gaylor, D.W. 1988. Quantitative risk estimation, in *Risk Assessment and Risk Management of Industrial and Environmental Chemicals* (C. R. Cothern, M. A. Mehlman, and W. L. Marcus, eds.), Princeton Scientific Publ. Co., Princeton, NJ. Vol. 15, pp. 23–43.

Graham, J. D. 1990. *The Weight of the Evidence on the Human Carcinogenicity of 2,4-D,* Harvard School of Public Health, Boston, MA.

Hayes, W. J. 1982. *Pesticides Studied in Man,* Williams and Wilkins, Baltimore, MD.

Johnson, W. W., and M. T. Finley. 1980. *Handbook of Acute Toxicity of Chemicals to Fish and Aquatic Life,* Fish and Wildlife Service, U.S.D.I., Washington, DC. Resource Publication 137.

Maron, D. M., and B. N. Ames. 1983. Revised methods for the *Salmonella* mutagenicity test. *Mutat. Res.* **113:** 173–215.

Marty, M. A., S. V. Dawson, M. A. Bradman, M. E. Harnly, and M. J. Dibartolomeis. 1994. Assessment of exposure to malathion and malaoxon due to aerial application over urban areas of Southern California. *J. Expos. Anal. Environ. Epidem.* **4:** 65–81.

Marx, J. L. 1981. Malathion threat debunked. *Science* **213:** 526–27.

McKim, J. M. 1985. Early life stage toxicity tests, in *Aquatic Toxicology* (G. M. Rand and S.R. Petrocelli, eds.), Hemisphere Publ. Corp., Washington, DC, pp. 58–95.

Mount, D. I. 1989. *Methods for Aquatic Toxicity Identification Evaluation: Phase III Toxicity Confirmation Procedures,* U.S. Environmental Protection Agency, Duluth, MN. EPA-600/3-88-036.

NARA. 1990. *Data Requirements for Registration,* Code of Federal Regulations, CFR 40, Office of the Federal Register, National Archive and Registration Agency, Washington, DC, Part 158.

Suter, G. W. 1993. *Ecological Risk Assessment,* Lewis Publishers, Boca Raton, FL.

Tucker, R. K., and D. G. Crabtree. 1970. *Handbook of Toxicity of Pesticides to Wildlife,* Fish and Wildlife Service, U.S.D.I., Washington, DC. Resource Publication 84.

USPHS. 1993.*Vital Statistics of the United States 1989,* U.S. Public Health Service,National Center for Health Statistics, Hyattsville, MD. Vol. 2: Mortality Part A.

Wolfe, H. R., W. F. Durham, and J. F. Armstrong.1967. Exposure of workers to pesticides. *Arch. Environ. Health* **14**: 622–33.

Special Topic 8. Epidemiology

Epidemiology is a rigorous, quantitative science requiring a series of basic steps derived from Koch's postulates of over a century ago:

1. An effect must be observed in a population (no one-person epidemics);

2. Extensive data must be collected on the effect and possible causes;

3. A statistically valid association must be established between the effect and possible causative factors in relation to no-effect controls;

4. The causal factor(s) must be clearly identified; and

5. The factor(s) must produce the effect when a new group of humans or other appropriate living organisms becomes exposed.

Data gathering and interpretation are conducted in a number of ways, the most common being case-control and cohort studies. The former assumes that a disease is associated with a specific environmental exposure; "cases" are already known to have the disease, while controls from the same population do not. Based on information collected from both, the cases should have received the greater exposure. The quantitative measure of relative risk is the odds ratio (OR), and if both groups show the same incidence of exposure, OR = 1.0. Cohort studies are based on the incidence of exposures compared between those known to have been exposed and a separate group not exposed. Results rely on preexisting data, never on experiments.

A typical epidemiological investigation (Armstrong et al., 1969) took place at a small maternity hospital, where, over a period of 4 months, 20 newborn infants suffered symptoms of excessive sweating, rapid heartbeat, fever, and enlarged liver. Nine became severely ill, and two died as a result, but 63 others remained healthy. The healthy ones had stayed at the hospital for significantly shorter periods than had those who became ill. Microbial disease was suspected, but thorough disinfection did not alleviate the problem. The symptoms eventually were associated with phenol toxicity, and blood and tissue analyses identified residues of pentachlorophenol (PCP) in the affected children but not in the others. The contaminant was traced to a laundry product used to wash diapers, shirts, blankets, and other nursery items from which it was absorbed through the baby's skin upon prolonged contact. The symptoms were consistent with those of known PCP poisoning in humans. Note that all 5 of the required epidemiological steps were taken.

Another example concerns the possible carcinogenicity of the herbicide 2,4-D (2,4-dichlorophenoxyacetic acid), whose butyl ester was a major component of the Agent Orange defoliant used in the Vietnam war. Three cohort studies involving 25,582 factory workers and agricultural applicators over as long as 37 years revealed no statistically significant correlation between exposure to the herbicide and the incidence of tumors, with special attention to the Hodgkin's disease (HD), soft-tissue sarcoma (STS), and non-Hodgkin's lymphoma (NHL) reported earlier by Swedish scientists. Of 6 major case-control studies, the two that received the most attention involved 424 Kansas males diagnosed with HD, STS, or NHL, plus 1005 controls, and 385 diagnosed cases and 1432 controls in Nebraska.

Neither the Kansas nor Nebraska investigation could correlate herbicide use with HD or STS, but both suggested a weak association with NHL (OR 2.5 and 1.5, respectively). However, the results were ***confounded;*** that is, factors were identified that cast doubt on the validity of the results. For example, as NHL is a rare type of cancer, the number of affected farm workers actually was small; exposure to 2,4-D could not be separated from exposure to other chemicals; and answers to questionnaires and interviews about the occurrence and extent of exposure relied entirely on the memory of individuals or even next of kin, often years after the fact. An expert panel eventually concluded that 2,4-D exposure could not be associated with NHL at present (Graham, 1990).

Although originally designed for applications in human health, the principles of epidemiology are equally applicable to the ecological effects of environmental chemicals (Suter, 1993). Like human epidemiology, "ecological epidemiology" is based in the real world and in seeking, observing, and measuring *actual* impacts on individuals, populations, and communities. Laboratory tests can suggest causality but not necessarily prove it. However, unlike human epidemiology, there seldom are health records to go on, a staggering number of species may be involved rather than just the one, and a diverse selection of measurement endpoints may be required. Although epidemiological methods often are used in ecotoxicology, there are at present few who would consider themselves to be Ecological Epidemiologists.

The predilections of the popular press notwithstanding, statistical arguments by themselves are never sufficient to draw conclusions from either type of epidemiological studies (Feinstein, 1988). One is reminded of the perfect correlation that can be drawn between the decline in the number of human births in Holland in earlier years and the simultaneous decline in the stork population. Epidemiology is much more complex than that.

APPENDIX 8.1.
Basic Toxicity Tests for Pesticide Registration[a]

Acute tests
 Oral toxicity, in rats
 Dermal toxicity, usually in rabbits
 Inhalation toxicity, in rats
 Primary eye irritation, in rabbits
 Primary skin irritation
 Skin sensitization
 Delayed neurotoxicity (required only for organophosphates)

Subchronic tests
 90-Day feeding, in rodents
 90-Day feeding, in a nonrodent
 21-Day dermal or 90-day dermal
 90-Day inhalation, in rats
 90-Day neurotoxicity, in hens or mammals

Chronic tests
 Chronic feeding, in rodents
 Chronic feeding, in a nonrodent
 Oncogenicity, in rats
 Oncogenicity, in mice
 Teratogenicity, in two species
 Reproduction, over two generations
 Mutagenicity, gene mutation

APPENDIX 8.1. *(continued)*

Mutagenicity, structural chromosome aberration
Mutation, other genotoxic effects such as mammalian cell transformation
General metabolism
Dermal penetration

---

[a] NARA (1990).

APPENDIX 8.2.
Ecotoxicology Tests for Pesticide Registration[a]

Wildlife
  Avian oral LD_{50}, in mallard duck or bobwhite quail
  Avian dietary LC_{50}, in mallard duck and bobwhite quail
  Avian reproduction, in mallard duck and bobwhite quail
  Wild mammal acute toxicity
  Simulated or actual field tests, in mammals and birds

Aquatic organisms
  Acute freshwater fish LC_{50}, in rainbow trout and bluegill sunfish
  Acute freshwater invertebrate LC_{50}, in *Daphnia*
  Acute LC_{50} in estuarine and marine organisms
  Fish ELS
  Fish life-cycle
  Aquatic invertebrate life-cycle
  Bioaccumulation in aquatic organisms
  Simulated or actual field testing in aquatic organisms

Other Tests
  Target area phytotoxicity
  Nontarget seed germination and shoot emergence
  Nontarget vegetative vigor
  Nontarget aquatic plant growth
  Acute contact LD_{50}, in honey bee
  Toxicity of foliage residues, in honey bee
  Nontarget toxicity to aquatic insects, predators, and parasites[b]

---

[a] NARA (1990).
[b] Not currently required, pending test protocols.

APPENDIX 8.3.
Threshold Limit Values for Environmental Chemicals[a]

Chemical	TLV (ppm)	TLV (mg/m^3)	STEL[b] (ppm)	STEL[b] (mg/m^3)	TLVC[c] (ppm)	TLVC[c] (mg/m^3)
Ammonia	25	17.4	35	24		
Benzene	0.3	1.0	0.2	0.64		
Carbon monoxide	50	57.3				
Chlorine	0.5	1.5	1	2.9		
Chloroform	10	49				
Ethanol	1000	1885				
Formaldehyde	2	2.5			0.3	0.37
Gasoline	300	890	500	1480		
Hexane	50	176				
Hydrogen cyanide					4.7	5
Lindane	0.04	0.5				
Methanol	200	262	250	328		
Ozone	0.05	0.1	0.2	0.4		
Trichloroethylene	50	269	100	536		
Vinyl chloride	5	13				

[a] ACGIH (1994).
[b] Short-term exposure limit (15 minute).
[c] TLV ceiling.

Intoxication Mechanisms

9

9.1 | THE BIOCHEMICAL LESION

Intoxication does not just happen. It is the result of a long chain of events (Section 7.1) leading to a chemical reaction or other binding between the proximate toxicant and some biochemical receptor or target to produce a "biochemical lesion". Intoxication and toxicology have their basis in chemistry.

Most organisms share a common biochemistry (Appendix 9.1). Energy initially arises from plant photosynthesis *via* the breakdown of sugars (glycolysis) coupled to the tricarboxylic acid (TCA) cycle, and the necessary reducing power (NADH) is produced from water. Proteins, nucleic acids, and other essential cell components are built, in all kinds of organisms, from the same basic building blocks by the same routes. However, each kind of living creature exhibits its own unique characteristics: flowering plants harvest solar energy to build the required metabolic intermediates, animals rely on a sensitive nervous system for information and action, bacteria build unusual cell walls, and fungi and arthropods use the aminopolysaccharide chitin for structural strength. Interference with any of these processes can spell disaster for the individual, and xenobiotics *do* interfere (Appendix 9.1). The way they do it is called a "mechanism."

Considering the vast numbers of environmental chemicals and the complexity of even the simplest organisms, a thorough discussion of intoxication mechanisms is not possible here. Many mechanisms remain obscure, and the details of even the best known are only now becoming clear. However, limited generalities can be made (Appendix 9.2), and a few important examples will illustrate other possibilities.

9.2 | RECEPTORS AND TARGETS

A dictionary defines a **receptor** as "a chemical group or molecule in a plasma membrane or cell interior that shows an affinity for another specific chemical group or molecule." This once-controversial concept was first detailed in the 1908 Nobel laureate address of German biochemist Paul Ehrlich, but it was long considered to be only an abstraction. However, several lines of evidence developed in support of it. First, the effectiveness of many toxic chemicals and drugs at concentrations down to 10^{-12} M requires defined and very specific sites with which the toxicant can associate. Second, some toxic effects require a specific type of chemical structure, that is, a certain molecular shape at the site of action, illustrated by the frequent difference in toxicity between one optical isomer and its antipode. Third, the high degree of enzymatic specificity in toxic action, such as the inhibition of primarily a single enzymatic process (AChE hydrolysis) by organophosphorus esters, suggests that a reaction occurs selectively with some very specific biochemical entity.

Although a biochemical receptor for the neurotransmitter acetylcholine (ACh) was postulated in the 1920s, it was not actually observed until half a century later. Figure 9.1 shows an electron micrograph of a group of ACh receptors in the postsynaptic membrane from the electric organs of the electric ray, *Torpedo,* the subject of much research (Giraudat and Changeux, 1980). At the center of each glycoprotein rosette can be seen the "pore" or channel through which sodium and potassium ions move during the transfer of a nerve impulse from one neuron to another or, in this case, to a muscle (Changeux et al., 1984). ACh receptors at neuromuscular junctions are called "nicotinic," as the tobacco alkaloid nicotine mimics ACh there; in another type, the "muscarinic receptors," ACh is mimicked by the quaternary amine musca-

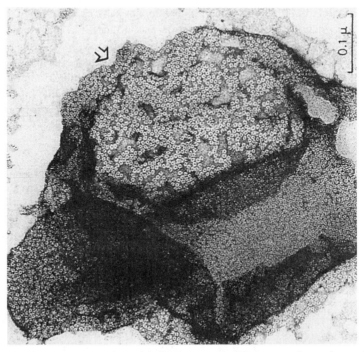

Figure 9.1. Electron micrograph of acetylcholine receptors in the postsynaptic membrane of *Torpedo mercato.* The Na$^+$–K$^+$ channel appears at the center of each rosette (Albert, 1985). © 1985 by Chapman and Hall, reproduced with permission.

rine from the toxic toadstool, *Amanita muscaria*. Figure 9.1 lays to rest any doubt about the existence of receptors (Changeux et al., 1984).

Most pharmacologists restrict the term **receptor** to only this kind of protein. However, other biochemical entities are more often involved in toxic action, including the active sites of enzymes such as the cholinesterases, coenzymes like the dihydrolipoamides required for respiration, cellular membranes, metabolic intermediates such as the plastoquinones, and macromolecules such as DNA (Appendix 9.2). Although these, too, are sometimes referred to as receptors, they will be differentiated here as **targets**.

9.3 | MECHANISMS OF GENERAL TOXICITY

Some types of chemicals are toxic to all forms of life. They affect biochemical processes shared by most living organisms, such as respiration and the production and utilization of energy. The halo- and nitrophenols provide an example, of which pentachlorophenol (PCP) is probably the best known. Such phenols are toxic to animals, plants, and microorganisms and thus have received wide use as pesticides (Section 15.6). They function by inhibition of oxidative phosphorylation.

Oxidative phosphorylation is the biochemical process by which the generation of the principal energy intermediate, adenosine triphosphate (ATP), is coupled to the electron transport system. Analogous to an automobile's power train, energy produced by the engine (electron transport) is coupled to the actual propulsion (ATP generation) by the clutch (oxidative phosphorylation). For chemiosmotic coupling, electron transport is accompanied by the pumping of hydrogen ions across the inner mitochondrial membrane to the outside, and the resulting electrochemical (proton) potential drives ATP synthesis:

$$2NADH + 2H^+ + O_2 + 6ADP + 6P_{inorg} \longrightarrow 2NAD^+ + 2H_2O + 6ATP \quad (9.1)$$

Impermeability of the membrane to protons is essential for maintenance of that potential. The lipophilic phenols, which are relatively strong acids (proton donors), penetrate the membranes and uncouple ATP synthesis by removing the restriction to proton flow. Electron transport and oxygen utilization continue, but no ATP is produced; the engine is running, but, with the clutch not engaged, the wheels fail to move and cellular activity grinds to a halt. Not surprisingly, inhibitors such as dinitrophenol fungicides and PCP are broadly toxic (rat oral $LD_{50} \sim 20$ mg/kg), and mammalian symptoms include lack of oxygen (cyanosis), rapid respiration and increased heart rate as the body struggles for more, overheating as the metabolic rate increases, and eventual collapse, coma, and death.

Nonspecific toxicity also is shown by compounds of arsenic. As mentioned in Section 1.1, the extreme toxicity of arsenicals once made them the poisons of choice for elimination of bothersome enemies or relatives, and many insects, weeds, and fungi as well. Large quantities of sodium arsenite were used in ant baits, lead arsenate was a major insecticide, calcium arsenate was used against snails, and copper aceto-arsenite (Paris green) was a major fungicide. Arsenic's high toxicity is due to the stability of its bonds with sulfur.

The pyruvate produced during glycolysis is converted into the metabolic building block, acetylcoenzyme A (acetyl-CoA), by the pyruvate dehydrogenase (PDH) system (Fig. 9.2). Pyruvate condenses with thiamine pyrophosphate (TPP) to form hydroxyethyl-TPP, which acetylates dihydrolipoyl transacetylase (DLT); DLT then converts coenzyme A into acetyl-CoA. The active site of the transacetylase is the 1,3-dithiol, dihydrolipoamide (DHL, Eq. 9.2), with which trivalent arsenic bonds covalently. The driving force for the reaction is unusually great because the As–S bond angles allow formation of a stable six-membered ring. In the equation, R represents the four-carbon lysine sidechain on the transacetylase protein.

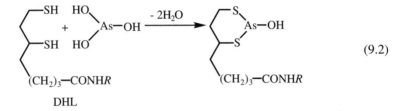

$$(9.2)$$

As DHL mediates the transfer of electrons as well as the activated acyl groups generated by the oxidation of glucose, its deactivation by arsenic results in loss of the energy derived from glycolysis; that is, the engine, clutch, and power train are functional, but the fuel has run out. Respiration is inhibited, and the cells—whether of animals, plants, or microorganisms, die through progressive inactivity. The amount

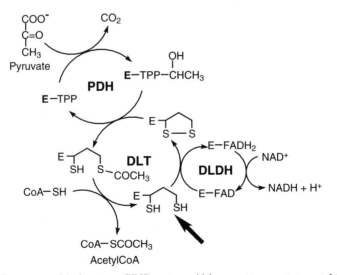

Figure 9.2. The pyruvate dehydrogenase (PDH) system, which converts puruvate to acetyl Coenzyme A. DLDH is dihydrolipoyl dehydrogenase that generates the dihydrolipoamide 1,3-dithiol (arrow) responsible for binding metals. E = enzymeprotein.

of DHL in an organism is very limited, so very little arsenic is required for poisoning. As only trivalent arsenic reacts, pentavalent arsenates require prior reduction (Section 11.4.2).

Organoarsenic war gases such as Lewisite (2-chlorovinyl dichloroarsine, $ClCH = CH\text{-}AsCl_2$) killed or injured many people in World War I. At the start of World War II, Rudolph Peters in England used arsenic biochemistry to devise an **antidote,** that is, a treatment to counteract intoxication. Administration of massive amounts of nontoxic 2,3-dimercaptopropanol [$HSCH_2CH(SH)CH_2OH$] to war gas victims resulted in displacement of the DHL-bound As as its cyclic compound with the dimercapto-propanol and its excretion in water-soluble form in the urine. This antidote against arsenic poisoning, the most effective one in over 2000 years, was called BAL or British antilewisite and is still in use as the drug Dimercaprol. It won a knighthood for Peters (Peters, 1963). BAL later was found also to counteract poisoning by copper, mercury, thallium, and lead (Section 11.5), which also react with DHL.

9.4 | ANIMAL-SPECIFIC MECHANISMS

9.4.1. Nervous System

A unique feature of animal life is the existence of a nervous system (Section 7.4). The neurotransmitters are key players, as they convey nerve impulses across the synaptic cleft to other nerve cells and to muscles. For example, the arrival of an impulse at the presynaptic side of the cleft stimulates a rapid release of acetylcholine (ACh), which then diffuses to the postsynaptic membrane (Fig. 7.3), creates a Na–K ion imbalance at the ACh receptor, and thus generates a new impulse or an action.

However, excess ACh must be destroyed at the postsynaptic membrane to avoid continuous stimulation, and this is done by a hydrolytic enzyme, acetylcholinesterase (AChE), whose active site is diagrammed in Fig. 9.3A. The ACh forms an enzyme–substrate complex at the serine hydroxyl of amino acid residue 203 (Fig. 9.3B) and acetylates it (Fig. 9.3C), releasing nontransmitting choline; water rapidly hydrolyzes the acetylserine to prepare the site for another ACh. The reaction at this so-called "esteratic site" where Ser acetylation occurs is assisted by glutamic acid 334 and histidine 447, the electron-rich imidazole N making the serine oxygen more electronegative via H-bonding. The amino acid sequence at the active site is crucial (Whittaker, 1986): Glu 202 adjacent to the Ser electrostatically holds the ACh cation in place at the choline subsite shared with Trp 86, while the rings of phenylalanines 295 and 297 form an "acyl pocket" for the acetyl group (Taylor and Radić, 1994).

Certain other esters can substitute for ACh in the reaction with serine, notably reactive carbamates and phosphates (Section 15.3). Figure 9.3D shows phosphorylation of Ser 203 by a phosphate ester, paraoxon, and this phosphorylated serine (Fig. 9.3E) is quite stable to hydrolysis and effectively blocks further reaction with ACh. In fact, such hydrolysis of phosphorylated serine as does occur may eventually re-

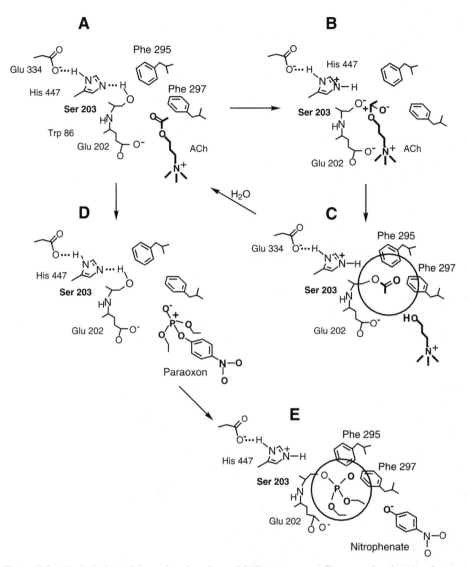

Figure 9.3. Vertical view of the active site of acetylcholinesterase as ACh approaches it (A) and reacts with Ser 203 (B) to acetylate it (C), circled. The acetylserine is hydrolyzed to reactivate AChE (A). Paraoxon likewise can approach Ser 203 (D) but irreversibly phosphorylates it (E), circled. Based on Taylor and Radić, 1994.

move one of the alkoxyl groups instead, a process referred to as "aging," which makes the block essentially permanent. Covering the serine results in interruption of nerve function. As ACh accumulates, muscarinic action at the smooth muscles and glands causes a feeling of tightness in the chest, constriction of pupils, and increased saliva, tears, and sweating, while the nicotinic effects include fatigue, weakness, and convulsions. Sulfates and borates apparently are too reactive to reach the active site unchanged, and although phosphate esters can phosphorylate the serine, the more stable thiophosphates require oxidation to the oxons (Section 15.4).

While water is not reactive enough to hydrolyze the phosphorylated serine, a strong organic base can react. The most widely used of these is N-methyl-2-pyridaldoxime iodide, called 2-PAM (Pralidoxime), whose anion displaces the blocking group and regenerates the serine OH (Eq. 9.3). The resulting phosphorylated 2-PAM is readily excreted in the urine.

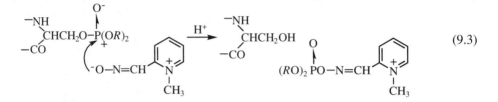

$$\text{(9.3)}$$

ACh receptors can also be inhibited directly. For example, the toxic component of the South American arrow poison, curare (D-tubocurarine), blocks nicotinic receptors by covering them, and the alkaloid atropine from Jimson Weed *(Datura stramonium)* similarly affects muscarinic receptors.

The chlorinated hydrocarbon insecticides, too, act upon the nervous system, but primarily on the axonal transmission of nerve impulses. DDT and its close relatives (Section 14.2) cause rapid and repetitive firing, which results in tremors, hyperexcitability, and eventual paralysis. One long-standing hypothesis (Holan, 1969) held that such wedge-shaped molecules pry and hold open the Na^+ channels so that ion movement is unimpeded and the axon is kept in a continual state of polarization. Other wedgelike substances with an apex diameter similar to that of sodium ion (6.3 Å), such as the natural pyrethrins (Section 15.7), also are insecticidal and were thought to behave similarly. Increased temperature unblocks the channels, so DDT exerts little effect above about 30°C. This served to explain why humans, with a body temperature of 37°C, were not readily poisoned by these insecticides, while the colder-blooded insects, fish, and aquatic invertebrates were extremely senstive.

However, more recent biochemical research has shown that DDT's toxicity is much more complex than that. The insecticide interferes with the movement of ions through neuronal membranes by delaying the closing of the Na^+ channels, slowing the opening of the K^+ gates, inhibiting the neuronal ATPase that controls repolarization and blocking calcium ion transport into the neuron as well. Release of aspartic and glutamic acids produces hyperexcitability that contributes to the tremors and convulsions seen in insects, birds, and mammals poisoned by substantial doses of DDT.

Lindane and the cyclodiene insecticides (Section 14.3) behave differently. As their toxicity is neither temperature sensitive nor moderated by rapid biodegradation, they are more toxic to mammals than is DDT. Lindane produces tremors and convulsions, but unlike DDT, it acts as an excitant that antagonizes the action of the neurotransmitter γ-aminobutyric acid (GABA) and the uptake of chloride ions necessary for neuron function (Matsumura, 1985). In insects, at least, it mimics the natural

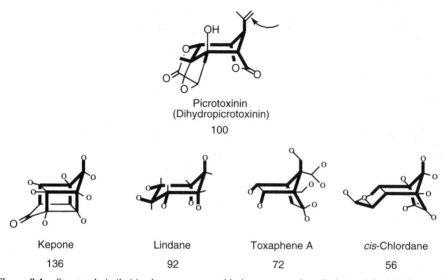

Figure 9.4. Structural similarities between organochlorine compounds and picrotoxinin. Numbers indicate relative binding to the picrotoxinin receptor of the American cockroach, *Periplaneta americana* (Tanaka et al., 1984). Dihydropicrotoxinin is saturated at the arrow. o = Cl.

excitant picrotoxinin (from the plant *Anamirta cocculus*) at the GABA receptor, and its clear structural similarity to other halogenated alicyclic insecticides and picrotoxinin is apparent from Fig. 9.4. Other BHC isomers display a similar action but much less toxicity, although all of them induce hepatic enzymes (IPCS, 1992).

A number of other intoxication mechanisms affect the nervous system. For example, the most powerful known mammalian poison, botulinus toxin released by the anaerobic bacterium, *Clostridium botulinum,* inhibits the presynaptic release of acetylcholine, and nicotine mimics acetylcholine at the ACh receptor (Section 12.2).

9.4.2. Liver

Another unique feature of most animal species is their liver or hepatopancreas (Section 7.4.5). As this organ is primarily responsible for detoxication and waste removal, its impairment has serious consequences. The processing of large flows of blood or hemocoel exposes it to unusually high levels of xenobiotics and the constant danger of hepatotoxicity (Zimmerman, 1978), which includes lipid accumulation, necrosis, lipid peroxidation, and cirrhosis. Chlorinated solvents, nitrosamines, aflatoxins, toadstool poisons, and pyrrolizidine alkaloids are among the worst offenders.

Perhaps the best understood hepatotoxicant is carbon tetrachloride. This familiar cleaning solvent is reductively dechlorinated by the cytochrome P450 enzyme system (Section 6.2.3) to produce highly reactive trichloromethyl radicals which cause peroxidative damage to membrane phospholipids, inhibit the glutathione so important in detoxication, and destroy the P450 itself (Equation 9.4).

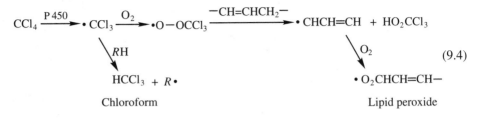

$$(9.4)$$

Chloroform   Lipid peroxide

Among the most interesting and controversial of the chemicals affecting the liver are the dibenzo-p-dioxins typified by 2,3,7,8-tetrachlorodibenzo-p-dioxin, TCDD (Section 13.3.4). TCDD is among the most toxic of all manmade substances, and its acute oral LD50 of <2.5 μg/kg in the guinea pig is widely cited. However, TCCD toxicity has other unusual features: After exposure to lethal doses, animals do not die quickly but instead stop eating, lose weight, and eventually just waste away. The weight loss can be prevented by intravenous feeding, but the lethality cannot. Other effects in experimental animals include immunotoxicity, embryotoxicity, and carcinogenicity (actually, tumor promotion), although none of these has been confirmed in humans.

The broad spectrum of dioxin effects is ascribed to an affinity for a stereospecific cytosolic *Ah* receptor in the liver (Fig. 9.5). After binding, the receptor–ligand complex is transported into the cell nucleus and induces or represses the expression of a battery of cellular proteins that account for the observed lesions (Whitlock, 1990). For example, hepatic monooxygenases (*e.g.,* CYP1A1) and aryl hydrocarbon hydroxylase (AHH) are strongly induced, and key detoxication enzymes including glucuronyl transferase, glutathione-*S*-transferase, and aldehyde dehydrogenase are affected adversely.

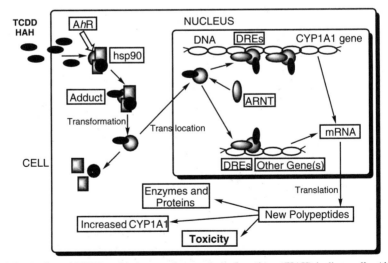

Figure 9.5. Action of TCDD and halogenated aromatic hydrocarbons (HAH) in liver cells. AhR is the *Ah* receptor, DRE a dioxin-responsive element (DNA recognition site), ARNT is the *Ah* receptor nuclear translocator, and mRNA denotes messenger RNA. Courtesy of M. S. Denison.

The dimensions of the receiver's binding site appear to be about 3×10 Å (Scheme 9.5), which accounts well for the structural specificity.

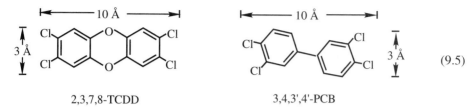

2,3,7,8-TCDD   3,4,3',4'-PCB

(9.5)

Other planar halogenated aromatic hydrocarbons (HAH) with these dimensions, including some PCBs and dibenzofurans, produce similar effects. Binding strength (EC_{50}) at the receptor depends upon lipophilicity (π), electronegativity (σ), and H bonding (Safe,1986), and in a series of fourteen 2,3-dichlorodibenzo-p-dioxins, log $(1/EC_{50}) = 1.24\pi + 6.11$, where π is Fujita's substituent constant (Section 16.5). The more lipophilic the compound, the tighter the binding.

Although TCDD is exceptionally toxic to young guinea pigs, responses vary widely among other species. The oral LD_{50} in C57BL mice is 284 μg/kg, that in an outbred strain of rat 297 μg/kg, beagle dogs 1000 μg/kg, and the Syrian hamster as high as 18,500 μg/kg, that is, 18 mg/kg (Geyer et al., 1990). One explanation for these species differences relates toxicity to the proportion of body fat; the leaner the animal, the greater the sensitivity (Geyer et al., 1990). However, despite thousands of man-years of research, the actual mechanism of lethality remains an intriguing mystery.

9.4.3. Intermediary Metabolism

Fluoroacetic acid, FCH_2COOH, has long been recognized as a natural poison found in plants of the genera *Acacia* and *Dichapetalum,* and its sodium salt finds wide use as a rodenticide. Rudolph Peters (1963), of BAL fame, found that it is converted into fluoroacetyl-coenzyme A and then, entering the normal TCA cycle, into fluorocitrate (see Appendix 9.1). As the fluorine is about the same size as an hydrogen atom, fluorocitrate readily binds to the enzyme, aconitase, that is responsible for the dehydration of citrate to *cis*-aconitate, and so shuts down mitochondrial energy production. The heart and CNS are especially sensitive to this loss of energy, and poisoning symptoms include erratic heartbeat and fibrillation, cyanosis, and convulsions. Peters termed this type of mechanism *lethal synthesis.*

Plant aconitase is not particularly sensitive to fluorocitrate and so is not seriously affected, but at about 0.2 mg/kg in rats, the acute oral LD_{50} of sodium fluoroacetate makes it a powerful rodenticide. The compound is even more toxic to canines (LD_{50} 50 μg/kg), and its use has resulted in the intentional or unintentional poisoning of large numbers of dogs, foxes, and wolves. Possible effects on aquatic animals also is a concern, as the replacement of persistent chlorofluorocarbon refrigerants (Freons) by more degradable ones has led to atmospheric deposition of fluoroacetic acid (Special Topic 5).

9.5 | PLANT-SPECIFIC MECHANISMS

9.5.1. Photosynthesis.

A unique feature shared by most plants is the ability to use sunlight to generate energy (ATP) and reducing power (NADPH) from water. Simultaneously, carbon dioxide is converted into organic metabolites by way of glyceraldehyde-3-phosphate, pyruvate, and finally glucose (Appendix 9.1). The capture of visible light energy is accomplished primarily by several closely related tetrapyrrole pigments: chlorophylls in green plants, phycoerythrin in red algae, phycocyanin in bluegreen algae, and bactcriochlorophyll in photosynthetic bacteria. Algae also utilize chlorophyll, and phycoerythrin and phycocyanin are known as "accessory pigments." In green plants, the process takes place in specialized organelles called chloroplasts.

Electrons released by the oxidation of water are boosted to a higher energy level by 680 nm light at a chlorophyll (Chl) reaction center and there reduce the pigment, phaeophytin (Fig. 9.6). They then move through a series of reducible intermediates, receive another energy boost at the 700 nm reaction center, and finally generate NADPH which is used to reduce CO_2 via phosphoglycerate and phosphoglyceraldehyde. The intermediates in the photosynthetic process are reminiscent of those in respiration, but the movement of electrons from water to CO_2 is just the reverse. If respiration is like burning fossil fuel to generate electricity, photosynthesis generates the electricity via two different solar panels.

A wide variety of chemicals can inhibit photosynthesis (Table 9.1), and many

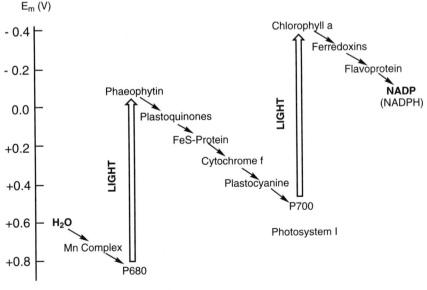

Figure 9.6.   Electron flow in plant photosynthesis.

TABLE 9.1
Inhibition of Photosynthesis by Herbicides[a]

Type	Structure type	Example	Structure	I_{50} (μM)[b]
sym-Triazine		Simazine		2.2
Anilide	—NHCOR	Propanil		0.76
Uracil		Bromacil		0.49
asym-Triazinone		Metribuzin		0.2
Phenylurea	—NHCON<	Diuron		0.18

[a] Corbett et al., 1984.
[b] I_{50} is the molar concentration providing 50% inhibition of oxygen release in the Hill reaction.

find use as herbicides ("weed killers"). Most have been discovered by screening candidate compounds for their effect on the Hill reaction:

$$2H_2O + 4Fe^{3+} \xrightarrow[\text{Light}]{\text{Chloroplasts}} O_2 + 4H^+ + 4Fe^{2+} \qquad (9.6)$$

Isolated chloroplasts are illuminated in the presence of an electron acceptor, such as ferricyanide, and oxygen evolution is measured. Inhibition of the Hill reaction occurs close to the plastoquinone step that follows Photosystem II (Fig. 9.6), but although the steric fit of inhibitors onto the plastoquinone molecule has received extensive study, the inhibitory mechanism remains unclear. Almost without exception, chemicals slowing the Hill reaction prove to be toxic to intact plants, although low activity and a susceptibility to biodegradation often limit their practical application.

A group of herbicidal dipyridinium salts inhibit photosynthesis by competing with chlorophyll a as the electron acceptor in Photosystem I. Paraquat (1,1-dimethyl-4,4'-bipyridylium chloride) has long been known for its high reduction potential (E_0 -446 mV) and used as an oxidation–reduction indicator called methyl viologen. Accepting an electron generates a remarkably stable free radical that reduces molecular oxygen to the powerful oxidant, superoxide anion ($\bullet O_2^-$), which reacts with and damages cell constituents. As shown in Eq. 9.7, paraquat (P) can be considered as just a catalyst for the reduction of oxygen, and so a small amount can produce a lot of very destructive superoxide.

$$P^{2+} + e^- \longrightarrow P^{+\bullet} \xrightarrow{O_2} \bullet O{-}O^- + P^{2+} \qquad (9.7)$$

9.5.2. Amino Acid Biosynthesis

Glyphosate (Roundup®) has become a widely used herbicide, especially in home gardens. It kills or stunts herbaceous plants but does not seem to affect most woody species. Whether in bacteria, algae, or higher plants, it blocks formation of the aromatic amino acids phenylalanine, tyrosine, and tryptophan, which share the shikimic acid biosynthetic pathway through the intermediate phosphoenolpyruvate (Fig. 9.7). The enzyme affected is 5-enolpyruvylshikimate-3-phosphate (EPSP) synthase, also called 3-phosphoshikimate-1-carboxyvinyl (PSCV) transferase, which occurs widely in plants (Duke, 1988). Biosynthesis of both protein and lignin ceases. Animals share the shikimate pathway with plants but fail to absorb this highly polar herbicide.

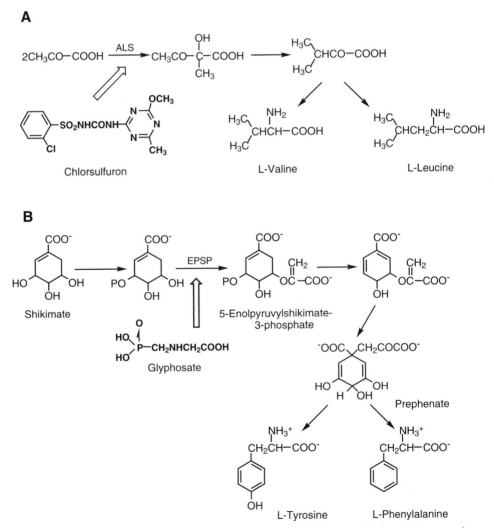

Figure 9.7. Biosynthesis of amino acids inhibited (A) by sulfonylureas and (B) by glyphosate at the steps shown. ALS = acetolactate synthase, EPSP = 5-enolpyruvylshikimate-3-phosphate synthase.

However, some plants contain a form of EPSP synthase that is relatively insensitive to glyphosate. The gene that encodes for this enzyme can be transferred to susceptible plants to confer resistance toward the herbicide, and this technique has now been put into agricultural practice so that an entire field can be sprayed with Roundup®, but only the weeds are affected.

A relatively new group of herbicides, the sulfonylureas, also act on amino acid biosynthesis. In this case, the branched amino acids valine, leucine, and isoleucine fail to form because of inhibition of acetolactate synthase (ALS). This key enzyme catalyzes the condensation of a pyruvate molecule with a second pyruvate to form α-acetolactate and, eventually, valine and leucine (Fig. 9.7); it also reacts with α-ketobutyrate to give α-aceto-α-hydroxybutyrate and finally isoleucine (Beyer et al., 1988). Resistant crop species such as wheat were found to be up to 4000 times less sensitive to chlorsulfuron than were weeds such as mustard, because they detoxify the herbicide efficiently by ring oxidation followed by glucosylation (Section 6.3.2).

9.6 | MICROBE-SPECIFIC MECHANISMS

9.6.1. Cell Wall Synthesis

Bacteria and fungi possess unique cell walls. Where animal cells lack a discrete wall, and plant cells have walls composed of cellulose and other carbohydrates, bacterial cells have an interior membrane held within a more-or-less rigid cage of cross-linked polymer called murein. Gram-negative bacteria (those not stained by a mixture of crystal violet and iodine) have a thin outer membrane in addition. The heavier walls of Gram-positive bacteria, 15–50 nm thick, are composed of up to 25 separate layers of murein and account for as much as 20% of the cell's weight. As the cell grows, new wall must be synthesized, and it is here that many antibiotics act.

Murein consists of strands of a poly(peptidoglycan) composed of repeating units of N-acetylglucosamine and N-acetylmuramic acid linked to the tetrapeptide L-alanyl-D-glutamyl-L-lysyl-D-alanine (Structure 9.8) and cross-linked via a chain of 5 glycines between the D-alanine carboxyl and the side-chain amino group of a lysine on another strand.

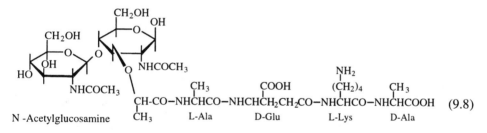

(9.8)

N-Acetylglucosamine   N-Acetylmuramic acid   Tetrapeptide

The cell may be thought of as a bag of protoplasm held within one giant polymer molecule (Fig. 9.8). Construction of the cage starts with biosynthesis of UDP-*N*-acetylglucosamine and proceeds through several steps to UDP-*N*-acetylmuramic acid, adds the first three amino acids, and completes the unit with a terminal D-alanyl-D-alanine. The second *N*-acetylglucosamine is then introduced, followed by attachment of the glycine pentapeptide to the lysine and polymerization of the peptidoglycan monomer to form a high-molecular-weight strand. The antibiotic D-cycloserine inhibits formation of the D-alanyl-D-alanine fragment, and vancomycin inhibits the polymerization.

In the final transpeptidation step, the strands are cross-linked by displacement of the terminal D-alanines by the amino groups of the pentaglycine chains, the reaction taking place outside the cell membrane. It is the transpeptidase "zipper" that is inhibited by penicillin, among several other complex steps. Steric models show that penicillin looks remarkably like a conformation of the D-alanyl-D-alanine, and the anti-

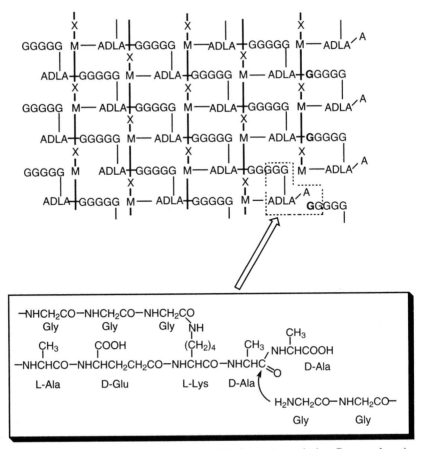

Figure 9.8. Structure and synthesis of bacterial peptidoglycan. A = alanine, D = D-glutamic acid, G = glycine, L = lysine, *M* = *N*-acetylmuramic acid, X = *N*-acetyl-*D*-glucosamine. At lower right, terminal glycine (boldface) and D-alanyl-D-alanine are ready to be cross-linked.

biotic is thought to acylate the active site of the enzyme through its very reactive β-lactam ring:

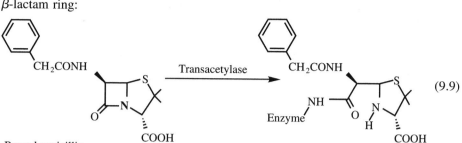

$$(9.9)$$

Benzyl penicillin

Bacteria are under a high internal pressure, and cell walls weakened by the action of penicillin actually burst. As only certain prokaryotes have this type of biochemistry, antibiotics such as penicillin are nontoxic to other kinds of organisms, although allergic reactions can occur in humans. A good discussion of the cell wall biosynthesis is provided by Albert (1985) and Zubay (1986).

9.6.2. Sulfhydryl Groups

Copper compounds have been used since Roman times to control fungal diseases of crops, and such classical treatments as Bordeaux mixture (cupric hydroxide stabilized with calcium sulfate) and Burgundy mixture (cupric hydroxycarbonate) are still used for this purpose in much of the world. Like arsenic, the copper combines tightly with the SH groups of dihydrolipoyl transferase (Section 9.3) and stops respiration. Yet, as these fungicides are insoluble in fatty solvents, how does the Cu enter the fungal cell?

Fungi readily absorb lipophilic substances from their immediate surroundings. Copper forms cyclic, lipid-soluble coordination compounds (chelates) with natural amino- and hydroxy-acids. For example, cupric ions combine with glycine to give first a 1:1 monoadduct and then, with excess glycine, a 1:2 diadduct.

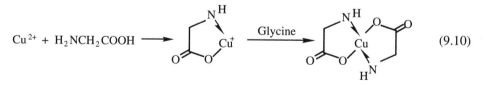

$$(9.10)$$

The fungi secrete amino acids, which dissolve the fungicide and carry its copper into the cell where the –SH groups of respiratory enzymes react to form stable S–Cu bonds. The equilibrium constants for chelate formation (Table 9.2) show that copper binding can be very strong and that the binding constant β for a sulfur ligand always is higher than that for amino acids, meaning that an –SH on an enzyme will displace copper from them.

Among the world's most popular synthetic fungicides are the dithiocarbamates, typified by sodium dimethyldithiocarbamate, $(CH_3)_2N$–CSSNa. They are easy and cheap to make and are effective against a broad range of pathogenic fungi while

TABLE 9.2
Relative Stability (log β) of Some Coordination Compounds^a

	Cu^{2+}		Ni^{2+}		Zn^{2+}		Fe^{2+}		Mn^{2+}	
Ligand	1:1	1:2	1:1	1:2	1:1	1:2	1:1	1:2	1:1	1:2
Glycine	8.5	15	6	11	5	9	4	8	3	5.5
Histidine	10.5	19	9	16	7	12	5	9	3.5	
Ethylene diamine	11	20	8	18	6	12	4	9.5	3	5
EDTA^b	19		18		16		14		13	
8-Hydroxyquinoline	12	23	10	18	8.5	16	8	15	7	12
Cysteine	c		10	19	10	18	6		4	
Dimethyldithiocarbamic acid	11	22				9				

^{a}Albert, 1985.
^{b}Ethylenediaminetetraacetic acid.
^{c}Oxidized by Cu^{2+}.

safe for animals and plants, which are able to detoxify them. Like the amino acids, dithiocarbamates form lipid-soluble coordination compounds with traces of natural copper present in the environment. The chelate is transported into the thin-walled fungal cells, where its copper combines with the DHL thiols and may have other disruptive action as well. The great stability of the 1:2 adducts (Table 9.2) suggests that the metal would not be easily displaced by DHL, and so the toxic action must be due to the monoadduct.

Ethylene-*bis*-dithiocarbamates are a related type of fungicide, of which nabam is typical. Although the dithiocarbamate salts are too polar to penetrate the fungal cell membrane, they are air-oxidized in solution to the more lipophilic ethylenethiuram disulfide (ETDS) and then converted to a related heterocyclic compound, ETEM:

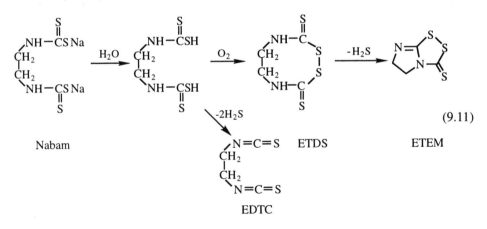

(9.11)

The free dithiocarbamic acid, formed by hydrolysis, also eliminates two molecules of H_2S to form ethylene diisothiocyanate (EDTC), and these three decomposition products all react with thiols and are fungitoxic. It seems likely that each is in some way implicated in the *bis*-dithiocarbamate mechanism of action. Other fungicides of this group, such as zineb, already contain chelated zinc, iron, or manganese, but the metal ligand only confers lipophilicity and does not take part in the toxicity (Corbett et al., 1984).

Fungicides such as captan, which contain the *N*-trichloromethylthio group, are decomposed *in vivo* to thiophosgene, $S = CCl_2$, which reacts nonspecifically with thiols. Alternatively, the $-SCCl_3$ group is displaced from the fungicide by the thiol. However, chlorothalonil is quite specific in its reaction with the thiol of glyceraldehyde-3–phosphate dehydrogenase.

(9.12)

Captan   Chlorothalonil

9.7 | PERSPECTIVE

Clearly, even the early steps of intoxication are chemical. A xenobiotic reacts with some specific site on a receptor, enzyme, membrane, or cofactor to alter or block its normal action, followed later by a physiological response such as loss of respiration or tremors. van der Waals, ionic, or covalent forces may be involved, but the bonds are chemical nevertheless.

Different kinds of organisms may respond differently to a particular substance, depending on their style of biochemistry, or they may not respond at all if the necessary target is absent. However, even individuals of the same species may differ in their response. Take their acetylcholinesterase for example. This large molecule is composed of four identical polypeptide chains, each of molecular weight 84,000 and containing about 400 amino acid units. With four catalytic sites per molecule, where a difference of even a single amino acid in the sequence can alter specificity, the great potential for variation is evident (Augustinsson, 1961). A human possesses either the normal form or an "atypical" sequence, –Gly–His–Ser–Ala–Gly–Ala–Ser, at the active site.

However, a frequent reason for interspecies differences in intoxication is the inability of a xenobiotic even to reach its target. Variations in absorption, distribution, and biotransformation between species and individuals may spell the difference between intoxication and safety, but *exposure* always is the necessary and dominant factor. Exposure is the subject of the next chapter.

9.8 | REFERENCES

Albert, A. 1985. *Selective Toxicity*, 7th Ed., Chapman and Hall, London, U.K.

Augustinsson, K.-B. 1961. Multiple forms of esterases in vertebrate blood plasma. *Ann. N.Y. Acad. Sci.* **94:** 844–60.

Beyer, E. M., M. J. Duffy, J. V. Hay, and D. D. Schlueter. 1988. Sulfonylureas, in *Herbicides: Chemistry, Degradation, and Mode of Action* (P. C. Kearney and D. D. Kaufman, eds.), Vol. 3, Marcel Dekker, New York, NY, pp. 117–89.

Changeux, J.-P., A. Devillers-Thiéry, and P. Chemeuilli. 1984. The acetylcholine receptor: An allosteric protein. *Science* **25:** 1335–45.

Corbett, J. R., K. Wright, and A. C. Baille. 1984. *The Biochemical Mode of Action of Pesticides*, 2nd Ed., Academic Press, New York.

Duke, S. O. 1988. Glyphosate, in *Herbicides: Chemistry, Degradation, and Mode of Action* (P. C. Kearney and D. D. Kaufman, eds.), Vol. 3, Marcel Dekker, New York, NY, pp. 1–116.

Geyer, H. J., I. Scheuntert, K. Rapp, A. Kettrup, F. Korte, H. Greim, and K. Rozman. 1990. Correlation between acute toxicity of 2,3,7,8-tetrachlorodibenzo-p-dioxin (TCDD) and total body fat content in mammals. *Toxicology* **65:** 97–107.

Giraudat, J., and J.-P. Changeux. 1980. The acetylcholine receptor. *Trends Pharm. Sci.* **1:** 198–202.

Holan, G. 1969. New halocyclopropane insecticides and the mode of action of DDT. *Nature* **221:** 1025–29.

IPCS. 1992. *Alpha- and Beta-hexachlorocyclohexanes*, WHO, Geneva, Switzerland. Environmental Health Criteria 123.

Josephy, P. D. 1997. *Molecular Toxicology*, Oxford University Press, New York, NY.

Matsumura, F. 1985. *Toxicology of Insecticides*, 2nd Ed., Plenum Press, New York, NY.

Peters, R. A. 1963. *Biochemical Lesions and Lethal Synthesis*, Macmillan, New York, NY.

Pitot, H. C., and Y. P. Dragan. 1996. Chemical carcinogenesis, in *Casarett and Doull's Toxicology: The Basic Science of Poisons*, 5th Ed. (K. D. Klaassen, ed.), McGraw-Hill, New York, NY, pp. 201–67.

Safe, S. H. 1986. Comparative toxicology and mechanism of action of polychlorinated dibenzo-p-dioxins and dibenzofurans. *Ann. Rev. Pharmacol. Toxicol.* **26:** 371–99.

Tanaka, K., J. G. Scott, and F. Matsumura. 1984. Picrotoxinin receptor in the central nervous system of the American cockroach: Its role in the action of cyclodiene-type insecticides. *Pestic. Biochem. Physiol.* **22:** 117–27.

Taylor, P., and Z. Radić. 1994. The cholinesterases: From genes to proteins. *Ann. Rev. Pharmacol. Toxicol.* **34:** 281–329.

Whitlock, J. P. 1990. Genetic and molecular aspects of 2,3,7,8-tetrachloro-p-dioxin action. *Ann. Rev. Pharmacol. Toxicol.* **30:** 251–77.

Whittaker, M. 1986. *Cholinesterase*, Karger, Basel, Switzerland.

Zimmerman, H. J. 1978. *Hepatotoxicity*, Appleton-Century-Crofts, New York, NY.

Zubay, G. 1986. *Biochemistry*, Addison-Wesley, Reading, MA.

Special Topic 9. Chemical Carcinogenesis

The very word *cancer* brings fear to the hearts of most people. The dictionary defines a cancer as "a malignant tumor of potentially unlimited growth that expands locally by invasion and systematically by metastasis." This describes a group of diseases that share characteristics of uncontrolled cell proliferation, association with nuclear DNA *(genotoxicity),* and a substantial latency period before becoming evident. Tumors have been observed in all tissues and in most animal species, including fossils of a 125 million-year-old dinosaur and Java Man *(Pithecanthropus)* from 1 million years ago.

Contrary to public perception, only 2–3% of human cancer can be ascribed to environmental pollutants, and the natural compounds in food and, especially, tobacco appear to be the principal culprits (Section 7.4.7). Tumors can be initiated by irritant solids such as asbestos, compounds of certain metals such as chromium, ionizing radiation including x-rays and ultraviolet light, and by *chemical carcinogens*—the focus of this Special Topic.

As most tumors arise in some way from chemicals, cancer may be assumed to start with a reaction in DNA. Indeed, the proximate carcinogen commonly is a highly reactive electrophile, an *initiator,* that combines with electron-rich sites on DNA, often a nitrogen atom on a purine or pyrimidine base. Methylating agents such as methyl iodide act primarily on the N^7 of guanine, although other N and O atoms also react (Eq. 9.13). PAH epoxides act on the guanine N^2, and nitrenium ions generated from amines and nitro compounds react exclusively at the guanine C^8 (Josephy, 1997). Other genotoxic electrophiles of environmental interest include methyl sulfate, ethylene dibromide, and ethylene oxide, all of which can *alkylate* DNA at a site such as the oxygen indicated by the arrow and so disrupt DNA structure.

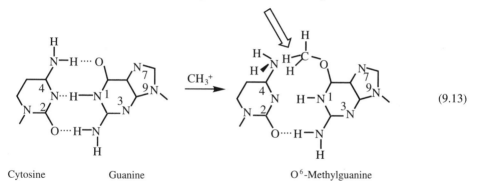

| Cytosine | Guanine | | O^6-Methylguanine |

$$(9.13)$$

However, most electrophiles are too reactive to survive long in the environment. The proximate carcinogen most often is generated metabolically within the cell, usually by enzymatic oxidation such as in the conversion of dimethylnitrosamine to the carcinogenic alkylating agent, diazomethane:

$$\underset{DMN}{\overset{H_3C}{\underset{H_3C}{>}}N-N=O} \xrightarrow[-HCHO]{[O]} CH_3NH-N=O \rightleftharpoons CH_3N=N-OH \xrightarrow{-H_2O} \underset{Diazomethane}{CH_2N_2} \qquad (9.14)$$

Other examples include the epoxidation of vinyl chloride (Section 13.3), aflatoxin (Section 12.6), and polycyclic aromatic hydrocarbons (Section 13.4), as well as the *N*-oxidation of aromatic amines and amides to phenylhydroxylamines, with subsequent conjugation to form electrophilic sulfates (Section 15.7).

Equation 9.13 shows that alkylation of a DNA nitrogen affects the H-bonding that provides structure to the double helix. If the carcinogen is a *bifunctional alkylating agent* such as ethylene dibromide or *bis*(chloromethyl) ether, N atoms on one strand can be cross-linked to those on another strand or even to an adjacent protein molecule. Aflatoxins may function in this way, via an epoxide at one end of the molecule and a lactone at the other. If the damaged cell manages to replicate, such changes can lead to *neoplastic conversion* in which the genome is permanently altered *(mutated)* through replication errors, base mispairing, and/or codon transposition. The altered cells multiply, with or without the assistance of an external chemical *promoter;* promoters (cocarcinogens) such as chlorodioxins are not themselves carcinogenic. If the alteration leads to changes in growth-regulating proteins, the cells may become *neoplastic* and proliferate out of control to develop into a tumor. The multiple steps in the development of neoplasms are classified as *initiation, promotion,* and *progression,* often resulting in *metastasis,* where the malignant cells are transported to other sites.

Not all carcinogens are genotoxic (mutagenic), that is, react with DNA. An increasing number of chemicals are being shown to produce tumors via other mechanisms, including benzene, chloroform, di(2-ethylhexyl) phthalate, lindane, and saccharin. Most are classified as promoters or progressors. One alternate mechanism is peroxisome proliferation, in which the carcinogen stimulates production of these subcellular bodies which, in turn, cause the oxidative damage to the DNA. A good general discussion of chemical carcinogenesis in animals is presented by Pitot and Dragan (1996).

It seems almost impossible that a reactive electrophile could survive the gauntlet provided by cellular nucleophiles and metabolic detoxication. Indeed, most of the methyl bromide used to fumigate grain becomes bound to S and O rather than to N. Once reaction with DNA has occurred, natural repair mechanisms also remove virtually all damaged components, and usually any remaining crippled cells die without replication. All this explains why most organisms live out a normal lifespan cancer-free despite being surrounded by a host of environmental carcinogens. Increasing knowledge of the whole process, better detection of mutagens and carcinogens (Section 8.4.5), and strict laws are all to the good, but can we *ever* identify and regulate *all* carcinogenic chemicals? Considering the thousands of untested natural products and the increasing annual introduction of new synthetic substances, the answer has to be "probably not."

APPENDIX 9.1
A Brief Summary of Intermediary Metabolism[a]

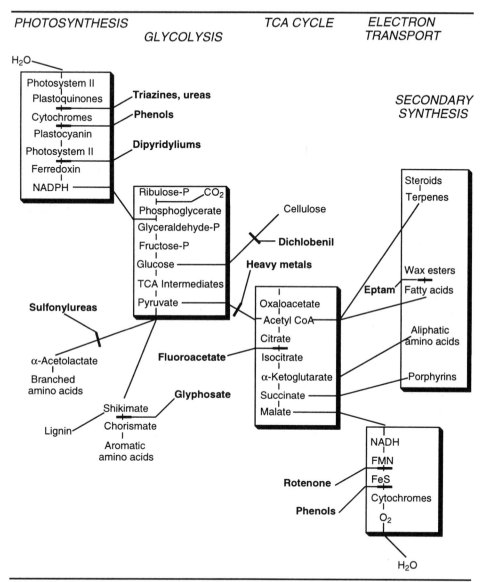

[a]Toxicants are shown in **bold.** P = phosphate.

APPENDIX 9.2.
Some Intoxication Mechanisms

Toxicant	Target	Typical Effects	Section
Aflatoxins	DNA	Liver cancer	12.6
Anagyrine	Unknown	Teratogenesis	7.4
Aromatic amines	DNA	Liver/kidney cancer	15.6
Arsenic	Sulfhydryl	Respiration inhibition	9.3
Bromoalkanes	DNA; sulfhydryls	Cancers, CNS damage	15.2
Carbon monoxide	Heme Fe	Respiration inhibition	7.4, 15.9
Carbon tetrachloride	Lipid	Liver degeneration	9.4
Cardiac glycosides	Na^+-K^+ ATPase	Heart arrhythmia	12.3
Dithiocarbamates	Sulfhydryls	Fungitoxicity	9.3
Ethanol	Membrane?	CNS depression	7.4
Ethylene glycol	Forms oxalate	Kidney damage	13.2
Fluoroacetate	Aconitase	Energy disruption	9.4
Halophenols	H^+ gradient	Energy disruption	9.3
Heavy metals	Sulfhydryls	Respiration inhibition	9.3, 9.6
Methanol	Forms formate	Blindness	13.2
Nicotine	ACh receptor	Muscular effects	12.2
Organochlorines	Na^+/K^+ balance	Convulsions	9.4
Organophosphates	AChE 34	Nervous system effects	7.4, 9.4
Penicillin	Peptide transferase	Bacterial cell disruption	9.6
Polycyclic hydrocarbons	DNA	Cancers	13.3
Snake venoms	ACh receptors	Nerve disruption	12.5
Toadstool toxins	Membrane	Liver necrosis	7.4
Triazines herbicides	Plastoquinones	Photosynthesis inhibition	9.5
Urea herbicides	Plastoquinones	Photosynthesis inhibition	9.5
Urushiol	T-cells	Skin rash	7.4

Exposure and Risk **10**

10.1 | HAZARD AND RISK

Hazard and risk are what most people really imply when they talk about "toxicity." *Hazard* is the potential to produce harm (Section 1.6). Its close relative, *risk,* measures the magnitude of the hazard by the probability of its occurrence. Hazard is qualitative; risk is quantitative. *Risk assessment,* the process defining just how dangerous a particular substance is to ourselves and our environment, involves four overlapping components: (1) hazard identification or evaluation, (2) dose–response evaluation, (3) exposure evaluation, and (4) risk characterization. This chapter shows how risks are assessed, managed, and communicated.

10.2 | EXPOSURE

Exposure is the key to intoxication and risk. The intoxication process, toxicity measurements, and toxic mechanisms described in previous chapters are meaningless unless the organism is exposed. Exposure ordinarily is defined simply as contact of an organism's exterior with a chemical, but toxicologically, it is the means by which the organism acquires a *dose* (Suter, 1993). The dose is the total amount of chemical received through any body areas that contact the outside world: skin or integument, mouth, nose, gills, eyes.

Exposure can take a number of forms: respiration of air, drinking or imbibing water, contacting soil, and taking in nutrients (Table 10.1). Although people tend to think of exposure only in human terms, it is universal among animals, plants, and microorganisms. It may come from such unexpected places as a hot shower or a

TABLE 10.1
Some Sources of Exposure to Toxicants

Source	Examples	Population Exposed	Analysis
Manufacturing	Factory	Workers, close residents	Air, surfaces
Transportation	Spills, storage, engine exhaust, waste oil	Workers, close residents, motorists	Air, water, surfaces
Consumer products	New cars, clothing	Consumers	Air, surfaces
Food	Natural constituents, pesticide residues, packaging	Consumers, pets, other animals	Food
Water	Drinking water, bathing, swimming, laundry	General	Water, air
Ambient air	Smog, indoor air	General, office workers, hospital patients	Air
Contaminated soil	Fields, gardens	AG workers, gardeners	Soil, air
Waste disposal	Incinerators, waste collection, landfills	Workers, homeowners	Air, soil, surfaces

bird's perch. *External exposure,* the kind usually measured, must be differentiated from *internal exposure,* the amount of chemical absorbed into the systemic circulation, and from the *biologically effective dose* available at the site of action.

Because external exposure requires contact with a chemical, it is linked closely to the chemical's environmental dissipation. The form and extent of dissipation govern contact. Volatility and dissolution quickly reduce a substance's availability, while adsorption and bioconcentration often prolong it (see Chapter 3). Once released into the environment, most chemicals eventually "disappear," so exposure level varies with time. Figure 10.1 charts the dissipation of parathion insecticide from the surface of citrus leaves and hence the decreasing potential for exposure. Chemical reactivity forms an important part of dissipation, in that most substances are transformed into less toxic products (Chapters 5 and 6). However, toxicity occasionally does increase, for example as parathion is converted to the much more poisonous paraoxon:

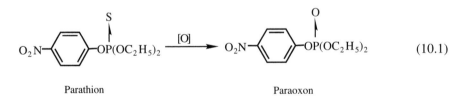

$$\tag{10.1}$$

Parathion   Paraoxon

External exposure is directly related to *contact rate,* the quantity of chemical per unit area and time. Factors that influence contact rate include surface areas, respiration rates, and food and fluid intake, and they vary widely among individuals of a population depending on age, size, health, activity, and behavior. Table 10.2 shows a set of physiological standard values developed for humans, although these are not necessarily realized in detail. For example, food consumption, the main route of total human exposure, has been standardized at 1500 g/day, although Appendix 10.1 shows an actual intake less than that except in adult males (USDA, 1980) Yes, food is a mixture of chemicals.

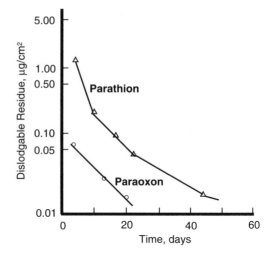

Figure 10.1. Dissipation of parathion and paraoxon from orange leaves. Parathion applied as 25% WP at 10 lbs/acre in August. Adapted from Gunther et al. (1977).

Even the values in the Appendix are misleading, as only the ***maximum*** value is presented, irrespective of age. Typically, the food intake of a 19–22-year-old man actually averages only 1398 g/day (2090 g/day if soft drinks, coffee, and beer are included), while his female counterpart averages 1000 g/day (1570 with the beverages). It is important to realize that each individual is truly different from his neighbors, and this individuality has to be considered in any precise measure of an organism's true exposure. The advantage of the default values of Appendix 10.1 is that they represent the average (although hypothetical) member of the general population.

Skin is the next most frequent route of exposure after food, and the Appendix shows the aveage male and female human skin area to be 1.94 and 1.69 m², respectively. However, except in instances such as bathing, few people ever expose their entire skin area to contaminants. Generally all but the hands, arms, head, and neck are covered, and even a person dressed in a short-sleeved shirt, short pants, and shoes exposes less than 0.3 m² of skin; over 90% of the skin area is covered when wearing a long-sleeved shirt and gloves. Actually, a small amount of surface contaminant can

TABLE 10.2
Standard Exposure Factors[a]

	Human		Mouse		Rat	
Factor	M	F	M	F	M	F
Lifespan, yrs	70	78	2	2	2	2
Body weight, g	75,000	60,000	30	25	500	350
Surface area, cm²	19,400	16,900	NA[b]		NA	
Food intake, g/day	1,500	1,500	5	5	20	18
Water intake, mL/day	2,500	2,500	5	5	25	20
Air intake, L/day	20,000	20,000	40	40	200	200

[a]ICRP (1975).
[b]NA = not available.

penetrate clothing, but the proportion that reaches the skin is less than 10% of that on the cloth surface.

Respiratory exposure data seem to be the least complete and consistent, probably because breathing is the most variable form of exposure. There is agreement that adult humans respire a total of about 20 m^3 of air a day, but this must vary greatly from day to day depending on lifestyle and activity level. However, multiple exposures are common, and it is the *total* intake of a chemical that counts. For example, a person typically might receive a skin exposure of 60 μg/day and respiratory exposure of 2 μg/day of a particular chemical at work, a residue of 2 μg/kg of the same chemical in the food he eats, 4 μg more in the water he drinks, and still another 50 μg via skin and lungs from a 10-minute shower for a total exposure of perhaps 120 μg/day. Thus the usual occupational exposure estimate might reveal only half the total exposure.

10.3 | RISK ASSESSMENTS

10.3.1. Occupational Exposure

Many people do receive their principal exposure to chemicals while at work in an industrial setting. Workers usually are tested for exposure in some noninvasive way, and specific tests are performed routinely where exposure is expected to be significant. For example, wood often is treated with pesticides such as pentachlorophenol (PCP) to preserve it against the attacks of insects and wood-rotting fungi, and so the urine of factory workers in an Idaho treatment plant was analyzed monthly for PCP in an attempt to identify and roughly quantitate exposure (Wyllie et al., 1975). Urine is easy to collect, readily analyzed by spectrophotometry or gas chromatography, and reflects the internal (absorbed) dose. A constant PCP level signifies a steady exposure, while a high level that declines with time due to metabolism and excretion would signal an acute exposure incident.

All the Idaho employees showed evidence of steady, elevated PCP exposure. The average urine content of a pressure treatment operator was 296 μg/L and that of the office manager 64 μg/L, although no one showed symptoms of intoxication. The urine of a person from outside the plant averaged 3.4 μg/L, within the range of 6.3$\pm$4.3 μg/L reported for the U.S. population as a whole. Everyone is exposed daily to traces (1–170 μg) of PCP via drinking water, food, clothes, and treated lumber, but occupational exposure may reach 100 to 1000 times that amount (see Special Topic 8). While symptoms of intoxication start as low as 3–10 mg/kg/day, workers as well as the rest of us rapidly develop a tolerance through increased degradative ability.

Exposure to toxic substances is commonplace among agricultural workers. In addition to toxic plants and animals, pesticides and their formulating agents are a continual cause for concern (Section 2.5). Prior to application, pesticide active ingredients are diluted with carriers, which ordinarily help to reduce exposure, but solid formulations such as dusts and wettable powders form a ***dislodgable deposit*** of residue on treated leaves. Consequently, fruit pickers in orchards may work in a cloud of pesticide particles stirred up by their own activities. A study of orange harvesters (Spear et al., 1977) showed that their dermal and respiratory exposure to cholinesterase-inhibiting parathion and paraoxon averaged a total of 390 and 740 μg, respectively, during the first week of work. By the second week, exposure had risen to 255 and 1080 μg, while the third week reached 1040 and 3140 μg as more parathion was oxidized to oxon and more deposit was loosened (Table 10.3).

Although such exposure can be estimated by chemical analysis of the pesticide or its metabolites in urine, a more meaningful measure might be changes in blood cholinesterase levels that directly indicate the biologically effective dose presented to the receptor. While not as convenient as urine to collect, a drop of blood will suffice, and red blood cell (RBC) acetylcholinesterase (AChE) can be measured colorimetrically with an accuracy of 80–95%. The fractional change in AChE level, ΔAChE, is indicative of the degree of inhibition by the organophosphate (Section 9.4) and is related to insecticide dose:

$$\Delta\text{AChE} = 1 - e^{-k_e \Sigma (D/\text{LD}_{50})} \tag{10.1}$$

where D represents the exposure dose normalized to body weight (mg/kg), LD_{50} is the rat oral median lethal dose in mg/kg, Σ means total, and k_e is an empirical constant specific for the enzyme, AChE (Popendorf and Leffingwell, 1982). AChE was significantly depressed in the orange pickers (Table 10.3), and this depression could be traced back to the total exposure, E mg, related to the leaf residue R at any time t:

$$E = k_D R t \tag{10.2}$$

The k_D is a crop-specific dosing coefficient determined empirically for the specific situation. Considering RBC AChE to be a surrogate for neuronal AChE, and recognizing the high degree of individual variability in levels, a roughly 40% decrease in

TABLE 10.3
Calculated Dose–Response for Parathion in Orange Pickers[a]

Parameter	Week 1		Week 2		Week 3	
	PS[b]	PO	PS	PO	PS	PO
Dermal dose, μg	382	705	246	1020	1030	3080
Respired dose, μg	8	37	9	58	11	57
Total dose D, μg	390	240	755	1080	1040	3140
D, mg/kg ($\times 10^3$)[c]	5.6	10.6	3.6	15.4	14.9	44.9
$D/\text{LD}_{50}(\times 10^3)$[d]	0.6	10.6	0.4	15.4	1.5	44.9
$\Sigma D/\text{LD}_{50}$		0.011		0.016		0.046
ΔAChE		−4.0%		−12%		−23%
k_e		4.0		8.5		5.7

[a] Spear et al., 1977.
[b] PS = parathion, PO = paraoxon.
[c] Assumes standard 70 kg weight.
[d] Assumes a parathion LD_{50} of 10 mg/kg, paraoxon 1 mg/kg.

cholinesterase activity is necessary to see clear symptoms of poisoning. Measurements are accurate to within 5–15%.

One value of such estimates is that they can be used by regulatory agencies to set *reentry periods,* that is, the length of time required for environmental dissipation to reduce exposure to a level that permits workers to reenter the field or orchard safely (Popendorf and Leffingwell, 1982). In orange pickers, possible scenarios calculated for an average parathion dissipation rate showed that AChE levels would be reduced 20% if workers started picking only 5 days after pesticide application. This loss could be held to 8% by waiting 17 days, to 4% by waiting 27 days, and to a negligible 2% after 35 days. The period established for reentry into California orange groves treated with parathion was set at 21 days if less than 4 lbs/acre/year of active ingredient had been applied, 30 days with 4–10 lbs/acre/year, and 45 days for more than 10 lbs.

10.3.2. Exposure at Home

Everyone is exposed to toxic chemicals around the home. Tabulation of telephone calls to poison control centers in 1992 (Litovitz et al., 1993) showed that detergents and cleaners headed the list with 10.5% of acute exposures, followed in decreasing order by analgesics such as aspirin (9.6%); cosmetics including perfumes, cologne, and toilet water (8.2%); cold medications (5.8%); toxic plants (5.7%); and bites or stings from insects and snakes (4.0%). Pesticides, including rodenticides, accounted for only 3.8%, although this still represented over 70,000 calls. Surprisingly, some causes of serious home poisoning, such as liquor, carbon monoxide, and lead-based paints, were not specified, perhaps because so many resulted in immediate visits to the emergency room.

However, many home exposures are more subtle. They include exposure to phenols and PAH in wood smoke, nicotine alkaloids in tobacco smoke, and chloramine gas generated by mixing bleach and ammonia in a toilet bowl. Household water usually contains traces of the weakly carcinogenic chloroform and trichloroethylene (Section 2.4), commonly at <0.1–1.5 mg/L, and exposure occurs orally by drinking it, and via skin and lungs during showers or baths. McKone (1987) estimated that concentrations of chloroform or TCE in the air of the shower stall, bathroom, and the rest of the house averaged 20, 4, and 0.2 mg/m^3, respectively, from a 1 mg/L (1 ppm) concentration in the shower water.

Jo et al. (1990a,b) were able to relate chloroform levels in the exhaled breath of volunteers (4 men and 1 woman) to those in the shower water and show that their internal dose was about equally divided between dermal and respiratory routes. Analysis of exhaled breath is familiar to both police and drinkers as a reliable and noninvasive indicator of volatile compounds such as alcohol in blood. McKone (1993) later developed a PBPK model (Special Topic 10) to determine a person's total internal exposure to chloroform and its distribution in the body. The calculated chloroform intake via skin and lungs minus the amount lost by metabolism, storage, and binding to organs predicted about the same exhaled amount as was actually found. None of these investigations considered the further halocarbon exposure that would come from doing the laundry, washing dishes, and cooking.

10.3.3. Risk Characterization

Having identified and evaluated hazards, exposures, and biological responses, the last step in the risk assessment process is generally to compile and evaluate information to estimate the types, magnitudes, and probabilities of adverse effects. A common feature of risk characterization is **uncertainty**, and it is here that the probabilistic nature of risk is seen most clearly. Where possible, a numerical estimate of risk is desirable, the most common of which for humans is the **individual lifetime risk** (See Table 1.3), which is the probability of a person experiencing a specific adverse effect in his or her lifetime due to exposure to the toxic agent or some other hazard. In the case of carcinogens, it is simply the slope of the **carcinogenic potency** (dose–response) curve multiplied by the dose in mg/kg/day.

Carcinogenic potency can be thought of as the relative effectiveness of a chemical for tumor generation at a given dose. It is estimated by mathematical extrapolation of experimental data into the region approaching zero dose and zero response (Section 8.4), a very uncertain process. In the widely used linear multistage model, the lifetime probability of tumors *(P)* is plotted as a function of internal dose *d:*

$$P = 1 - e^{-(\gamma_0 + \gamma_1 d + \gamma_2 d^2 + \ldots \gamma_x d^x)} \tag{10.3}$$

where each γ is a different constant and x is the number of (hypothetical) cellular stages affected by the chemical. At $d = 0$, $P = 1 - e^{-\gamma_0}$; so γ_0 represents the background level of spontaneous tumors. A typical graph based on Eq. 10.3 is shown in Fig. 10.2 (Cothern, 1988); remember, points on the curve are only a mathematical construct and do not represent actual data. The error bars indicating an upper 95% confidence limit show this model to have a relatively low variability compared to the familiar probit plot (see Fig. 8.1C), but the variability in any calculated probability depends greatly on the model used. The probit model places the level generating a one per million risk of contracting cancer from TCE at 150 μg/L, while the more conservative multistage model places it at only 4 μg/L. The slope of the line represents the carcinogenic potency, q^*, so that $P = q^* d$.

Section 10.3.2 showed that a 10-minute daily shower in 40°C tapwater containing 25 μg/L of chloroform provided a whole-body dose of 0.46 μg/kg, and drinking 2 L of the water daily added another 0.70 μg/kg/day. A recognized carcinogen, chloroform has a q^* of 0.26 (mg/kg/day)$^{-1}$, so the dose from the shower corresponds to an excess cancer risk of 1.20×10^{-4}, or 122 in a million and that from drinking the water adds another 180 per million, for a total of at least 300 per million. Compare that with the values for other cancer risks shown in Table 1.3—a 20 per million risk from a medical x-ray, for example. Such "excess risks" are merely superimposed upon the existing U.S. lifetime cancer risk of 0.25 (1 in 4, or 250,000 per million).

Before anyone gets too excited, the weaknesses of such oversimplified estimates must be recognized. They are illustrated by the widely used dithiocarbamate fungicides, maneb and zineb (Section 9.6.2), which contain an impurity, ethylenethiourea (ETU), which causes thyroid tumors in rats. Based on a Q^* of 1.76×10^{-2}, which is the q^* calculated from the upper 95% confidence limit, the National Research Council estimated a dietary cancer risk of 1.61×10^{-3}, or 1610 per million, from these chemicals in a normal diet. However, thyroid tumors are rare in the United States. Why the discrepancy?

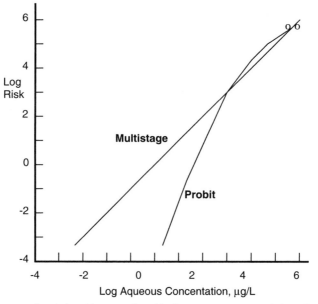

Figure 10.2. A comparison of multistage and probit dose–response extrapolations for lifetime cancer risk, per million, from exposure to TCE. Adapted from Cothern (1988). Error bars show upper 95% confidence limit; two data points at top right show LOEL.

First, q^* is based on the assumptions that humans respond to a given chemical in the same way as test animals, that P is linear with **d** at low levels, and that the upper 95% confidence limit for Q^* accurately represents the potency. Second, the NRC estimate assumes that all dithiocarbamate residues (and hence ETU) are at the tolerance level, that all treated food commodities contain this maximum level, that the foods with this level will be eaten maximally every day for a lifetime, and that ETU always is present and none is lost between spraying and eating. The first set of assumptions probably is erroneous, and the second certainly is. Dithiocarbamate residues seldom appear in U.S. food, and then at only a small fraction of the legal tolerance; residue-containing food is seldom consumed; and most ETU would be removed through environmental breakdown, washing, and cooking. The problem with the well-meaning policy aimed at "safety" is that the error in each worst-case scenario is *multiplied* by, rather than added to, the error in the others.

10.4 | ECOLOGICAL RISK

10.4.1. Toxicity Data

The assessment of ecological risk follows the same steps as that for humans: hazard identification, dose–response, exposure assessment, and risk characterization.

However, until recently, detailed ecological risk assessment was seldom undertaken, partly because of public apathy, partly because of a lack of suitable methods, and partly because of the diverse species and situations involved.

The toxicity information upon which hazard determination is based usually comes from bioassays. As toxicity tests in all the species comprising a natural community would be impractical, one must settle for data from only a few *indicator species*. But which species? The EPA prefers very sensitive organisms, whether actual members of the at-risk community or not. Should economic importance be a factor, using young salmon, for example? What about biological relevance, species such as the small copepods at the base of marine foodchains whose loss might most affect an entire ecosystem? Societal relevance (popularity) is problematic, as it may tend toward dolphins and koala bears, and test organisms must be accessible to measurement, for example, not whales. And what toxicity endpoints should be used? Death (often too late)? AChE level (too variable)? These are questions risk analysts struggle with.

Because of such difficulties, dose–response in wild animals usually has been extrapolated from that in common laboratory animals. However, the demonstrated differences in response between species usually make this inadvisable. Who could have predicted the egg shell thinning by DDT or the extreme toxicity of organotin compounds to molluscs? Quantitative structure–activity relations (QSAR), a few of which are shown in Eq. 10.4–10.6, have become a popular way of predicting a wide variety of effects (Section 16.5), but they, too, are only models.

Freshwater fish (96 h LC_{50}):
$$\log LC_{50} = -0.94 \log K_{ow} + 0.94(\log 0.000068\, K_{ow} + 1) - 1.2 \quad (10.4)$$

Marine fish (96 h LC_{50}): $\log 1/LC_{50} = 0.73 \log K_{ow} - 3.69$ \quad (10.5)

Daphnid (48 h LC_{50}): $\log 1/LC_{50} = 0.91 \log K_{50} - 4.72$ \quad (10.6)

Similarly, the toxicity of chemicals to 44 species of freshwater fish, amphibians, and invertebrates is represented by Eq. 10.7 (Holcombe et al., 1988) based on the acute LC_{50} in the fathead minnow (**fm**, *Pimephales promelas*):

$$LC_{50} = 0.91(LC_{50}\text{fm}) + 0.42, \quad n\ 309,\ r^2\ 0.86 \quad (10.7)$$

The relation of acute toxicity to size was demonstrated by Patin (1982), who was able to correlate the LC_{50} of oil, PCBs, and heavy metals in marine organisms with the animals' average length (Fig. 10.3).

Unlike humans, many kinds of organisms such as fish, aquatic plants, and earthworms may be completely and continually exposed to contaminants. Their external exposure generally is estimated by chemical analysis of the surrounding environment, assuming that the measured concentrations adequately reflect those available at the body surface. Cumulative exposure may be roughly estimated from carcass analysis if K_{ow} for the chemical is known, as BCF may then be calculated. For example, if 1.00 ppm of chlorobenzene (K_{ow} 830) were found in a fish, the BCF from Eq. 3.7 would be 39.8, and the aqueous concentration of chlorobenzene 0.025 mg/L according to Eq. 3.5. However, environmental analysis may not reveal the proportion of chemical actually available for absorption, as discussed in Section 7.3.

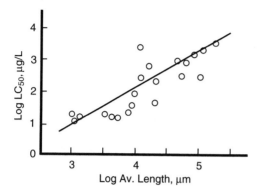

Figure 10.3 Relation of LC_{50} to body length in adults of 19 species of marine animals. Adapted from Patin (1982).

An extensive compilation of exposure factors for wild mammals, birds, reptiles, amphibians, and other vertebrates has been published by the U.S. EPA (1993).

10.4.2. Scaling

Comparative toxicology has assumed that normalizing toxicity values on a dose per body mass basis, such as mg/kg, accounts for anatomical differences among species. This may be true in a very general way, but more careful examination shows that physiological relationships tend to be exponential and conform to the equation $p = \mathbf{a}M^b$, where p is some physiological parameter, M is the body mass, and $\mathbf{a}$ and b are species-independent constants. For many functions, including cardiac output, respiration rate, clearance rate, and metabolic rate, b is close to 0.75 (Travis, 1993). Figure 10.4 illustrates the linearity of this relationship for metabolic rates in 28 animal species ranging from mice and small birds weighing less than 20 g to elephants weighing 4000 kg.

Another important consideration is the duration of exposure and dissipation as viewed by other species. Why should we assume that toxicological events take place

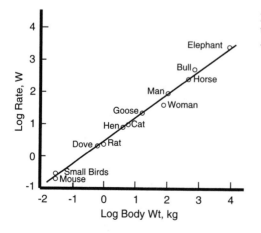

Figure 10.4. Metabolic rates (watts) in relation to body weight M, where rate $= kM^{0.75}$. Adapted from Kleiber (1947).

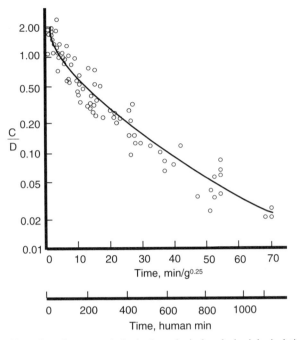

Figure 10.5. Disposition of methotrexate, in both chronological and physiological time, in mice, rats, dog, monkeys, and humans. The y axis shows plasma concentration C (μg/mL) divided by dose D (μg/g). Adapted from Dedrick et al. (1970).

on a human time scale in, say, a mouse or *Daphnia?* While it is convenient for us to measure chronological time t by *our* clocks, biological processes actually occur according to a physiological time, $\mathbf{t'}$ related to ours by:

$$\mathbf{t'} \; = \; t/M^{0.25} \tag{10.8}$$

Although no environmental example is available, Figure 10.5 illustrates this relationship for the disposition of the anticancer drug methotrexate in 5 mammalian species, including humans, according to the physiological time scale of each (Dedrick et al., 1970). While the metabolic half-life of a chemical varies sharply among these same species when measured in our chronological time (minutes), it is the same in all when shown in physiological time. That is, the various animals metabolize a given *proportion* of chemical in the same number of heartbeats! Good discussions of scaling and of toxicokinetics in physiological time are offered by Mordenti (1986) and Travis (1993).

10.4.3. Ecosystem Risk

Most toxicology has been concerned with effects in individual organisms or small populations. However, in the real environment, populations of many species interact to form organized communities. Ecosystems are considered to be made up of these individuals, populations, and communities, but unlike them, an ecosystems is only a human invention defined by the flow of energy and materials through its parts rather than by physical boundaries (Special Topic 1). Primary producers (green plants) use solar energy to convert inorganic nutrients into living matter that is consumed, in turn, by herbivorous animals and they by carnivores. After death, all these consumers are assimilated by scavengers whose substance is eventually returned to mineral form by microbial decomposers. and the cycle is repeated (Fig. 1.1).

Ecosystems may be thought to possess structure, including composition, trophic levels, species diversity, resistance to change, and resiliency to return to equilibrium after disturbance (Special Topic 1). They also possess important functions, including energy and material flow, biogeochemical cycling, nutrient retention, and undoubtedly a generalized ability to sequester and detoxify chemicals (self-cleaning). According to Suter (1993), an ecosystem's only purpose is to capture, distribute, transform, and dissipate biochemical potential energy. This is accomplished through primary production at one end and respiration at the other, and all the complex food-web transfers of energy and materials are just there to provide the means.

Can such a process be influenced directly by intoxication? Again, Suter (1993) suggests four possibilities: (1) altered ability of a population to interact with populations of other species, such as avoidance of predation, (2) altered predation, as by lack of prey, (3) altered ecosystem structure, such as by reduction in the number of species or trophic levels, and (4) alteration of an ecosystem function, such as interruption of primary production. Damage may be temporary, as when susceptible insects are removed from your back yard ecosystem after you apply insecticide, or more extensive if you destroy all of the plants and animals in the yard with a soil sterilant (Special Topic 1). Toxic chemicals can be hazardous even at an ecosystem level.

Models that might provide data for ecosystem risk assessment include those for energy flow, material cycling, and prediction of natural changes, as well as the standard microcosms and mesocosms (Section 16.6). Despite the traditional view that an ecosystem is greater than the sum of its parts, an interesting beginning has been made toward the extrapolation of toxic effects in individuals and populations to ecosystems. Regression of the No-Observable-Effect-Level from some of the few ecosystem-level chemical tests, $NOEL_E$, against the lowest reported acute LC_{50} in a single species gives Eq. 10.9, while regression against the lowest reported single-species *chronic* NOEL gives Eq. 10.10 (Sloof et al., 1986). An analysis of variants shows that single-species tests may apply to ecosystems after all.

$$\log NOEL_E = 0.81 \log LC_{50} - 0.55, \quad r^2\ 0.59 \qquad (10.9)$$

$$= 0.85 \log NOEL + 0.63, \quad r^2\ 0.72 \qquad (10.10)$$

So far, relatively little attention has been paid to risk assessmant at an ecosystem level. The reasons include a lack of suitable methods such as those that incorporate physical transport and chemical reactivity with biological factors, difficulty of the identifying suitable endpoints, and the differences in the very definition of an ecosystem. Disagreement also remains about what ecosystem properties need to be preserved. Is *any* deviation from a "normal" state acceptable? Apparently it is, considering society's acceptance of present logging, farming, and fishing practices right down to mosquito abatement. The need for improved ecosystem-level risk assessment seems obvious and urgent.

10.5 | RISK MANAGEMENT

10.5.1. Regulations

Once a true risk has been identified, evaluated, and characterized, something normally should be done to bring it to an acceptable level. That is the function of risk management, which combines the scientific data from risk assessment with economic, political, and social values to determine what course of action to take. Some risk management decisions are easy: A person's exposure may be reduced by running away, fish may simply swim out of range, or one can practice better personal hygiene, wear protective clothing, and even drink bottled water if necessary.

On a broader scale, federal and state laws long have served to protect health and the environment from the worst problems caused by toxic chemicals (Appendix 10.3). The oldest such law still in existence, the Federal Food, Drug, and Cosmetics Act of 1938, has prohibited the adulteration of food, cosmetics, or drugs by foreign objects or chemicals. It was followed in 1947 by the Federal Insecticide, Fungicide, and Rodenticide Act (FIFRA), which set composition standards for commercial pesticides and required registration of all new pesticide formulations based on extensive tests. Most of the federal laws have been amended and reauthorized periodically, although the process has become increasingly politicized. An example is CERCLA, the Comprehensive Environmental Response, Compensation, and Liability Act of 1980, whose purpose was to reduce toxic chemical releases from storage, treatment, and disposal sites (Special Topic 15). Despite heavy criticism of its implementation, an improved CERCLA was reauthorized in 1986 as SARA, the Superfund Amendments and Reauthorization Act, up for debate again in 1997. A concise summary of toxic chemical legislation is provided by Harte et al. (1991).

A 1954 amendment to the Food, Drug, and Cosmetics Act required safety evaluation of any pesticide residues remaining on raw agricultural commodities, and a 1958 amendment required similar evaluation of any intentional additive to food and limited the amount. Another, the still-controversial "Delaney Amendment" prohibited any additive "found to induce cancer in man or animal."

One result of such legal requirements has been the establishment of pesticide residue *tolerances,* called *maximum residue levels* (MRLs) in other countries (Table 2.8). A tolerance must be set by EPA and FDA for each pesticide on the food commodity on which it is used. It is calculated from the NOEL for that pesticide, in the most sensitive laboratory test species, by dividing by a safety factor of at least 100 to account for variability in responses in both the test animals and humans. However, this legal limit may vary, depending on the regulatory agency. For example, the U.S. tolerance for the insecticide diazinon is 500 ppb on almonds, while the MRL in

Europe is 100 ppb; the largest residues actually found are about 2 ppb. Tolerances for suspected carcinogens are determined from quantitative risk assessment models (Section 10.3).

In actuality, washing and cooking largely eliminate any remaining pesticide. However, a curious regulatory gap has emerged: Natural toxicants in food go largely unregulated. While limits have been set on the levels of aflatoxins and cyanide, the PAHs that result from both broiling and normal biosynthesis generally are not regulated. The same is true of most natural carcinogens (Ames, 1983), although less toxic synthetic additives have been banished by the Delaney amendment.

Risk management must balance dangers against benefits. For example, the risks from pesticide use are balanced against increased food quality and yields. In view of the relatively limited resources available, EPA experts have prioritized a number of environmental risks for possible action. High priorities were given to atmospheric pollution, indoor air pollution, stratospheric ozone depletion, and pesticide residues. Superfund sites (Special Topic 15), underground storage tanks, and groundwater contamination ranked relatively low. Interestingly, later surveys showed that the public was most concerned over hazardous waste disposal, atmospheric pollution, and pesticides, while they gave indoor air, consumer products, and radiation a low priority.

10.5.2. Risk Communication

Risk communication stands at the confluence of science, psychology, and public relations. As people's exposure to toxic chemicals becomes better defined, an increasingly important consideration in a democracy is that the public and their representatives adequately understand and evaluate risks so that appropriate action can be taken. This turns out to be more difficult than one might think. Some key difficulties from the communicator's side include disagreements among scientific experts on the significance of data, poor understanding of the interests and concerns of the audience, and over-reliance on technical terms and jargon. On the audience's side, difficulties include inadequate technical background, demand for scientific certainty, diminished faith in science, and poor perception of relative risk.

One attempt to alleviate the latter has been to compare toxic risks to more familiar ones, as in Table 1.3. The basis of many comparisons is the annual risk of being struck by lightning, about one in a million in the United States, a number to which most people who have never been struck seem to relate. However, such comparisons often suffer from a failure to convey the large uncertainties in any risk estimate, and this is also true in presenting the estimates themselves: "Omigod, did you read that we can get cancer by just taking a shower?" Perception is everything.

People's perception of risk can be colored by such emotions as familiarity (pesticide residues are feared, bacterial food poisoning is not), personal stake (living near a waste dump is feared, taking waste to the dump is not), victim identity (your own child vs 100 nameless people poisoned), and voluntariness (pollution of drinking water by pesticides vs. spraying your own garden) (Covello, 1989). Which brings us to the subject of *acceptable risk.* Risk acceptable to one person may not be acceptable to another. While the risk of cancer from taking a shower or eating vegetables actually is small, it is not zero, and Americans' fear of toxic risk may make it unacceptable to many. One practical approach to this is *de minimis* or trivial risk, a level such as the one in a million (1×10^{-6}) often invoked today, below which the risk is officially ignored.

Why are modern-day Americans so afraid of chemicals? The answer is complex and may have something to do with our national character, as citizens of other countries seem less concerned. One suggestion is that chemicals offer a simple explanation for dreaded maladies, such as cancer and birth defects, which relieves victims and their families from responsibility for their personal lifestyle. As with our increasing demand for litigation and compensation for "wrongs," trouble is always someone else's fault. The uneasiness may also have to do with our increasing reliance on chemicals for everything material in our lives, from food to medicine to household conveniences, while society is less and less knowledgeable about science and technology. Some of these psychological aspects are discussed by Simon et al. (1990).

And what are these toxic chemicals we are so regularly reminded of? The next few chapters provide detailed examples.

10.6 | REFERENCES

Ames, B. N. 1983. Dietary carcinogens and anticarcinogens. *Science* **221:** 1249–64.

Chen, C. W., and K.-C. Hoang. 1993. Incorporating biological information into the assessment of cancer risk to humans under various exposure conditions and issues related to high background tumor incidence rates, in *Health Risk Assessment* (R. G. M. Wang, J. B. Knaak, and H. I. Maibach, eds.), CRC Press, Boca Raton, FL, pp. 309–29.

Cothern, C. R. 1988. Uncertainties in quantitative risk assessments—two examples: Trichloroethylene and radon in drinking water, in *Risk Assesment and Risk Management of Industrial and Environmental Chemicals* (C. R. Cothern, M.A. Mehlman, and W. L. Marcus, eds.), Princeton Scientific Publ. Co., Princeton, NJ. Vol. 15, pp. 159–80.

Covello, V. T. 1989. Communicating right-to-know information on chemical risks. *Environ. Sci. Technol.* **23:** 1444–49.

Dedrick, R. L., K. B. Bischoff, and D. S. Zaharko. 1970. Interspecies correlation of plasma concentration history of methotrexate (NSC-740). *Cancer Chemotherapy Rep.* **54** (Part 1): 95–101.

Gunther, F. A., Y. Iwata, G. E. Carman, and C.A. Smith. 1977. The citrus reentry problem: Research on its causes and effects, and approaches to its minimization. *Residue Rev.* **67:** 1–132.

Harte, J., C. Holdren, R. Schneider, and C. Shirley. 1991. *Toxics A to Z*, UC Press, Berkeley, CA.

Holcombe, G. W., G. L. Phipps, and G. D. Veith. 1988. Use of acute lethality tests to estimate safe chronic concentrations of chemicals in initial ecological risk assessments, in *Aquatic Toxicity and Hazard Assessment* (G. W. Suter and M. Lewis, eds.), ASTM, Philadelphia, PA, pp. 442–67.

ICRP. 1975. *Report of the Task Force on Reference Man,* International Commission on Radiological Protection, Report No. 23, Pergamon Press, Elmsford, NY.

Jo, W. K., C. P. Weisel, and P. J. Lioy. 1990a. Routes of chloroform exposure and body burden from showering with chlorinated tap water. *Risk Anal.* **10:** 575–80.

Jo, W. K., C. P. Weisel, and P. J. Lioy. 1990b. Chloroform exposure and the health risk associated with multiple uses of chlorinated tap water. *Risk Anal.* **10:** 581–85.

Kleiber, M. 1947. Body size and metabolic rate. *Physiol. Rev.* **27:** 511–41.

Litovitz, T. L., K. C. Holm, C. Clancy, B. F. Schmitz, L. R. Clark, and G. M. Oderda. 1993. 1992 Annual report of the American Association of Poison Control Centers Toxic Exposure Surveillance System. *Amer. J. Emerg. Medicine* **11:** 494–555.

McKone, T. E. 1987. Human exposure to volatile organic compounds in household tap water: The indoor inhalation pathway. *Environ. Sci. Technol.* **21:** 1194–201.

McKone, T. E. 1993. Linking a PBPK model for chloroform with measured breath concentrations in showers: Implications for dermal exposure models. *J. Expos. Anal. Environ. Epidem.* **3:** 339–65.

Mordenti, J. 1986. Man versus beast: Pharmacological scaling in mammals. *J. Pharm. Sci.* **75:** 1028–40.

Patin, S. A. 1982. *Pollution and the Biological Resources of the Ocean,* Butterworth Scientific Publishers, London, U.K.

Popendorf, W. J., and J. T. Leffingwell. 1982. Regulating OP pesticide residues for farmworker protection. *Residue Rev.* **82:** 125–201.

Simon, G. E., W. J. Katon, and P. J. Sparks. 1990. Allergic to life: Psychological factors in environmental illness. *Amer. J. Psychiatry* **147:** 901–906.

Sloof,W., J. A. M. van Oers, and D. de Zwart. 1986. Margins of uncertainty in ecotoxicological hazard assessment. *Environ. Toxicol. Chem.* **5:** 841–52.

Spear, R. C., W. J. Popendorf, J. T. Leffingwell, and T. H. Milby. 1977. Worker poisoning due to paraoxon. *J. Occup. Med.* **19:** 411–14.

Suter, G. W. 1993. *Ecological Risk Assessment,* Lewis Publishers, Boca Raton, FL.

Travis, C. C. 1993. Interspecies extrapolation of toxicological data, in *Health Risk Assessment* (R. G. M. Wang, J. B. Knaak, and H. I. Maibach, eds.), CRC Press, Boca Raton, FL, pp. 387–410.

USDA, 1980. *Food Intakes: Individuals in 48 States, Year 1977–78.* U.S. Dept. of Agriculture, Washington, DC. Nationwide Food Consumption Survey, 1977–78.

USEPA. 1989. *Exposure Factors Handbook,* Office of Health and Environmental Assessment, U.S. Environmental Protection Agency, Washington, DC. EPA/600/8-89/043.

USEPA. 1993. *Wildlife Exposure Factors Handbook,* Office of Health and Environmental Assessment, U.S. Environmental Protection Agency, Washington, DC, 2 vols. EPA/600/R93/187a.

Wyllie, J. A., J. Gabica, W. W. Benson, and J. Yoder. 1975. Exposure and contamination of the air and employees of a pentachlorophenol plant, Idaho—1972. *Pestic. Monit. J.* **9:** 150–53.

Special Topic 10. PBPK Models

Interest in physiologically-based pharmacokinetic (PBPK) models is increasing (Section 8.5.3). While two-compartment physiological models, generally representing blood and lipid, are still widely used, the recognized complexity of the intoxication process has led to more general and flexible representations of the internal functions of an organism (Fig. 10.6). The models are constructed very logically.

The idea is to balance the total amount of a chemical entering, inside, and leaving the body and its parts in relation to time. The process ideally combines respiratory, dermal, and oral exposures, although often only one or two routes actually are specified. Typically, C stands for concentration in mg/L, t for time in minutes, Q for flow rate in L/min, V for volume in L, and P for a partition coefficient (Appendix 10.2). For example, pulmonary uptake balances the amount of chemical in the lungs (ventilation rate times the concentration difference between outside air and alveolar air) against that in the blood, that is, total blood flow times the concentration difference between arterial (**a**) and venous (**v**) blood:

$$Q_{vent}(C_{air} - C_{alv}) = Q_{blood}(C_\mathbf{a} - C_\mathbf{v}), \qquad \text{or} \tag{10.11}$$

$$C_\mathbf{a} = (Q_{vent}C_{air} + Q_{blood}C_\mathbf{v})/Q_{blood} + Q_{vent}/P_{\mathbf{a}-air}, \text{ where} \tag{10.12}$$

$$C_\mathbf{v} = C_lC_{vl}/P_l + Q_fC_{vf}/P_f + Q_{org}C_{vorg}/P_{org} + Q_mC_{vm}/P_m \tag{10.13}$$

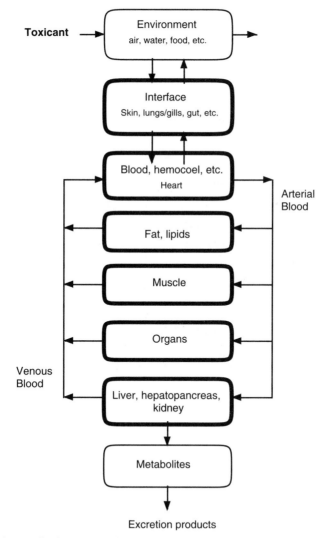

Figure 10.6.   A generalized PBPK model. For discussion, see Chen and Hoang (1993).

In Eq. 10.13, the venous blood concentration reflects the distribution of chemical among other body compartments such as liver (l), fat (f), internal organs (org), skin (s), and muscle (m). For example, C_{vl} denotes the concentration of chemical in venous blood in the liver. Note that arterial blood carries the chemical out of the lungs and *into* body compartments, and venous blood removes it *from* the compartments (Fig. 10.6).

The rate at which the chemical leaves the skin compartment, s, reflects the difference between dermal absorption (exposed skin area A and skin permeability constant K_{perm}) and systemic distribution in the body:

$$V_v \left(\frac{dC_{vs}}{dt} \right) = K_{perm} A (C_{Air} - C_{vs}/P_{s-air}) + Q_s (C_a - C_{vs}/P) \qquad (10.14)$$

The dose D of ingested chemical is assumed to be eliminated, at an exponential rate, entirely by enzymatic metabolism in the liver as reflected by $\mathbf{V}_{max}$ (where $\mathbf{V}$ is velocity) and the Michaelis–Menten constant, K_m. In Eq. 10.15, the amount of chemical in the liver at any given time represents the net difference between ingestion [A] and combined the systemic distribution [B] and metabolism [C], where k_g *is* the rate constant for intestinal absorption:

$$\qquad\qquad [A] \qquad\qquad [B] \qquad\qquad [C]$$

$$V_l(dC_{vl}/dt) = Dk_g e^{-k_g t} + Q_l (C_a - C_{vl}/P_l) - \frac{\mathbf{V}_{max}\ C_{vl}/P_l}{K_m + C_{vl}/P_l} \qquad (10.15)$$

Within any tissue group x, the mass balance equations represent only systemic distribution:

$$V_x (dC_{vx}/dt) = Q_x (C_a - C_{vx}/P_x) \qquad (10.16)$$

Use of the model requires analytical data on chemical concentrations in arterial and venous blood, inhaled and exhaled air, and the skin, as well as extensive physiological information about the exposed species (Appendix 10.2). Tissue groups, or compartments in the model, may also include brain, testes, nerve, and any other for which the necessary information can be gathered. Armed with this, it should be possible to estimate the half-lives of chemicals in the body, concentrations in each organ and at the site of action, and, by back calculation, the original exposure levels (Section 8.5.3).

PBPK analysis has now been applied to mammals, fish, and plants, but apparently not yet to invertebrates. The method should be applicable to any kind of organism, once the necessary data are developed. Theoretically, eventual extension of the concept to populations, communities and even to an ecosystem level should be possible.

APPENDIX 10.1.
Exposure Factors for Humans[a]

	Men	Women	Children[b]	Babies[c]
Weight, kg	81 **(70)**	68 **(60)**	25 **(20)**	11.6 **(10)**
lbs	178 **(154)**	150 **(132)**	55 **(44)**	25.5 **(22)**
Skin area, m^2	1.94 **(1.94)**	1.69 **(1.69)**	0.92 **(0.88?)**	**0.35**
Respiration rate				
Resting, L/min	11.7 **(7.5)**	5.0 **(6.0)**	6.7 (4.8)	**1.5**
Active, L/min	41.7**(20.0)**			
Total, m^3/day	**23.0**	**21.0**	**15.0**	**3.8**
Food consumption, maximum g/day[d]				
Meat, fish, poultry	292	191	154	103
Milk, dairy products	569	466	483	439
Grain products	304	241	227	65
Fats, oils	19	14	9	5
Vegetables, nuts	309	249	170	100
Fruit and fruit juice	256	254	211	204
Sugar and sweets	29	36	17	29
[Other beverages]	1012	832	232	153
Total food, kg/day[e]	1.72 **(1.50)**	1.45 **(1.50)**	1.27 **(1.50)**	0.95 **(1.50)**
Fluid consumption, L/day	1.73 **(2.0)**	1.73 **(2.0)**	0.86 **(2.0)**	0.74 **(2.0)** ?
Soil ingestion, mg/day	61 **(50)**	61 **(50)**	24 **(100)**	165[f] **(50)**

[a] USEPA (1989); USDA (1980) standard values in **bold.**
[b] Age 6–8 years.
[c] Age 1–2 years.
[d] Maximum consumption of edible portion for any age range.
[e] Excludes "other beverages" (coffee, tea, soft drinks, beer).
[f] Age 2–5 years.

APPENDIX 10.2.
Parameters for PBPK Models[a]

Parameter	Symbol	Rats	Humans	Symbol	Rats	Humans
Body weight, kg	M	0.35	70.0			
Alveolar vent., L/min	Q_{alv}	0.083	7.5			
Blood flow, L/min				Tissue volume, L		
Total	Q_t	0.104	6.20	V_t	0.316	62.6
Fat compartment	Q_f	0.0092	0.31	V_f	0.032	14.0
Organs[b]	Q_{org}	0.0434	2.76	V_{org}	0.015	3.5
Muscle[c]	Q_m	0.0074	1.26	V_m	0.220	36.4
Liver[d]	Q_l	0.0389	1.55	V_l	0.014	1.7
Skin	Q_s	0.0052	0.31	V_s	0.035	7.0
Partition coefficients[e]						
Blood/air	$P_{a\text{-}air}$	18.9	10.3			
Skin/air	$P_{s\text{-}air}$	—	505.4			
Fat/blood	P_f	109.0	109.0			
Organ/blood	P_{org}	3.18	3.72			
Muscle/blood	P_m	1.06	3.72			
Liver/blood	P_l	3.72	3.72			

APPENDIX 10.2. *(continued)*

Parameter	Symbol	Rats	Humans	Symbol	Rats	Humans
Skin/blood	P_s	—	505.4			
Metabolic constants[e]						
Metabolic rate, mg/min	V_{max}	0.00586	0.703			
Michaelis, mg/L	K_m	2.938	32.043			
Skin permeability, cm/h	K_{perm}	0.668	0.17			

[a] Chen and Hoang, 1993.
[b] Rapidly perfused tissue group.
[c] Slowly perfused tissue group.
[d] Metabolic tissue group.
[e] For dichloromethane.

APPENDIX 10.3.
U.S. Laws Governing Toxic Chemicals

Title	Acronym	Year	Control
Food, Drug, and Cosmetics Act	FDCA	1938	Adulterants
(Pesticide Residues Amendment)		1954	Pesticides
(Food Additive Amendment)		1958	Food additives
(Color Additive Amendment)		1960	Food colors
Federal Insecticide, Fungicide, and Rodenticide Act	FIFRA	1947	Pesticides
Federal Hazardous Substances Act	FHSA	1960	Toxic chemicals
Clean Air Act	CAA	1970	Exhaust emissions
Occupational Safety and Health Act	OSHA	1970	Workplace chemicals
Lead-Based Paint Poisoning Prevention Act		1971	Household lead
Consumer Product Safety Act	CPSA	1972	Household chemicals
Safe Drinking Water Act	SDWA	1972	Water pollutants
Federal Water Pollution Control Act	FWPCA	1972	Water pollutants
(Clean Water Act)	CWA	1987	Water pollutants
Resource Conservation and Recovery Act	RCRA	1976	Hazardous waste
Toxic Substances Control Act	TSCA	1976	Industrial chemicals
Comprehensive Environmental Response, Compensation, and Liability Act	CERCLA	1980	Hazardous waste
(Superfund Amendments and Reauthorization Act)	SARA	1986	Hazardous waste

CHAPTER

Inorganic Toxicants 11

11.1 | TOXIC INORGANIC CHEMICALS

Some of the earliest recognized toxic substances were inorganic. The toxicity of mercury compounds was known and used in ancient Egypt, the Romans recognized that lead acetate ("sugar of lead") was poisonous, and medieval alchemists gave arsenic the symbol of a snake. "Inorganic" implies a mineral rather than biological origin and largely excludes compounds containing carbon, although carbon dioxide is deemed inorganic while carbon tetrachloride and carbon disulfide are "organic."

The term *organic* is a holdover from the time when carbon compounds were thought to derive only from living things. However, as will be shown later, many inorganic compounds are not fundamentally different from organics in reactivity or physical properties, so the distinction is largely historical. A real difference is that inorganic raw materials come directly from ores, brines, or air (Chapter 13). The top three chemicals in U.S. production volume consistently include salt (NaCl), phosphate rock [$Ca_3(PO_4)_2$], and lime (CaO), and the huge annual production of ferrous and nonferrous metals comes entirely from mineral ores (Table 13.2).

Although any chemical can be poisonous under the right circumstances, many inorganic substances show only a low degree of toxicity. Certain anions such as cyanide and fluoride generally do confer toxicity, and many barium compounds are poisonous, but otherwise the salts of alkali and alkaline earth elements are toxicologically uninteresting. Compounds of elements such as the lanthanides ("rare earths") may have toxic properties but are so uncommon and provide such limited exposure that little toxicity information exists for them. However, in terms of persistence, many inorganic substances *never* go away.

On the other hand, certain elements produce compounds so universally poisonous that they have even entered some of our more gruesome lore, witness Snow White and the poisoned apple and the stage play *Arsenic and Old Lace*. These elements are highlighted on the periodic table shown in Fig. 11.1 and form natural groups based

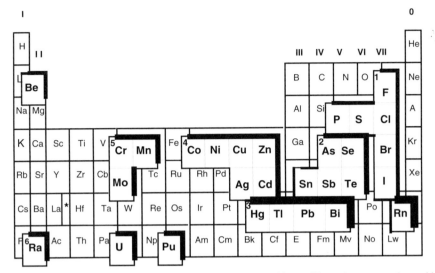

Figure 11.1. The periodic table of elements. Those elements with notably toxic compounds are highlighted. The lanthanide series is shown by an asterisk.

on chemical properties: heavy metals, metalloids, transition metals, etc. Before discussing each group in more detail, a review of some fundamental chemistry is in order.

11.2 | SOME BASIC CHEMISTRY

The periodic classification groups elements according to common chemical and structural properties based on electronic configuration. Elements that share the same column as nitrogen all possess two *s* electrons and three *p* electrons in their outer shells, show a tetrahedral geometry, and form tri- and pentachlorides. Phosphorus trichloride boils at 76°C and reacts violently with water, while carbon tetrachloride boils at 77°C and is unreactive and virtually insoluble, and the familiar sodium chloride is a water-soluble crystalline solid boiling at 1413°C. What causes such a wide range of properties among seemingly related compounds?

The answer lies in electronegativity (Fig. 11.2). The electronegativity scale, originally compiled by Linus Pauling based on bond lengths, is a measure of how strongly an element's nucleus attracts and holds electrons. Electronegativity (E_n) values conform closely to Coulomb's law, in which the force acting on an orbiting electron, distant a covalent radius r from the atom's nucleus, can be calculated. Screening of the atom's nuclear charge by inner electrons results in an effective nuclear charge (Z_{eff}) that controls how tightly the outer electrons, its own or those of another atom, are held:

$$E_n = \frac{Z_{eff}}{r^2} \tag{11.1}$$

The E_n difference between two atoms determines the degree of charge separation, or polarity. In NaCl, the difference of 1.9 between the elements is almost twice the entire E_n of 1.0 for the sodium; the chlorine claims the bulk of the electrons, causing it to be negatively charged and the sodium's positive nuclear charge to be expressed. The electrons are virtually not shared, the forces holding Na^+ and Cl^- together in the crystal lattice being dispersed among a number of adjacent ions and readily overcome by the attraction of water molecules. Thus sodium chloride is water soluble and is said to be "ionic."

On the other hand, the E_n difference in CCl_4 is only 0.4, meaning that electrons are largely shared between the elements, there is little charge separation, and the bonds are "covalent." The attraction of water is slight (a saturated aqueous solution of carbon tetrachloride is only about 5 mM), and the molecules likewise hold very little attraction for each other and so separate (volatilize) easily. PCl_3 is intermediate (E_n difference 0.8), more "organic" than "inorganic," an oily liquid with a moderate boiling point and soluble in CCl_4 and benzene. However, the attraction of electrons by the Cl nucleus allows the positive charge of the P nucleus to show through and become vulnerable to attack by nucleophiles such as the hydroxide ion of water.

Electronegativity increases as one moves to the right across the periodic table. Alkali and alkaline earth elements have a low E_n, transition elements such as Fe, Cu, Mo have an E_n of 1.3–1.7, nonmetals and metalloids such as C and As have an E_n of 1.9–2.9, and elements in the right-hand columns all have an E_n above 3.0. Although heavy elements such as mercury, thallium, lead, and bismuth have E_ns similar to

H 2.1						
	Be 1.5	**B** 2.0	**C** 2.5	**N** 3.0	**O** 3.5	**F** 4.0
Na 1.0		**Al** 1.5		**P** 2.1	**S** 2.5	**Cl** 2.9
	Cu 1.8			**As** 2.1	**Se** 2.4	**Br** 2.8
	Ag 1.6	**Cd** 1.6	**Sn** 1.8			**I** 2.4
		Hg 1.7	**Pb** 1.7			

Electronegativity ⟶

Figure 11.2. Pauling electronegativity of selected elements.

those of transition metals, they stand out by forming covalent compounds with high toxicity, relatively low melting points, and solubility in organic solvents. Whether inorganic or organic, two features of any chemical can always be associated with toxicity: *lipophilicity* **and** *electrophilic reactivity* (reactivity toward electron-rich nucleophiles such as SH groups). By contrast, common alkali metal salts generally lack both features and so show low toxicity.

11.3 | NONMETALLIC ELEMENTS

11.3.1. Halogens

The group of elements labeled **1** on Fig. 11.1 are nonmetals with high E_ns, low electrical conductivity, and no metallic luster. Although many of their compounds are poisonous, the toxicity of the elements themselves is of particular interest. The common *halogens* fluorine, chlorine, bromine, and iodine are the best examples. Older people recall using iodine to disinfect minor wounds, and deep-red liquid bromine has occasionally been used to disinfect water. However, chlorine has largely taken over as a disinfectant and now is almost universally added to public water supplies, swimming pools, and bacterially contaminated wastewater. Other swimming pool disinfectants include compounds, such as dichlorohydantoin, that readily release chlorine or hypochlorite in contact with water. Today, there are few communities that do not keep a large tank of liquid chlorine for their local pool, and treated water is a common source of chlorine exposure for most people.

However, chlorine, bromine, and iodine are toxic to all life forms, not just microorganisms. Gaseous chlorine is especially dangerous, as it is heavy, corrosive, fat soluble, extremely reactive, and very widely available. Exposure in mammals may result in suffocation due to reflex closing of the air passages, and low levels result in an often-fatal water accumulation in the lungs known as pulmonary edema (WHO, 1982). This combination of properties resulted in many human injuries and deaths during the World War I use of chlorine for chemical warfare, and fatalities still occur due to spills of liquid Cl_2. In contact with water, chlorine forms hypochlorous acid, and the HOCl is bactericidal as well as highly toxic to aquatic life.

$$Cl_2 + H_2O \longrightarrow HOCl + HCl \tag{11.2}$$

Although human or environmental exposure to the very reactive but uncommon elemental fluorine almost never happens, exposure to fluoride ion is commonplace. Fluoride minerals are widespread in the environment, and many are ionic and occur in natural waters. As the element is vital to healthy teeth, people in fluoride-deficient areas such as Polynesia tend to suffer from dental caries, and fluoride often is added artificially to drinking water and toothpaste (as tin fluoride) even in other parts of the world.

"Fluoridation" is continually enveloped in controversy, as inorganic fluorides are toxic and long were used as insecticides and rodenticides. Sodium fluoride, sodium fluorophosphate, and cryolite (sodium aluminum fluoride) are common forms. Acute toxicity arises from kidney damage, but the mechanism is unclear. Chronic exposure causes mottled teeth and eventually various bone defects as F^- replaces OH^- in the structural mineral, apatite. Although most damage to human and environmental health results from industrial activity, a daily human exposure of as little as 8 mg of fluoride in water can be hazardous (WHO, 1984). The corresponding hydrogen fluoride, b.p. 19.5°C, is the most important industrial fluoride, and, together with its 38% aqueous solution (hydrofluoric acid), is extremely toxic and corrosive. Being both highly lipophilic and reactive, these compounds are readily absorbed into skin and mucosa and later produce painful burns from the inside.

11.3.2. Phosphorus

Elemental phosphorus exists in two common forms, yellow (or "white") and red. Yellow phosphorus possibly was known to the alchemists but certainly was exhibited in Germany in the late 1600s for the cold green light (phosphorescence) it emitted. With its tetrahedral structure and covalent character, its reactions often resemble those of carbon more than those of nitrogen. Yellow P_4 ignites spontaneously in air, is lipophilic and readily absorbed through skin to produce painful burns, and is very poisonous—dangerous stuff. Not surprisingly, its use is common in chemical warfare. An allotropic form, red phosphorus (P_x), a cross-linked polymer that is less reactive and toxic, has replaced P_4 in matches.

Phosphorus burns in air to P_4O_{10} ("phosphorus pentoxide") and in chlorine to PCl_3, both of which react violently with water to form phosphoric and phosphorous acids, respectively. As expected, its reactions are entirely analogous to those of arsenic (see Fig. 11.4), the halides reacting with nucleophiles such as alcohols to provide common solvents, dielectric fluids for electrical transformers, fire retardants, and pesticides. The highly volatile and toxic phosphine, PH_3, generated *in situ* from metal phosphides, is widely used for fumigating grain and storage areas to kill insects and rodents.

$$AlP + 3H_2O \longrightarrow PH_3 + Al(OH)_3 \qquad (11.3)$$

11.3.3. Sulfur

Sulfur provided the ancients with medicine, bleach, and gunpowder. Although it is still used for these purposes today, the major applications now include the manufacture of sulfuric acid, vulcanization of rubber, agricultural pest control, and soil amendment. Elemental sulfur occurs in Nature as yellow, solid cyclooctasulfur (Fig. 11.3), S_8, which can exist in several crystal forms. Rhombic S_α, the stable form often seen in large yellow piles in agricultural areas, melts to a yellow liquid at 112.8°C and is soluble in organic solvents. Sublimation provides a common microcrystalline form known as "flowers of sulfur." Further heat produces monoclinic S_β, m.p. 119°C, and even higher temperatures generate a long-chain helical polymer,

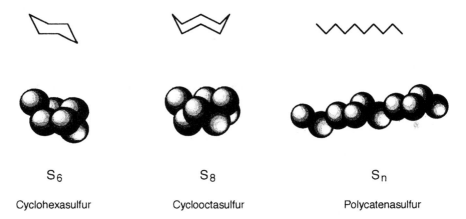

S_6 S_8 S_n

Cyclohexasulfur   Cyclooctasulfur   Polycatenasulfur

Figure 11.3. Three stable allotropic forms of sulfur. See Meyer, 1977.

polycatenasulfur, containing up to 8×10^5 S atoms and insoluble in both water and organic solvents (Meyer, 1977). When acidified, aqueous sodium thiosulfate forms a series of polythionates and the highly reactive cyclohexasulfur, S_6 (Fig. 11.4):

$$2S_2O_3^{2-} + H^+ \longrightarrow HS{-}S{-}SO_3^- + SO_3^{2-} \tag{11.4}$$

$$6S_2O_3^{2-} + H^+ \longrightarrow S_6 + 5SO_3^{2-} + HSO_3^- \tag{11.5}$$

Several forms of sulfur still find use as insecticides and fungicides, especially "lime-sulfur" and the colloidal sulfur prepared by acidification of sodium thiosulfate. To make lime sulfur, calcium oxide or hydroxide and powdered sulfur are mixed into water, heat is evolved, and the solids dissolve to form a red solution containing calcium polysulfide and calcium thiosulfate, ostensibly via the commonly seen Equation 11.6.

$$6Ca(OH)_2 + 21S \longrightarrow 6Ca^{2+} + 2S_2O_3^{2-} + 3S_4^{2-} + S_5^{2-} + 6H_2O \tag{11.6}$$

However, this traditional equation disregards the facts that the sulfur starts as S_8 and the predicted proportion of thiosulfate to calcium in the product is smaller than that actually observed. The demonstrated rapid attack of nucleophiles on the S_8 ring (Schmidt, 1965) suggests instead that hydroxide ion from $Ca(OH)_2$ opens the ring, and the resulting polysulfide disproportionates.

The mechanism of the fungitoxic action remains unkown. Various intermediates have been proposed, including oxidized (polythionate) and reduced (hydrogen sulfide), and finally abandoned. One possible biochemical route involves absorption of the lipophilic S_8 into the cell, insertion into the electron transport chain after FMN (Appendix 9.1), and subsequent generation of the observed hydrogen sulfide (Tweedy, 1969). Another is that S_6 is generated from thiosulfate or polysulfide by environmental carbonic acid and reacts with cellular nucleophiles such as glutathione. Hopefully, we will not have to wait another 3000 years for this mystery to be solved.

11.4 | THE METALLOIDS

11.4.1. Metalloids

The elements marked **2** on Fig. 11.1—arsenic, selenium, tin, antimony, and tellurium—have some of the properties we associate with metals, such as luster and at least limited conductivity, but they also differ in many ways. They have relatively high E_ns, provide primarily covalent compounds, and form acidic (amphoteric) hydroxides. Arsenic is typical.

11.4.2. Arsenic

Like sulfur, the medicinal and toxic properties of arsenic compounds have been recognized and put to use for thousands of years (Section 1.1). Most behave very much like organic compounds, with a tetrahedral configuration and covalent centers (arsenic trichloride boils at 130°C) that react readily with nucleophiles (Fig. 11.4). The surprisingly volatile arsenic trioxide ("white arsenic"), long formulated as As_2O_3, eventually was recognized to be the ball-like As_4O_6 which is easily oxidized to As_4O_{10} and rapidly attacked by hydroxide ion to form arsenious acid and arsenites.

Although elemental arsenic does occur in Nature, the principal commercial product, As_4O_6, sublimes as a byproduct during the smelting of certain ores. Orpiment, As_4S_6, and realgar, As_4S_4, minerals highly coveted as pigments by the ancients and mined by slaves who frequently became poisoned, demonstrate the stability of the As–S bond. The 110° S–As–S bond angle is close to the tetrahedral angle of 109.5°, allowing the formation of a stable six-membered ring containing the arsenic and two sulfur atoms of the respiratory enzyme, dihydrolipoyl transacetylase (Section 9.3).

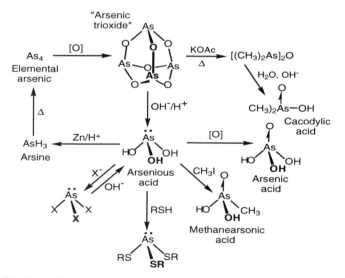

Figure 11.4.   Reactions of "arsenic trioxide" (As_4O_6). X = Cl, Br, I; R = H or organic.

Arsenic hydride, or arsine, is a toxic gas generated under a wide variety of reducing conditions. It forms the basis of the century-old Marsh test, where, heated as it passes through a glass tube, it deposits a black mirror of elemental As_4 that provides a semiquantitative measure of arsenic. Methylarsines are important environmental transformation products of As (Special Topic 11), and small doses of arsenic compounds are excreted by mammals in methylated form, primarily as dimethylarsinate, also known as cacodylate (Fig. 11.4).

11.4.3. Selenium

Selenium shares many of the properties of sulfur and arsenic. Its compounds are covalent (Se_2Cl_2 boils at 130°C), it exists in several allotropic forms including Se_8, and the oxide dissolves in dilute base to give selenites (*e.g.*, Na_2SeO_3). Although it is an essential nutrient in small amounts, selenium and its compounds are toxic at only slightly higher levels. Elemental selenium is widely used in electronic semiconductors, as it conducts electricity in the light but not in the dark, and hexavalent Se occurs widely as selenate in natural waters.

The acute oral LD_{50} of sodium selenite in rats is 7 mg/kg, and that of sodium selenate 4 mg/kg, the principal action being on the nervous system. Natural selenate is readily absorbed into plants from soil, the Se replacing S in the essential amino acids cysteine and methionine and making the forage toxic to anmals (Section 12.5). Perhaps the most spectacular example of the environmental toxicity of selenium emerged at the Kesterson Wildlife Sanctuary in Northern California, where agricultural drainage water from the dry west side of the San Joaquin Valley was collected in a reservoir in lieu of transporting it into San Francisco Bay (Ohlendorf et al., 1986). There, the water evaporated, concentrating its natural selenate, which was subsequently taken up by aquatic animals and plants with devastating effects in waterfowl that fed upon them.

Although found widely in food, the corresponding compounds of *tellurium* are not particularly toxic, the most notable result of excessive intake being a very offensive body odor, called "tellurium breath," which results from biochemical reduction and methylation to the volatile garlic-odored dimethyl telluride.

11.4.4. Tin

Tin is the most metallic of this group of elements. Although it is lustrous, malleable, and conductive, its compounds are largely covalent ($SnCl_4$ boils at 114°C), and its oxide, the mineral cassiterite, dissolves in hot base to form stannates such as Na_2SnO_4. Elemental tin is safe enough to have been used as coating on the inside of steel cans ("tin cans"), and stannous fluoride is approved for use in toothpaste, so tin and its compounds previously have held relatively little interest for toxicologists. Triphenyltin hydroxide has been applied to combat fungal grain diseases in Europe, and tricyclohexyltin hydroxide (cyhexatin) is used as a fungicide on tree fruits such as almonds.

However, tributyltin compounds (TBT) now are known to be extremely toxic to aquatic animals. Tributyltin oxide and chloride have long been used in antifouling paints to keep barnacles and other sessile aquatic organisms from settling on boat hulls. If barnacles appear on the bottom of your sailboat, they are a nuisance; if they grow on the hull of an aircraft carrier, they slow the ship appreciably, require enormous amounts of extra fuel, and are very expensive to remove. Organotin compounds proved highly effective in reducing the growths, and it was not until the late 1980s that their extreme toxicity to nearby molluscs was discovered—less than a part per billion in the seawater often was lethal. The compounds are somewhat less toxic to fish but could still be dangerous, as TBT concentrations in mussels from Humboldt Bay, California, were found by the California Mussel Watch Program to have bioconcentrated Sn to as much as 15 ppm (SWRCB, 1988).

Butyl- and methyltin species also are very mobile in the environment and may be detectable far from their probable source (Clark et al., 1988). Usually, maximum concentrations in the water column are 100–200 ng/L (ppt), about 2 μglL (ppb) in the surface microlayer, and up to 1 *milligram per liter* (1 ppm) at the surface of sediments. As sublethal effects can be observed in molluscs at 10–20 ng/L, the alkyltin compounds pose a real threat to coastal aquatic life. Many states in the United States have now limited their use, although no other antifouling materials have equaled their effectiveness.

11.5 | HEAVY ELEMENTS

11.5.1. The Heavy Metals

The term *heavy metal* now is familiar to almost everyone, and is even the name of a rock-music magazine. However, the definitions often conflict and may include such elements as cadmium, copper, zinc, and nickel. For us, only the truly *heavy* metals—mercury, thallium, lead, and bismuth—will be considered here, Group **3** on Fig. 11.1. They have atomic weights between 200 and 210, are true metals in appearance and conductivity, but often form covalent compounds. Although mercuric chloride melts at 277°C, it is appreciably volatile at ambient temperatures and dissolves easily in benzene. Heavy metal compounds usually are highly toxic, and all appear to share the same mechanism of toxic action as arsenic. (Section 9.3).

11.5.2. Mercury

This liquid metal has fascinated people throughout the ages. Already known to the ancients, it sometimes occurs naturally but more often is obtained simply by heating the widely found sulfide mineral, cinnabar (HgS). The Hg–Hg bond is very weak, 10 kcal/g-atom, and the metal dissolves slightly in water, ether, and benzene. It also dissolves many other metals to give solid solutions, called **amalgams**, used widely for fillings in dentistry. It reacts with nonmetals such as sulfur or chlorine even at room temperature.

Mercuric chloride, "corrosive sublimate," has been used in medicine for more than 3000 years. The solubility in organic solvents such as benzene reveals its covalent character, and >99% exists as undissociated $HgCl_2$ in distilled water. However, it readily forms coordination complexes, so that molecular $HgCl_2$ amounts to only 4.3% in a 0.5 M salt solution (seawater), while $HgCl_3^-$ represents 16%, and over 80% of the mercury is present as poorly absorbed $HgCl_4^{2-}$ (Fig. 11.5A). Due to the presence of bromide in seawater, bromine-containing complex ions also appear, but mercuric ions are almost completely absent. Thus the exact *speciation* of an element becomes very important in environmental toxicology.

Like the metalloids but unlike most other metals, mercury forms environmentally stable organomercury compounds. Phenylmercuric acetate, prepared by heating mercuric acetate with benzene, was once used extensively as an agricultural fungicide, and methoxyethylmercuric acetate and chloride (called Ceresans) prepared from ethylene and mercuric acetate in methanol long were used to control major seed-borne fungal diseases of crops.

$$AcO\text{-}Hg\text{-}OAc \xrightarrow{CH_2=CH_2} AcOCH_2CH_2\text{-}Hg\text{-}OAc \xrightarrow[AcOH]{CH_3OH} CH_3OCH_2CH_2Hg\text{-}OAc$$

Mercuric acetate   Ceresan

(11.7)

However, methylmercury pesticides are no longer registered and have received such environmental notoriety that they are included in this chapter's Special Topic 11 on inorganic methylation.

Natural methylation by microorganisms is a major contributor to the cycling of

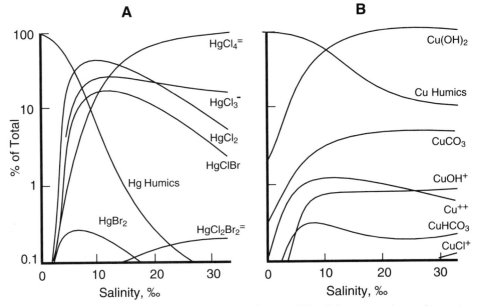

Figure 11.5. Proportions of dissolved Hg and Cu species as calculated for an estuarine environment. Adapted from Mantoura et al., 1978. Humics = humic acid complexes.

mercury in the environment (Fig. 11.6). Mercuric ion, released from a mercury ore such as cinnabar or from other sources, is methylated to methylmercury species that can be absorbed into organisms or eventually converted to volatile dimethylmercury. Photodegradation in water or the atmosphere removes the alkyl groups and returns the Hg to inorganic form.

Mercury is not especially poisonous in liquid form, but it is easily absorbed when finely divided or vaporized. However, most mercury compounds are very toxic, and a single gram of mercuric chloride provides a lethal dose for an adult human. They often are readily absorbed through the skin, making a microbiologist's traditional disinfection of his lab bench with mercuric chloride solution a very dangerous practice. A general mechanism of toxicity, as for many heavy metals, is reaction with the –SH groups of respiratory enzymes in a manner similar to arsenic (Section 9.3), except that the large size of the mercury atom requires the formation of S–Hg–S polymers rather than rings.

The lipophilic methylmercury compounds readily penetrate nerves and bind to cysteines on ACh receptors, resulting in neurological dysfunction. The ability to pass lipid membranes leads to degeneration and necrosis in the small sensory neurons of the cerebral cortex in animals, adversely affecting vision, hearing, speech, movement, sensation in hands and feet, and intelligence and mental ability. Brain size and development are affected in the young.

Mercuric and methylmercuric ions react with sulfhydryls in general, and in fact are used to precipitate enzymes out of dilute aqueous solution. They bind to SH groups on the surface of red blood cells, and although only about 1% of the groups react, that is enough to severely inhibit sugar transport and consequent energy pro-

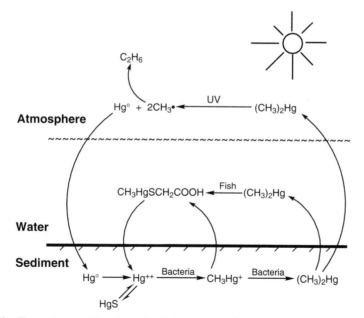

Figure 11.6.   The environmental cycling of methylmercury species.

duction. Thus, many of the bewildering variety of recorded poisoning symptoms may turn out to be due to a single type of chemical reaction after all.

11.5.3. Lead

Lead is the most widely distributed of the toxic elements (Table 11.1), although not the most abundant (0.002% of the earth's crust). It enters the environment by escape during smelting of its sulfide ore, galena, as well as through use in storage batteries, pipe and conduit, solder and pewter, and especially the addition of tetraethyl lead to gasoline. Lead tetroxide is used in corrosion-resistant paint ("red lead"), lead monoxide is added to fine glass ("leaded glass"), and lead hydroxycarbonate and sulfate were the principal white paint pigments until replaced in recent years by titanium dioxide. Millions of pounds of lead arsenate were applied for insect control in the first four decades of this century, and it contaminates large areas of once-agricultural land now occupied by housing developments. We are surrounded by lead.

Although lead has about the same E_n as mercury, divalent lead compounds such as $PbCl_2$ tend to be ionic and water soluble. Lead acetate and other aliphatic acid salts are fat soluble, and lead tetraethyl and lead tetrachloride are covalent, lipid-soluble liquids. However, water solubility does not preclude toxicity, as lead ions form lipophilic coordination complexes in water. In alkaline solution, lead concentrations are limited by precipitation of the hydroxycarbonate, but even metallic lead has an aqueous solubility of up to 500 $\mu g/L$ under acidic conditions.

Lead intoxication takes several forms. Like other heavy metals, lead binds to and inactivates SH-containing substrates such as dihydrolipoyl transacetylase (Section 9.3), inhibits heme biosynthesis, and can replace Ca in bone and in biochemical

TABLE 11.1
Distribution of Lead in the Environment[a]

Reservoir	Average Conc. ($\mu g/kg$)	Total Pb (10^6 kg)
Atmosphere	0.0035	18
Lithosphere		
Soils	16000	4.8×10^6
Sediments	47000	4.8×10^{10}
Hydrosphere		
Oceans	0.02	2.7×10^4
Interstitial water	36	1.2×10^7
Lakes and rivers	2	61
Glaciers	0.003	61
Groundwater	20	82
Biosphere[b]		
Terrestrial	100 (3000)	83 (2100)
Marine	500 (2500)	0.8 (2500)
Freshwater[c]	2500	825

[a] Adapted from Eisler, 1988.
[b] Living biota (dead biota).
[c] Living and dead.

processes. However, even low doses cause neurotoxicity, apparently due to its ability to compete with Ca^{2+} in nerve function. This can drastically affect the brain (lead encephalopathy), but more subtle effects include hyperactivity, decreased attention span, and as much as a 9-point reduction in IQ in exposed children (WHO, 1995). Lead levels in plasma often are correlated with antisocial behavior, and many juvenile criminals show elevated lead in their blood. The health-effect threshold currently is placed at 100 μg/L in blood, so the U.S. average of 28 μg/L may signal danger, as lead measurements are notoriously inaccurate. Excessive lead levels in both children and workers have been lowered satisfactorily by administration of BAL (Section 9.3) or, more recently, *meso*-dimercaptosuccinic acid (Aposhian and Aposhian, 1990) which operates by the same mechanism.

Lead specifically blocks several steps in the biosynthesis of heme pigments, including inhibition of ALA dehydratase (ALAD), the enzyme that catalyzes condensation of two δ-aminolevulinic acid molecules on the way to porphyrins. Among the results (WHO, 1995) are anemia and cardiovascular symptoms due to lack of hemoglobin, disturbed Ca metabolism leading to impaired bone and tooth development, and lowered cytochrome P450 levels which affect detoxication ability and alter brain neurotransmitter function.

The exposure of inner city kids comes primarily from old lead-based paints, but environmental lead is ubiquitous. It is found in most foods, in water, and in air due to vehicle exhaust (Table 11.2). It can enter humans from glazed pottery, from liquids

TABLE 11.2
Lead in the Human Environment[a]

Route	Concentration	Daily Exposure	Absorption Factor	Absorbed Pb (μg/day) (%)
Urban adult smoker				
Air	0.75 μg/m^3	20 m^3	0.4	6 (13.3)
Water	10 μg/L	2 L	0.1	2 (4.4)
Food	150 μg/kg	2 kg	0.1	30 (66.7)
Dust/soil	500 μg/g	0.02 g	0.1	1 (2.2)
Tobacco	5 μg/cigarette	30	0.4	6 (13.3)
Total				**45 (100)**
Rural adult nonsmoker				
Air	0.08 μg/m^3	20 m^3	0.4	0.7 (2)
Water	10 μg/L	2 L	0.1	2 (6)
Food	150 μg/kg	2 kg	0.1	30 (91.7)
Dust/soil	50 μg/g	0.02 g	0.1	0.1 (0.3)
Total				**32.8 (100)**
Urban child exposed to Pb paint				
Air	0.75 μg/m^3	10 m^3	0.4	3 (0.5)
Water	10 μg/L	1.4 L	0.5	7 (1.3)
Food	100 μg/kg	1 kg	0.5	50 (9.1)
Dust/soil	500 μg/g	1 g	0.3	150 (27.3)
Paint[b]	10^4 μg/g	0.2 g	0.17	340 (61.8)
Total				**550 (100)**

[a] Adapted from NRC, 1980.
[b] Oral exposure.

stored in pewter containers, or even from the lead seals on wine bottles. The decline of the Roman Empire has been attributed to the widespread use of lead vessels, but lead poisoning was more likely due to the common practice of spiking sour wine with sweet-tasting lead acetate ("sugar of lead"). Freshly cut or melted lead is not poisonous but soon becomes coated with a layer of toxic lead oxide and carbonate, a problem until the use of lead pipe to carry drinking water was recently discontinued. Birds still are poisoned by mistaking lead shot for seeds or from carrying shotgun pellets in their bodies.

In addition to the effects on mammals and birds, lead is bioconcentrated by and uniformly toxic to all aquatic organisms. Impaired reproduction, reduced survival, and restricted growth have been observed at Pb levels as low as the 1.0–5.1 μg/L common in natural waters (Eisler, 1988). Efforts to restrict lead-based paints, leaded gasoline, and lead shot already have been effective in reducing the hazard in the United States, but the use of lead worldwide is not declining. The 300,000,000,000 kg released into the environment over the past 5000 years is not going to disappear.

11.5.4. Thallium and Bismuth

Although most people have never heard of thallium, it actually is more prevalent in the earth's crust (0.7 ppm) than is mercury (0.5 ppm). It is used in electronic equipment, and its sulfate was once a common rat poison. The symptoms of poisoning are those of the other heavy metals, especially polyneuritis in the extremities, and the resulting hair loss (alopecia) was put to use in depilatories at one time. Perhaps the most notable physiological feature of thallium toxicity is its competition with potassium, which is of similar size and charge, at key sites such as nerves and Na^+–K^+ ATPase. Reversal of Tl poisoning by BAL indicates that SH groups also are involved.

Although bismuth compounds such as $BiBr_3$ and $BiCl_3$ are soluble in organic solvents and obviously possess covalent character, they do not express any outstanding toxicity. This may be due partly to the large size of the bismuth atom and partly to rapid hydrolysis of the compounds in water to give insoluble and infusible oxyhalides. Organic complexes of Bi are still used in antacids such as Pepto-Bismol®.

11.6 | TRANSITION ELEMENTS

11.6.1. Transition Metals

The transition elements are so called because they represent the "transition" from sequential filling of p orbitals to the filling of d orbitals. As seen on the periodic table (Fig. 11.1), the first transition, $4p$ to $3d$, includes such toxicologically interesting elements as chromium, manganese, cobalt, nickel, copper, and zinc, while the second transition is represented by molybdenum, silver, and cadmium. Transition elements are metallic and exhibit other common properties including primary valences of 2 or

3, a tendency to accept electron pairs from nucleophilic ligands, brightly colored salts, and limited covalent character. Copper is typical.

11.6.2. Copper

Natural metallic copper was found and used by the ancients; the "Bronze Age" refers to an early alloy of copper, tin, and zinc. Copper compounds have long been used in medicine and agriculture, for example, verdigris (basic copper acetate) as a pesticide in Roman vineyards, Bordeaux mixture (cupric hydroxycarbonate) used for over a century to combat mildew on grapes, and the "bluestone" (cupric sulfate pentahydrate) still employed worldwide as an algicide.

The electron configuration of copper is $1s^2 2s^2 2p^6 3s^2 3p^6 3d^{10} 4s^1$, so its principal valence should be $+1$. However, the $3d$ orbitals actually are of lower energy than the $4p$, and dsp^2 hybridization allows divalent cupric ion to predominate and assume a modified octahedral structure in solution despite the preference of copper for square planar. Absorption of 800 nm (red) light boosts a $3d$ electron into a $4p$ orbital, and so cupric salts appear blue.

Copper derivatives are uniformly toxic, the Cu absorbed *via* natural chelates as described in Section 9.6 (Phinney and Bruland, 1994). This general toxicity has been employed variously to kill or repel bacteria, fungi, algae, aquatic invertebrates, and undesirable fish. The LC_{50} of Cu^{2+} in the bluegill sunfish *(Lepomis macrochirus)* is 240 μg/L, but respiration is affected even at 10 μg/L. Mammals are not excepted: The oral LD_{50} of copper sulfate is 900 mg/kg in rats, and even <15 mg/kg orally leads to stomach upset and vomiting in humans. Symptoms include tremors, labored respiration, and hemolysis, all signs of an SH binding which is reversible by BAL (Section 9.3). A mysterious Monday morning illness in an office building was explained by the discovery that, over weekends, copper was leached from the feed lines in newly installed soft-drink machines and complexed into lipophilic form by chelate-forming food additives such as glycine and citric acid (Section 9.6) present in the liquid.

Such copper tubes and pipes exposed to water and carbon dioxide quickly become coated with a layer of cupric hydroxycarbonate, the mineral malachite. Used as an algicide, bluestone added to a natural water quickly forms a precipitate of insoluble malachite, but the exact speciation of copper depends on pH and the presence of ligands such as halide ions, ammonia, hydrogen sulfide, and dissolved organic matter. Figure 11.5B shows the changes in Cu speciation in an estuary as the proportion of seawater and pH increase. Unlike Hg, copper complexes only weakly with Cl^- but strongly with humic substances, so the high toxicity of Cu^{2+} seen in laboratory tests with aquatic animals may not apply equally in the natural environment because of the more limited bioavailability of the metal. The acute toxicity of metals observed in freshwater organisms is generally at least an order of magnitude greater than it is in marine species.

Dilute aqueous solutions of metal ions such as Cu^{2+} and Ni^{2+} can affect olfaction and behavior in aquatic animals (Rand and Petrocelli, 1985). For example, laboratory tests showed that rainbow trout *(Salmo gairdneri)* avoided copper at 70 ppb but were attracted to it at 460 and 760 ppb, and the aquatic amphipod *Gammarus lacustris* behaved similarly. However, the opposite response to nickel was observed in *Salmo,* while *G. pulex* avoided the copper wherever possible. The environmental significance of these observations is unclear.

11.6.3. Other Transition Elements

Zinc represents the last of the first-row transition elements, with all $3d$ orbitals filled. Although this element occurs everywhere and forms stable chelates, zinc is not particularly interesting toxicologically. However, compounds of its second-row analog, **cadmium,** are considered toxic pollutants. Cadmium is used primarily for protective plating, in solder, and in Nicad (nickel–cadmium) batteries. It occurs closely associated with zinc, and so is found anywhere that element is used. For example, vehicle tires contain a high proportion of zinc oxide for strength, and the powder formed as tires wear down is a major source of Cd pollution. Cadmium also occurs at relatively high levels in cigarette smoke and can be absorbed readily into the lungs.

Cd is bound and stored by high-sulfur proteins called *metallothioneines,* which occur in mammals primarily in kidneys. Metallothioneines and related low-molecular-weight proteins also bind Cu, Zn, and other elements and are considered to be an organism's means of detoxifying them. They are found widely among animal species, in molluscs, for instance (Roesijadi, 1992), and their levels increase in response to the presence of the metals. A similar protein, hemosiderin, specifically binds Hg, Pb, Bi, and Se, and ferritin binds Fe. Neither Co, Ni, Mn, nor Pb induce metallothioneines.

Molybdenum is a trace element essential for function of the key enzyme, xanthine oxidase, and is responsible for the bacterially derived ability of some plants to fix atmospheric nitrogen. It occurs extensively in food and water, is added to fertilizers, hardens steel, and provides important industrial catalysts. Ingestion of too much hexavalent Mo such as molybdate by grazing animals results in "teart" disease, characterized by anemia and poor growth, which can be reversed by supplemental copper and sulfate. Similarly, the hexavalent **chromium** in chromate (CrO_4^{2-}) and dichromate ($Cr_2O_7^{2-}$) causes kidney tubule necrosis and, *via* inhalation, lung cancer. Cr is used in chrome plating, leather tanning, textiles, and especially in wood preservation (Wolman salts), and environmental exposure occurs through waste releases by these industries.

Manganese is of comparatively recent environmental concern. Although a supply of Mn is essential to all forms of life, breathing airborne particles of manganese ores such as pyrolusite, MnO_2, causes acute respiratory disease and a severe chronic neurotoxicity ("manganism") resembling Parkinsonism (WHO, 1981). Manganism also resulted from drinking water that contained 16–18 ppm of dissolved Mn. A notable biochemical feature of manganism is inhibition of dopamine and serotonin synthesis in the brain.

Environmental exposure originally was largely limited to mining and smelting wastes, the dithiocarbamate fungicide maneb, and the occasional broken dry cell whose central core is black manganese dioxide. Acute respiratory illness in humans was associated with airborne Mn_3O_4 at levels as low as 1 $\mu g/m^3$ in communities near ferromanganese factories (WHO, 1981), but effects on the environment remain unexplored. However, a cutback in leaded fuels in the 1970s led to introduction of derivatives of manganese carbonyl, especially methylcyclopentadienylmanganese tricarbonyl (MMT), $C_6H_8Mn(CO)_3$, to boost octane ratings. Although this additive is used at only about 20 mg/L, the internal combustion engine converts it into finely divided airborne Mn_3O_4 particles, of which over 50% are respirable. Manganese pollution seems not to have led to general health effects so far, but a lot more research needs to be done—soon.

Nickel is a rather common element, representing 0.018% of the earth's crust compared to copper (0.007%) and lead (0.0015%). Its resistance to corrosion provides many industrial and consumer uses, stainless steel, for example, and U.S. and Canadian 5-cent pieces, "nickels," are 25% Ni and 75% Cu. Most nickel is mined as its sulfide, pentlandite, $(FeNi)_9S_8$.

Nickel compounds are unusually toxic and carcinogenic. Industrial exposure arises from processing of the ore which converts it to Ni_2S_3, and from refining and thermal plating with nickel carbonyl, $Ni(CO)_4$. The carbonyl is acutely very poisonous and volatile (b.p. 43°C), and breathing the vapor leads to pulmonary edema (fluid accumulation), pneumonia, and respiratory failure (Brown and Sunderman, 1980). It was originally suspected of being the prime cause of nasal and pulmonary cancer among smelter workers, now recognized to stem primarily from inhalation of finely divided Ni_3S_2, NiO, and even Ni itself.

Among inorganic compounds, nickel compounds are the most prevalent and potent causes of contact dermatitis (Section 7.4.3). Mere contact with nickel plate or coins can cause rash in sensitive individuals, and there is evidence that even elevated levels of Ni in food may cause sensitization. Considering the many possible sources of exposure to nickel, surprisingly little information exists on toxic effects in nonmammals, even though pollution from nickel mining and smelting are thought to be responsible for a severe decline in marine life in New Caledonia.

11.7 | RADIOACTIVE ELEMENTS

Uranium, radium, and radon are natural radioactive elements, but they also are toxic. *Uranium* is a starting point for nuclear energy, although its oxide also is used in specialty glass and ceramics. It is primarily (99.3%) ^{238}U, with small proportions of the 235 and 234 isotopes, and is produced from pitchblende and uraninite (UO_2) and from carnotite, $K_2(UO_2)_2(VO_4)_2$. The ore is extracted with nitric acid to provide fat-soluble and hexavalent uranyl nitrate, $UO_2(NO_3)_2$, a cause of kidney disease and skin lesions and considered very poisonous. Uranium deposits extensively in bone and lung, and although not highly radioactive ($t_{1/2}$ 4.5 billion years), it is in a position to cause long-term damage.

Through a series of steps, uranyl nitrate is converted to the covalent and highly volatile uranium hexafluoride, the source of **plutonium.** The principal natural isotope, ^{239}Pu, has a half-life of 24,390 years and emits highly energetic α particles, that is, He nuclei. This isotope is the fissionable material in nuclear bombs and the fuel in nuclear reactors, while ^{238}Pu provides heat for the thermoelectric power devices used on communications satellites and medical pacemakers and is planned to power future manned space stations. These isotopes also are the major contributors to the nuclear waste, which will be with us for thousands of years to come, and to nuclear fallout, along with ^{90}Sr and ^{131}I (Sections 4.4.3 and 4.6).

Plutonium toxicity is due almost entirely to its radioactivity (Bair and Thompson, 1974). The principal commercial form of the element, PuO_2, understandably has received the most toxicological attention. Very little Pu is absorbed by skin or GI tract, and intoxication is primarily pulmonary. In dogs, about 80% of inhaled PuO_2 initially in the alveoli was still present 10 years after exposure, the rest having migrated to liver, bone, and lymph nodes. Toxicity in mammals primarily takes the form of bone and lung cancer.

Radium (^{226}Ra, $t_{1/2}$ 1600 years) once was used to produce luminous numerals on clocks and instruments. It caused serious injuries among the mostly female dial painters, who sharpened their brush tips with lips or tongue. Medical followups provided a large store of knowledge about radium toxicity. Ra is chemically analogous to Ca; it is deposited in bones and leads to malignant tumors and blood disease there. It causes tumors in skin and air passages as well. The radioactive decay of radium produces a gaseous element, **radon,** and due to the wide distribution of Ra in soil and rocks, Rn is a common and dangerous pollutant (Section 2.3.3). It volatilizes from soil and building materials and collects in enclosed spaces such as basements and poorly ventilated rooms. Although a normal household level of this carcinogenic gas provides about 1.5 picocuries/liter (pCi/L) of radioactivity, that is, 56 atomic disintegrations/sec/m^3, many homes show 100 times that level (Nero, 1988).

11.8 | OTHER TOXIC ELEMENTS

Several other elements are toxic or produce toxic compounds. **Beryllium** is responsible for a chronic lung disease known as berylliosis. **Barium** salts are markedly poisonous, although a large quantity of very insoluble $BaSO_4$ must be ingested before fluoroscopic examinations. The covalent and toxic liquid tetrachlorides of **silicon** and **titanium** are often used to produce smoke for movies and for aerial advertising, where they hydrolyze in moist air to form fine particles of the corresponding oxides. More information on inorganic poisons is provided by Friberg et al. (1979). However, recall that inorganic substances generally are toxic because they are **lipophilic** and **electrophilic.**

11.9 | REFERENCES

Aposhian, H. V., and M. M. Aposhian. 1990. *meso*-Dimercaptosuccinic acid: Chemical, pharmacological, and toxicological properties of an orally-effective metal-chelating agent. *Ann. Rev. Pharmacol. Toxicol.* **30:** 279–306.

Bair, W. J., and R. C. Thompson. 1974. Plutonium: Biomedical research. *Science* **183:** 715–22.

Brown, S. S., and F. W. Sunderman. 1980. *Nickel Toxicology,* Academic Press, New York.

Clark, E. A., R. M. Sterritt, and J. N. Lester. 1988. The fate of tributyltin in the aquatic environment. *Environ. Sci. Technol.* **22:** 600–4.

Eisler, R. 1988. *Lead Hazards to Fish, Wildlife, and Invertebrates: A Synoptic Review,* U.S. Fish and Wildlife Service, Washington, D.C., Biol. Report. 85 (1.14).

Friberg, L., G. F. Nordberg, and V. B. Vouk. 1979. *Handbook of the Toxicology of Metals,* Elsevier–North Holland, Amsterdam, The Netherlands.

Mantoura, R. F. C., A. Dickson, and J. P. Riley. 1978. The complexation of metals with humic materials in natural waters. *Estuar. Coastal Mar. Sci.* **6:** 387–408.

Meyer, B. 1977. *Sulfur, Energy, and Environment,* Elsevier Scientific Publishing Co., Amsterdam, The Netherlands.

NRC. 1980. *Lead in the Human Environment,* National Research Council, National Academy of Sciences, Washington, DC.

Nero, A. V. 1988. Controlling indoor air pollution. *Sci. American* **258**(5): 42–48.

Ohlendorf, H. M., D. J. Hoffman, M. K. Saiki, and T. W. Aldrich. 1986. Embryonic mortality and abnormalities of aquatic birds: Apparent impacts of selenium from irrigation drain waters. *Sci. Total Environ.* **52:** 49–53.

Phinney, J. T., and K. W. Bruland. 1994. Uptake of lipophilic organic Cu, Cd, and Pb complexes in the coastal diatom *Thalassiosira weissflogii. Environ. Sci. Technol.* **28:** 1781–90.

Rand, G. M., and S. R. Petrocelli. 1985. *Fundamentals of Aquatic Toxicology,* Hemisphere Publishing Corp., Washington, DC, Chap. 9.

Roesijadi, G. 1992. Metallothioneines in metal regulation and toxicity in aquatic animals. *Aquatic Toxicol.* **22:** 81–114.

Schmidt, M. 1965. Reactions of the sulfur–sulfur bond, in *Elemental Sulfur: Chemistry and Physics* (B. Meyer, ed.), Interscience, New York, NY, pp. 301–36.

Sellers, P., C. A. Kelly, J. W. M. Rudd, and A. R. MacHutchon. 1996. Photodegradation of methylmercury in lakes. *Nature* **380:** 694–97.

SWRCB. 1988. *California State Mussel Watch, Ten Year Summary, 1977–1987,* Calif. State Water Resources Control Board, Sacramento, CA. Water Quality Monitoring Report 87-3.

Toribara, T. Y., T. W. Clarkson, and D. W. Nierenberg. 1997. More on working with dimethylmercury. *Chem. Eng. News* **75** (24):6.

Tweedy, B.G. 1969. Elemental sulfur, in *Fungicides: An Advanced Treatise* (D. C. Torgeson, ed.), Vol. II, Academic Press, New York, NY, pp. 119–46.

WHO. 1981. *Manganese,* World Health Organization, Geneva, Switzerland. Environ. Health Criteria 17.

WHO. 1982. *Chlorine and Hydrogen Chloride,* World Health Organization, Geneva, Switzerland. Environ. Health Criteria 21.

WHO. 1984. *Fluorine and Fluorides,* World Health Organization, Geneva, Switzerland. Environ. Health Criteria 36.

WHO. 1995. *Inorganic Lead,* World Health Organization, Geneva, Switzerland. Environ. Health Criteria 165.

Wood, J. M. 1974. Biological cycles for toxic elements in the environment. *Science* **183**: 1049–52.

Wuosmaa, A. M., and L. P. Hager. 1990. Methyl chloride transferase: A carbocation route for biosynthesis of halometabolites. *Science* **249**: 160–62.

Special Topic 11. Methylation of Metals and Metalloids

When Claire Booth Luce was ambassador to Italy, she and her family were lodged in an old villa. They soon complained of nausea and other symptoms consistent with arsenic poisoning, and an international incident seemed imminent. Some chemical sleuthing revealed that the damp, old wallpaper in the bedrooms had been colored with arsenic-containing dyes, and fungi were converting the arsenic to gaseous methylarsines. The Luces were inhaling the poison while they slept, and methylarsine's garlic odor did not seem unusual under the circumstances.

Other inorganic methyl compounds, too, are volatile, highly lipophilic, and very toxic. The metalloids Se, Sn, and Te also undergo biochemical methylation by *S*-adenosylmethionine in a manner similar to As, via methylcarbonium ions (Section 6.2). The methylation of arsenic also is easily carried out in the laboratory (Fig. 11.4), and even the alchemists knew how to prepare the malodorous and poisonous cacodyl oxide, known to them as "Cadet's fuming liquid," by distillation of a mixture of white arsenic and an acetate:

$$As_4O_6 + 8KOCOCH_3 \xrightarrow{\Delta} 2\,(CH_3)_2As\text{-}O\text{-}As(CH_3)_2 + 4K_2CO_3 + 4CO_2 \tag{11.8}$$
$$\text{Cacodyl oxide}$$

Air oxidation of cacodyl oxide leads to the herbicidal cacodylic acid, dimethylarsinic acid, which received controversial use as Agent Blue defoliant during the Vietnam war.

The heavy metals Hg, Tl, and Pb also become methylated in the environment, but this reaction requires a nucleophilic rather than electrophilic reagent (Wood, 1974). The source of this methyl carbanion is methylcobalamine, one of the forms of vitamin B_{12} (Eq. 11.9, R = dimethylbenzimidazole). The resulting methylmercuric ion can be methylated again to dimethylmercury, $CH_3\text{-}Hg\text{-}CH_3$, or if sufficiently concentrated, simply disproportionate to the same product. Methylation is carried out primarily by bacteria in sediment and water, and lipophilicity assures that the products will be bioconcentrated into aquatic organisms.

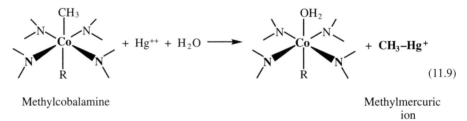

Methylcobalamine   Methylmercuric
ion

$$\tag{11.9}$$

The methylmercurys first received notoriety when a methylmercury catalyst used in the manufacture of acetaldehyde in Japan escaped into nearby Minamata Bay, and hundreds of people were poisoned by eating mercury-contaminated seafood. Once absorbed, methylmercuric ions react with sulfhydryl groups and eventually are excreted in urine as water-soluble CH_3HgSCH_2COOH or protein conjugates, while the more stable dimethylmercury quickly penetrates nerve cells and produces severe neurotoxicity, especially in the limbs.

Bromide and iodide, and to a smaller extent chloride, also are methylated via S-adenosylmethionine (Wuosmaa and Hager, 1990), and large quantities of methyl halides are released daily from the ocean into the atmosphere thanks to the metabolism of marine algae (Section 2.4). Methyl derivatives of elements farther to the left in the periodic table are too unstable to survive environmental conditions. For example, the methylmagnesium Grignard reagent familiar to generations of organic chemistry students reacts instantly with water:

$$CH_3MgI + H_2O \longrightarrow CH_4 + Mg(OH)_2 \qquad (11.10)$$

Methylated inorganics eventually break down in the environment, largely by oxidation and photolysis (Wood, 1974). Dimethylmercury forms metallic mercury upon UV irradiation in the atmosphere (Fig. 11.6), methylmercury behaves similarly in lake water (Sellers et al., 1996), and methylarsenic compounds are degraded microbially to arsenate in soil. The elements, however, persist and cycle forever. Along the way, methylation has provided many inorganic substances with the ability to become *lipophilic, reactive toward nucleophiles—and toxic.*

The fully methylated metals and -metalloids are unusually volatile, lipophilic, and dangerous, as shown by Ambassador Luce's experience. For example, dimethylmercury is a liquid with a vapor pressure of 58.8 torr at 23.7°C (b.p. 92°C) and a log K_{ow} of 2.59, values about like those of the cleaning solvent TCE. Consequently, it moves easily, is absorbed by human skin and lungs before the victim is even aware of exposure, and the resulting neurotoxicity is irreversible. In addition to the sufferers of Minamata, a number of people are known to have died from casual exposure to dimethylmercury in and around laboratories (Toribara et al., 1997), and any urge to experiment with this group of substances must be tempered with appropriate awareness and precaution.

12.1 | BIOTOXINS

"Toxic chemical" automatically means "synthetic" to most people. However, the chemicals one most frequently encounters are the natural poisons found in minerals, plants, animals, and microorganisms. Those derived from biota are sometimes called *toxins,* which a dictionary defines as "poisonous substances, especially proteins, produced by living cells or organisms and capable of causing disease when introduced into the body;" they also stimulate the production of antibodies (antitoxins) in their animal victims. The definition certainly describes the high-molecular-weight protein toxin from *Clostridium botulinum,* at present the most poisonous substance known.

However, many other kinds of natural substances are toxic to animals, plants, or microbes. Instead of referring repeatedly to "toxic organic natural products," we will call them **biotoxins,** a term coined by toxicologist B. W. Halstead that includes both protein and nonprotein poisons of organic origin. Biotoxins probably have existed since the emergence of living things and have evolved for defense, predation, nutrition, or possibly just for excretion of waste. Like synthetic chemicals, all natural products are at least potentially toxic, but biotoxins are exceptionally poisonous and can be distinctly hazardous. Microbial poisons such as aflatoxins, nicotine and other alkaloids from plants, and the venoms from bees and rattlesnakes are examples. The use of biotoxins by their producers against competitors, prey, and enemies is common enough to suggest that such chemicals are a force in ecology and evolution (Special Topic 12).

12.2 | THE ALKALOIDS

To be classed as an alkaloid, a substance must (1) be a natural organic compound, (2) contain a nitrogen atom in a heterocyclic ring, (3) provide an alkaline reaction in water, and (4) produce a pronounced physiological effect. A familiar example is ***nicotine,*** an alkaloid from *Nicotiana* (tobacco) species and a commercial byproduct of the tobacco industry. The name appears so often in cigarette ads, news headlines, and congressional reports that it probably is known to almost everyone in the United States. Although extensively applied at one time as an agricultural insecticide, the use of nicotine sulfate as a pesticide now is largely restricted to houseplants, primarily because of its high toxicity. Tobacco products represent by far the main source of human exposure today, as the average U.S. cigarette contains 15–20 mg of nicotine and many cigars twice that amount.

Nicotine is perhaps the most poisonous substance one routinely contacts. The oral LD_{50} in rats is 50 mg/kg and 10 mg/kg in the silkworm, *Bombyx mori,* but humans are much more sensitive. A lethal dose in an adult man was observed to be as little as 0.88 mg/kg, and an accidental dose of 0.04 mg/kg to a pregnant woman caused cardiovascular abnormalities in her unborn child. As small boys having their first smoke behind the barn could tell you, the early symptoms of nicotine poisoning include extensive salivation, nausea, and vomiting. Symptoms from greater exposure include muscular weakness, convulsions (the "nicotine fit"), and finally respiratory collapse, all signs of neurotoxicity.

Nicotine mimics and competes with the normal neurotransmitter, acetylcholine, in the central nervous system and acts at autonomous nerve ganglia and neuromuscular junctions. The initial response is excitation followed by a persistent depression. Despite nicotine's toxicity, serious poisoning from tobacco smoking is unusual, as the onset of illness precludes an overdose. Ingestion is another matter. Tobacco chewers are regularly poisoned by swallowing their "chaw," small children can be poisoned by chewing on cigarettes, and babies have died from drinking milk into which a cigarette or cigar had fallen accidentally.

Pure nicotine is a volatile, fat-soluble liquid easily absorbed through the skin, leading to occasional intoxication in tobacco handlers. Breast milk from smoking mothers may contain as much as 0.5 mg/L of nicotine, enough to affect nursing babies adversely. The half-life of nicotine in blood is only 30–60 min due to oxidative detoxication (Fig. 12.1A), and the induction of oxidative capacity is illustrated by the tolerance developed by long-term smokers. Thermal degradation during the burning of tobacco produces a large number of products of varying toxicity, one of which is myosmine (Fig. 12.1B). The strong, musty odor imparted to clothing, skin, and furnishings by tobacco smoke is largely due to this compound, and although smokers and nonmokers alike have significant exposure to it, the possible toxicological implications have received scant attention.

Many familiar plant species contain ***pyrrolizidine alkaloids*** (WHO, 1988). Among the principal troublemakers are *Senecio jacobaea* (tansy ragwort), *Heliotropium europaeum* (wild heliotrope), *Symphytum officinale* (common comfrey), *Amsinckia intermedia* (fiddleneck), and other members of these genera. Livestock poisoning was ascribed to these plants as early as 1800, and they remain a serious agricultural problem today. For example, tansey ragwort still kills or injures dozens of horses and cattle annually in the Pacific Northwest.

The two most important sources of human exposure are the pyrrolizidine-containing leaves incidentally harvested with wheat or rye and then baked into bread,

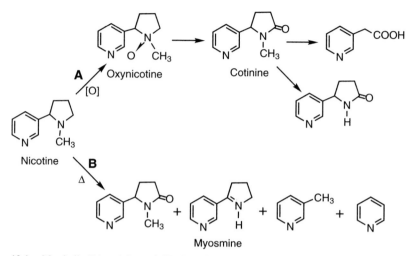

Figure 12.1.   Metabolic (A) and thermal (B) degradation of nicotine.

and the "herbal teas" concocted intentionally from wild plants. The herbal teas and medicines made from comfrey, heliotrope, and rattlebox *(Crotalaria)* have become increasingly important with growth of the back-to-nature movement in the United States, one typical poisoning victim dying after receiving an estimated 0.5–0.7 mg/kg/day of the alkaloids for several weeks during herbal treatment of psoriasis with *H. lasiocarpum.* An intake of more than 0.015 mg/kg/day of pyrrolizidines now is known to cause acute or chronic liver degeneration in humans, as both the alkaloids and their effects are cumulative.

Pyrrolizidine alkaloids derive their name from their shared *N*–heterocyclic ring system, shown by heavy lines in Fig. 12.2. A typical member, such as **symphytine**

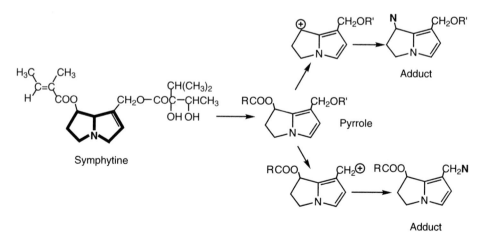

Figure 12.2.   Biotransformation of a pyrrolizidine alkaloid, symphytine, to a toxic pyrrole that alkylates DNA (**N**).

from *Symphytum,* must be metabolized to the corresponding pyrrole, as shown, to become toxic. The pyrrole then generates cationic species that alkylate DNA bases (Special Topic 9), but they are so reactive that they seldom escape the liver cells where they are formed. The site of acute intoxication is the veins of the liver, which become occluded (veno-occlusive disease), while longer exposure results in fat accumulation (cirrhosis). Both humans and domestic animals may still show signs of liver damage months after exposure, but pyrrolizidine poisoning has not been reported in wild animals except for deer.

Although the original definition required that they be derived from higher plants, a growing number of alkaloids have now been isolated from animals. For example, the small, bright-colored Colombian arrow poison frogs (*Phyllobates* species) secrete an unusual steroid alkaloid, batrachotoxin, which irreversibly increases the permeability of membranes to sodium ions and causes almost instant paralysis. A subcutaneous LD_{50} in mice of only 2 μg/kg has led to the age-old use of frog-skin extracts on arrow tips by South American aborigines.

Other well-known but toxic plant alkaloids include the antimalarial, quinine, atropine used to dilate the pupil during eye examinations, and narcotics such as morphine from opium poppies and cocaine from *Erythoxylon coca* (Section 7.4.2). Any interruption of narcotic use results in an intense craving for the drug, fatigue, and lassitude—withdrawal symptoms. Each of these reflects, in an exaggerated way, the situation with nicotine. Curiously, there has been a continuing public argument over whether or not nicotine is addictive, in which the clearest evidence probably comes from inveterate smokers who have tried unsuccessfully to quit!

12.3 | TOXIC GLYCOSIDES

A glycoside is a sugar to which a nonsugar group (aglycone) is bonded through an oxygen or nitrogen. An example is *solanine,* the glycoalkaloid of nightshades and other poisonous members of the family Solanaceae. Solanine also occurs in the common potato, *Solanum tuberosum,* but normally at levels below 10 ppm. However, when the tuber is exposed to sunlight ("green potato"), is injured, or sprouts, levels increase rapidly to as high as 100 ppm in the tuber and 500 ppm in the sprouts. Livestock have been poisoned by being fed cull or sprouting potatoes, and hungry people also have died this way. Poisoning symptoms include excessive salivation, trembling, and progressive weakness or paralysis, and in humans also may include loss of sensation (Kingsbury, 1964).

A toxic dose of solanine for an adult human is about 150 mg (2.8 mg/kg), but levels above 20 ppm in the potato are considered dangerous. Solanine and its relatives are cholinesterase inhibitors, like the organophosphorus insecticides, and dried potato skin was used by the ancient Incas to control body lice. The actual inhibitor is the aglycone, solanidine, which is released by acid or enzymatic hydrolysis (Eq. 12.1). The sugar portion, the unique trisaccharide solanose, simultaneously is hydrolyzed to its constituent D-glucose, D-galactose, and the unusual α-L-rhamnose. Many insect pests seem to attack potato sprouts with impunity, perhaps due to an inability to carry out the hydrolysis.

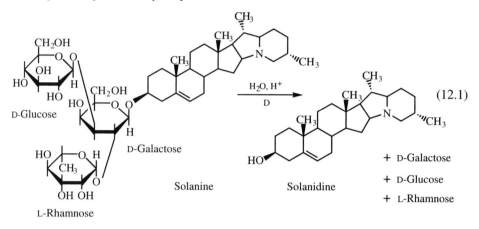

(12.1)

D-Glucose

D-Galactose

L-Rhamnose

Solanine   Solanidine

+ D-Galactose
+ D-Glucose
+ L-Rhamnose

Cyanogenic glycosides are widely distributed in the plant world, including human foods. Linamarin, from lima beans *(Phaseolus lunatus)*, is typical. Upon acid or enzymatic hydrolysis (Eq. 12.2), the beans release hydrogen cyanide (HCN), sometimes as much as 3500 mg from 1 kg (Liener, 1980).

$$\text{(12.2)}$$

Stone fruit kernels, apple seeds, and lima beans are especially rich in cyanogenic glycosides, and bird kills in Sacramento's Capitol Park, originally blamed on insecticides, were later found to be caused by the birds' consumption of immature fruit of ornamental almond trees. Another incident involved the fatal poisoning of a man who ate a handfull of prunasin-containing apple seeds he had saved for a treat. The U.S. government tests incoming shipments of cyanogenic foods and limits cyanide generation to 10 mg per 100 g portion, although the Canadian government allows twice this amount.

The release of cyanide requires two enzymes (Eq. 12.2), one a hydrolase to cleave sugar from aglycone, and the other a lyase to catalyze the breakdown of the resulting cyanohydrin into aldehyde or ketone and hydrogen cyanide. The cyanide, of course, is the actual poison (Section 15.9). The small amounts of cyanide that one is exposed to daily from food, air, and other sources are detoxified by conversion to thiocyanate (SCN^-) via the enzyme rhodanese (transsulfurase). However, many people, especially those in underdeveloped countries, have been poisoned by eating dark lima beans, cassava, or bamboo shoots. The HCN volatilizes if the food is boiled in an open pot, but a covered vessel causes it to condense and return at high concentrations to the pot liquor.

Many common garden plants, including foxglove, lily-of-the-valley, and olean-der, contain highly toxic *cardiac glycosides,* sugar derivatives of steroids that pos-sess an unsaturated lactone ring at the 17–position (Table 12.1). Those with a five–membered lactone ring are classified as *cardenolides.* Squill *(Scilla maritima)* was used by the ancient Romans as a heart stimulant, emetic, and rat poison, and digitalis from foxglove *(Digitalis purpurea)* was mentioned in Welsh writings of 1250 A.D. and is still in use today as a cardiac tonic. Typically, the active ingredients such as digitoxin act on the myocardium (heart muscle) to increase contractility and so are useful in restoring blood circulation after congestive heart failure and in the control of heart flutter (fibrillation).

Cardiac glycosides act by inhibiting $Na^+–K^+$-ATPase, thus increasing ATP and the energy it provides. Unfortunately, the therapeutic and toxic doses are very close, and overdose is common; a daily maintenance dose is 0.05–0.30 mg/day, while a toxic dose is 0.35 mg/day. Also, the drug is so widely used that children frequently gain access to it with serious results. In healthy young people, intoxication symptoms include bradycardia (slow heartbeat) and heart block (uncoordinated heartbeat), and older people also suffer fibrillation and cardiac arrest ("heart failure").

Oleander (Nerium oleander) is an ornamental shrub used extensively for land-scaping in temperate parts of the United States, such as on the median strips of freeways. All parts of the plant, both fresh and dry, are poisonous due to the steroid glycoside oleandrin, and ingestion of even a single leaf is considered dangerous for humans (Kingsbury, 1964). People have been poisoned accidentally by using the long straight stems as skewers for cooking meat, and grazing horses have been killed when tied even briefly to one of the shrubs. Oleandrin is a typical cardiac glycoside whose toxic aglycone (oleandrigenin) is generated, like solanidine, by hydrolytic re-

TABLE 12.1
Cardiac Glycosides[a]

Glycoside	Source	Sugar	Substituents	LD (mg/kg)[b]
Digitoxin	*Digitalis*	Digitoxose		0.33
Oleandrin	*Nerium*	Oleandrose	16β-OAc	0.20
Scillarin A	*Scilla*	Rhamnose, glucose	Δ^4; 6-ring[c]	0.15
Hellebrin	*Helleborus*	Rhamnose, glucose	5β-OH; 19CH=O, 6-ring	0.10
Convallatoxin	*Convallaria*	Rhamnose	5β-OH; 19CH=O	0.08
Thevetin A	*Thevetia*	Thevetose, gentiobiose	19CH=O	0.89
Cerberoside	*Thevetia*	Thevetose, gentiobiose	19CH=O	0.89

[a] Fieser and Fieser, 1959.
[b] Cat acute oral lethal dose.

moval of the unusual monosaccharide, oleandrose. Many cardiac glycosides have been identified, a few of which are listed in Table 12.1 together with an indication of their high toxicity.

The cardenolides of milkweeds (*Asclepias* spp.) have an interesting place in eco-toxicology. Monarch butterflies *(Daneus plexippus)* lay their eggs on milkweeds, on which the emerging caterpillars feed exclusively. The cardenolides are absorbed and bioconcentrated but have no effect on the larvae, as the insect heart differs from that of vertebrates. However, the insects now are toxic to predators, and young birds soon become painfully aware of the danger and so leave this particular food alone.

Glucosinolates are an interesting and widespread family of toxic natural organo-sulfur glycosides. In the presence of either acid or the enzyme thioglucosidase (myro-sinase), they are hydrolyzed to isothiocyanates, $RN = C = S$. These familiar "mustard oils" provide the pungent aroma and flavor of mustard, radish, and horseradish (Table 12.2). They are formed by a Lossen rearrangement of intermediate thiohydroxamic acids (Kjaer, 1960) which results in removal of a good leaving group, sulfate, and migration of the alkyl or aralkyl to a now-positive nitrogen in a manner reminiscent of the NIH shift (Section 6.2.2). The hydrolysis and rearrangement occur so easily that even a package of dry radish seed smells of mustard oil.

Mustard oils long have been recognized as bactericides and fungicides, but several also cause a thyroid disease (goiter) in animals. The disease is due to oxazolidi-nethiones (goitrins) formed by cyclization of hydroxyisothiocyanates, as demonstrated in lab animals and in domestic species foraging in fields where cole crops

TABLE 12.2
Some Glucosinolates and Their Isothiocyanates[a]

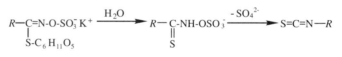

	Glucosinolate		Isothiocyanate

Name	Plant source	R
Sinigrin	Cabbage group *(Brassica oleraceae)*[b] Black mustard *(B. nigra)* White mustard *(Sinapis alba)* Horseradish *(Armoracia rusticana)*	$CH_2=CHCH_2-$
Glucoraphanin	Rape *(B. napus)*	$HO-C_6H_4CH_2-$
Gluconasturtiin	Horseradish *(Armoracia rusticana)* Turnip *(B. campestris)* Rape *(B. napus)* Papaya *(Carica papaya)* Nasturtium *(Tropaeolum majus)*	$C_6H_5CH_2CH_2-$
Glucorapiferin[c] (Progoitrin)	Cabbage group *(B. oleraceae)* Turnip *(B. campestris)* Rutabaga, rape *(B. napus)*	$CH_2=CHCH(OH)CH_2-$

[a] See Kjaer, 1960; Van Etten et al., 1969.
[b] Cabbage group consists of cabbage, Brussels sprouts, cauliflower, broccoli, kale.
[c] Cyclizes to form goitrin, *(S)*-5-ethenyl-2-oxazolidinethione.

have gone to seed. Symptoms have not been observed in humans, as people ordinarily are not exposed to sufficient concentrations, but the known transfer of goitrins from feed to milk is cause for concern (VanEtten et al., 1969).

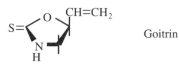

Goitrin

12.4 | PLANT PHENOLICS

Plants contain many kinds of phenols, often as glycosides, but most of them are devoid of notable toxicity. A prominent exception is the family of substituted cate-chols, called **urushiols,** that are the reactive allergens of poison oak, poison ivy, and other members of the Anacardiaceae. Poison oak (*Toxicodendron diversilobum,* Fig. 12.3) is a familiar West Coast shrub, while the very similar poison ivy *(T. radicans)* is found in all states east of the Cascade Mountains and the Great Basin. Poison sumac *(T. vernix)* grows east of the Mississippi River, most commonly in the Southeast. There is a less common Eastern poison oak, *T. quercifolium. T. radicans* is also found in Canada, throughout Mexico, and in Bermuda, the Bahamas, Japan, Taiwan,

Figure 12.3. Western poison oak, *Toxicodendron diversilobum.* Poison ivy, *T. radicans,* looks very similar. Photo by the author.

and western and central China. The older literature places all of these plants in the genus *Rhus*.

Contact of human skin with any part of these plants results in a delayed allergic contact dermatitis (Section 7.4.3). About two-thirds of the U.S. Caucasian population have been sensitized by previous contact and they "get poison oak," while the other third appear to possess an antigen-specific tolerance. The disease causes swelling and reddening of the affected area followed by itching and blisters, and crusting and scaling of the skin begin within a few days. This dermatitis is probably the most common cause of lost-time injury among the California workforce, hospitalization occasionally becomes necessary, and death can occur if smoke from the burning brush is inhaled.

While home "cures" such as yellow soap and potassium permanganate have been suggested, the best management seems to be administration of cortisone *after* symptoms appear. As the fat-soluble urushiols penetrate the skin within 10 minutes, washing with soap can remove any surface toxin not yet absorbed but seldom halts the toxic response in sensitized individuals. Over-the-counter preparations (petroleum emulsions) are available that literally solvent-extract unreacted urushiols out of the skin, but the best way to avoid the consequences of *Toxicodendron* poisoning is to avoid contact with the plants.

The 3-substituted catechols, such as **urushiol I** (Eq. 12.3), become oxidized to the corresponding *o*-quinones, and these react with skin proteins to form the antigens. Upon exposure to oxidants such as hydrogen peroxide or hypochlorite, or even air, urushiols polymerize to polyquinoid pigments, and contaminated clothing and tools quickly turn black when immersed in a little bleach.

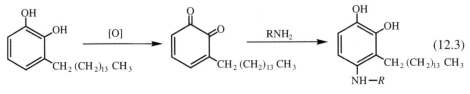

$$(12.3)$$

Urushiol I

Other plants of the Anacardiaceae contain urushiols, including cashew nut *(Anacardium occidentale),* and mango *(Mangifera indica),* and the shiny black lacquer that appears on some oriental bowls comes from the oxidized sap of the Asian sumac, *Rhus vernicipflua.* All are toxic.

12.5 | AMINO ACIDS, PEPTIDES, AND PROTEINS

Amino acids and peptides are the building blocks of proteins. Most natural amino acids have the amino and carboxyl groups bonded to the same asymmetric carbon atom (α-amino acids) in the L (and generally R) configuration. The protein amino acids show little mammalian toxicity, and even the most toxic, tryptophan, does not reveal its psychotoxicity below about 300 mg/kg. However, some very poisonous nonprotein amino acids occur in plants (Table 12.3), especially in the family Leguminoseae, as well as in fungi and bacteria.

Caribbean natives long have eaten fruit from a small tree they call *Akee (Blighia sapida)* that contains a very toxic amino acid, **hypoglycin A.** In undernourished individuals, especially children, the result can be violent vomiting, which may be followed by convulsions, coma, and often death (Liener, 1980). Mid-twentieth-century investigation showed the victims' blood to be extremely low in sugar (hypoglycemia), and there was significant loss of liver glycogen due to formation of a methylene-cyclopropylacetyl-CoA that blocked the long-chain fatty acid oxidation required for glucose biosynthesis (Tanaka et al., 1972). The vomiting is thought to be due to a buildup of branched fatty acids.

Another example is the poisoning seen in livestock that have eaten certain common legumes, such as milkvetch (*Astragalus* species), found in the Great Plains

TABLE 12.3
Some Toxic Amino Acids[a]

$$\underset{\underset{R-CH_2CH-COOH}{\quad}}{\overset{NH_2}{\mid}}$$

Amino acid	Source	R	Toxic action
α-Amino-β-(methylamino)-propionic acid	*Cycas cirialis* (cycad)	CH_3NH-	Neurotoxic
Azaserine	*Streptomyces* spp. (mold)	$N_2CHCOO-$	Fungicidal
β-Cyanoalanine	*Lathyrus* spp. (legume)	NC-	Neurotoxic
α,γ-Diaminobutyric acid	*Lathyrus* spp. (legume)	H_2NCH_2-	Neurotoxic
3,4-Dihydroxyphenylalanine	*Vicia fava* (legume)	HO, HO—⟨ ⟩—	Neurotoxic
Hypoglycin A	*Blighia sapida* (soapberry)	$CH_2=$◁	Hypoglycemic
Indospicine	*Indigofera spicata* (legume)	$H_2N-\overset{\overset{NH}{\parallel}}{C}CH_2(CH_2)_2-$	Hepatotoxic
Methylselenocysteine	*Astragalus spp.* (legume)	CH_3Se-	Neurotoxic
Mimosine	*Leucaena glauca* (legume)	HO, O=⟨ ⟩N-	Goitrogenic
Selenocystathionine	*Astragalus spp.* (legume)	CH_2SeCH_2-	Neurotoxic

[a]Liener, 1980.

states and Western deserts. These plants extract and concentrate natural selenium from the soil, sometimes to levels of thousands of ppm. They then pass it on to grazing animals in the form of Se-containing amino acids, such as selenocysteine and selenomethionine, in which S atoms have been replaced by Se. Incorporated into the victim's enzymes and other protein, these "unnatural" selenoamino acids produce severe disruption of biochemical processes.

Acute symptoms include appetite loss, labored breathing, coma, and death due to respiratory failure. Mild chronic exposure produces emaciation, lameness, and loss of hair and hoof, while greater exposure causes weakness, blindness, paralysis, and death (Kingsbury, 1964). Acute human poisoning has not been reported in the United States, but states such as Nebraska and South Dakota where grain is grown on selenium-rich soils report increased incidence of dermatitis, deteriorated teeth, and diseased nails (Liener, 1980). The chemistry of the selenoamino acids has not been reviewed since the work of Painter (1941).

Amino acids join through amide bonds to form peptides. Peptides with more than just a few amino acid units are referred to as polypeptides, and those with more than 20 units are called proteins. Many natural peptides are toxic, including important bacterial and fungal antibiotics (Kleinkauf and von Döhren, 1982). The toadstool poisons have both historical and practical significance and are typified by the amanitins and phalloidins from the Death's Cap, *Amanita phalloides* (Wieland, 1986). Common in Europe, Canada, and the United States, *A. phalloides* is easily mistaken for harmless edible mushrooms and so is responsible for over 90% of toadstool-related human deaths. It seems likely that this fungus, served at dinner by an ambitious wife, caused the death of Roman emperor Claudius in 54 A.D. The first symptoms, felt within a few hours, include nausea, vomiting, and diarrhea and are followed, after several days of recovery, by severe and often fatal damage to the liver and kidneys (Section 7.4.5). Contrary to folklore about a silver spoon turning dark, there is no simple test for *A. phalloides* toxins.

The ***phallotoxins*** are fast-acting seven-unit peptides with a mouse oral LD_{50} on the order of 2 mg/kg, and the slow-acting ***amatoxins*** are eight-unit peptides, mouse oral LD_{50} 0.3 mg/kg (Wieland,1968). The unusual bicyclic peptides (Fig. 12.4) were investigated thoroughly by the Wieland family starting in 1937, before modern spectrometric methods, and shown to contain the nonprotein amino acids L-*allo*hydroxyproline (Hyp) and L-dihydroxyisoleucine (Dhi), as well as a bridge in which the SH group of L-cysteine is attached at the 2-position of L-tryptophan. The structures have been confirmed by NMR and X-ray analysis (Wieland et al., 1983).

Among the many bioactive polypeptides of environmental interest, snake venoms hold a prominent place in the public's imagination. Most people hate snakes, and "poisonous" snakes in particular, although these animals are more correctly termed "venomous." A ***venom*** is a poison that is injected into the victim, as opposed to a poison taken internally or onto the skin. Many kinds of animals and a few plants produce venoms, among them spiders, bees and wasps, scorpions, and, of course, snakes. Snake venoms are complicated mixtures of neurotoxins, hydrolytic enzymes, and other chemicals intended to first immobilize the prey (or enemy) and then dissolve its tissues for easier digestion.

A   **B**

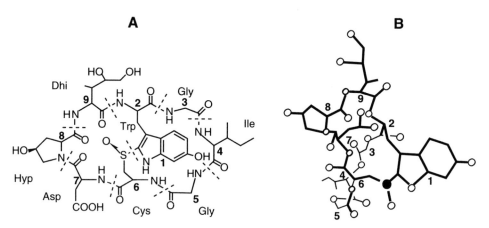

Figure 12.4. Structure of β-amanitin from *Amanita phalloides,* showing (A) amino acid composition and (B) steric configuration, S ●, O o, and N °. H atoms not shown. See Wieland et al., 1983.

Cobras belong to the family Elapidae that also includes the kraits and coral snakes. A lot is known about cobra venom. In addition to proteases, collagenase, phospholipases, and a powerful cardiotoxin, it contains a stable neurotoxin called **cobrotoxin** (Yang, 1965). Cobrotoxin from *Naja naja atra,* the Formosan cobra, contains 62 amino acid units (MW 6949), including 9 basic ones (arginine or lysine), but no alanine, phenylalanine, or methionine. The single peptide chain is cross-linked by 4 disulfide bonds (Fig. 12.5), and if any of these are modified by reduction or

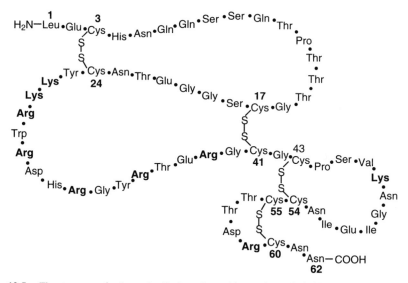

Figure 12.5. The structure of cobrotoxin. Basic amino acids are shown in bold type. Adapted from Yang and Chang, 1976.

oxidation, all toxicity is lost. Neurotoxins from the genus *Naja,* and indeed from the entire Elapidae, are very similar: All have cystine cross-links at positions 3–24, 17–41, 43–54, and 55–60; most have lysine at 27, 47, and 26 or 35; and most have arginine at 33 and either 36 or 39 (Table 12.5). Like the arrow poison curare, *Naja* toxins act post-synaptically to block nerve transmission at neuromuscular junctions by binding irreversibly to cholinergic receptors (Tu, 1977). They are exceptionally toxic, the subcutaneous LD_{50} of cobrotoxin in mice being only *65 ng/kg.*

Most bacterial toxins are true proteins. They cause serious human and animal diseases (Table 12.4), and a number of poisonous plants, such as castor bean *(Ricinus communis),* possess similar protein toxins. They are antigenic, have high molecular weights, and rely on configuration rather than unusual amino acid composition for toxicity. In recent years, the amino acid sequences of many protein toxins have been determined, usually by classical degradation reactions such as the cyanogen bromide and Edman thiohydantoin methods.

For example, Staphylococcal enterotoxin B, the cause of a nonfatal but very unpleasant type of food poisoning, is one of a number of related protein toxins produced by *Staphylococcus aureus.* It is a chain of 239 protein amino acids (MW 28,366), 33 of which are lysine but only 2 of which are cysteine (Huang and Bergdoll, 1970). The median emetic dose (the dose that makes a monkey vomit) is 5 μg/animal, so it is easy to see how the entire crowd at a family picnic could be felled by a single dish of unrefrigerated chicken salad. "Staph" food poisoning is the most common type in the United States.

While death from Staph poisoning is rare, the six protein toxins of *Clostridium botulinum* are among the most poisonous substances known. The disease they cause, known as botulism, has been recognized worldwide for centuries. Indeed, Emperor Leo VI, ruler of Byzantium from 886 to 911 A.D., outlawed blood sausage in his realm because it so often caused botulism. The toxins cause paralysis within 12–36 hours in humans who eat contaminated home-canned food such as olives, green beans, or home-prepared meats, and it also kills many kinds of domestic animals and migratory waterfowl. The disease usually is fatal in humans, with the 1281 known cases in the United States from 1899 to 1949 resulting in 833 deaths, but advances in diagnosis and treatment have reduced the mortality rate in recent years.

The crystalline Type A botulinus toxin, oral LD_{50} *1.5 ng/kg* in mice, was found to consist of an hemagglutinin (causing aggregation of red blood cells) and a neuro-

TABLE 12.4
Some Important Bacterial Toxins

Disease	Species	Toxin	Effect
Botulism	*Clostridium botulinum*	Type A	Neurotoxicity
Tetanus	*Clostridium tetani*	Tetanospasmin	Neurotoxicity
Diphtheria	*Corynebacterium diphtheriae*	Shick toxin	Dermonecrosis
Dysentery	*Shigella dysenteriae*	Hemorrhagin	Neurotoxicity
Food poisoning	*Staphylococcus aureus*	Enterotoxin A	Emesis
Cholera	*Vibrio cholerae*	Enterotoxin	Diarrhea
Tonsillitis	*Streptococcus pyogenes*	Streptolysin S	Hemolysis

TABLE 12.5
Partial Amino Acid Sequences in Elapidae Neurotoxins[a]

Name								Amino Acid Position										
	24	25	26	27	28	29	30	31	32	33	34	35	36	37	38	39	40	41
Formosan cobra (N. n. atra)[b]	Cys	Tyr	Lys	Lys	Arg	Trp	Arg	Asp	His	Arg	Gly	Tyr	Arg	Thr	Glu	Arg	Gly	Cys
Ringhals (N. haje haje)	Cys	Tyr	Lys	Lys	Arg	Trp	Arg	Asp	His	Arg	Gly	Ser	Ile	Thr	Glu	Arg	Gly	Cys
Yellow cobra (N. nivea)	Cys	Tyr	Lys	Lys	Arg	Trp	Arg	Asp	His	Arg	Gly	Thr	Ile	Ile	Glu	Arg	Gly	Cys
Black mamba (D. polylepis)	Cys	Tyr	Lys	Lys	Tyr	Trp	Arg	Asp	His	Arg	Gly	Thr	Ile	Ile	Glu	Arg	Gly	Cys
Spitting cobra (N. nigiricollis)	Cys	Tyr	Lys	Lys	Val	Trp	Arg	Asp	His	Arg	Gly	Thr	Ile	Ile	Glu	Arg	Gly	Cys
Black cobra (N. melanoleuca)	Cys	Tyr	Lys	Lys	Gln	Trp	Ser	Asp	His	Arg	Gly	Thr	Ile	Ile	Glu	Arg	Gly	Cys
Banded seasnake (H. cyanocinctus)	Cys	Tyr	Lys	Lys	Thr	Trp	Ser	Asp	His	Arg	Gly	Thr	Arg	Ile	Glu	Arg	Gly	Cys

[a]Ohsaka et al., 1976.
[b]Cobrotoxin.

toxin of MW ~158,000. However, the neurotoxin disaggregated in alkaline solution into smaller toxic units believed to contain the actual substance that prevents release of acetylcholine from cholinergic nerve endings (Boroff and DasGupta, 1971).

12.6 | LACTONE MYCOTOXINS

Mycotoxins are low-molecular-weight poisons produced by fungi. Many of them are lactones (cyclic esters), including tumorigenic penicillic acid found in corn and grain products, the vesicant and carcinogenic *patulin,* which occurs in apple juice, the anticoagulant *dicoumarol* from spoiled clover hay, and the extremely toxic *ochratoxin A* from moldy corn (Scheme 12.4).

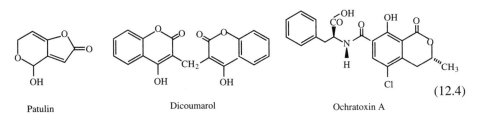

(12.4)

Patulin   Dicoumarol   Ochratoxin A

However, probably the most widely detected and best known are the **aflatoxins** produced by a mold, *Aspergillus flavus*. In 1961, over 100,000 turkeys died in England from an unknown disease eventually traced to feed containing moldy ground nuts (peanuts) imported from Africa. Chemical investigation revealed the presence of four related fungal toxins that could be separated by chromatography and detected by their intense blue or green fluorescence under UV light. They were named aflatoxins (from *A. flavus*) B_1, B_2, G_1, and G_2, and proved to be substituted furocoumarins, aromatic lactones possessing a pair of saturated furan rings (Fig. 12.6). They exhibited low acute oral LD_{50}s, for example, 5000 μg/kg for aflatoxin B_1 in the rat, 500 μg/kg in dogs and trout, and only 360 μg/kg in the duckling. Moreover, they were powerful hepatocarcinogens, producing liver tumors in rats at 75 μg/kg.

The carcinogenicity is due to *in vivo* formation of aflatoxin epoxides at the isolated 2,3-double bond that converts them into multifunctional alkylating agents capable of cross-linking strands of DNA. The aflatoxins are metabolized primarily by hydroxylation at the tertiary 4-position (to aflatoxin M_1), with subsequent sulfation and glucuronidation (Fig. 12.6). Although originally thought of as a detoxication product, the aflatoxin M congeners also turned out to be toxic and carcinogenic.

The aflatoxins have been found widely in human food as well as in animal feed. Sometimes the food becomes contaminated at the level of the primary producer, or sometimes later due to improper storage. Oily foods such as nuts are particularly prone to contamination, and a survey of peanut vending machines by the U. S. Public Health Service found that a majority served aflatoxin-containing nuts. One of the

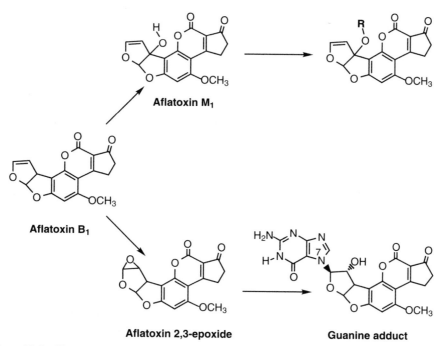

Figure 12.6.   The metabolism of aflatoxin B_1, R = sulfate/glucuronide.

favorite culture media for aflatoxin research has been moistened shredded wheat, which often did not even require further inoculation with fungus. Another survey found aflatoxin M_1 to be a frequent contaminant of milk and other dairy products, leading to more careful milk inspection programs. The widespread human liver cancer in tropical Africa, where food commonly is not refrigerated, has served as a wakeup call to developed nations that the same thing could happen to them in the absence of suitable precautions.

12.7 | PERSPECTIVE

To Lord Tennyson's 1850 description of "Nature, red in tooth and claw" we must now add "and rich in toxic chemicals." No synthetic pesticide can match the acute toxicity of many natural products, nor would it be allowed in our food if it did. Few people in the United States will ever be poisoned by a pesticide, but *everyone* will be poisoned sooner or later by a biotoxin from a toxic plant, animal, or microorganism. "Natural" is not necessarily synonymous with "good."

12.8 | REFERENCES

Boroff, D. A., and B. R. DasGupta. 1971. Botulinum toxin, in *Microbial Toxins* (S. Kadis, T. C. Montie, and S. J. Ajl, eds), Vol. IIA, Academic Press, New York, pp. 1–68.

Brower, L. P. 1969. Ecological chemistry. *Sci. Amer.* (2): 169–76.

Dumbacher, J. P., B. M. Beehler, T. F. Spande, H. M. Garraffo, and J. W. Daly. 1992. Homobatrachotoxin in the genus Pitohui: Chemical defense in birds? *Science* **258**: 799–801.

Eisner, T., and J. Meinwald. 1966. Defensive secretions of arthropods. *Science* **153**: 1341–50.

Fieser, L. F., and M. Fieser. 1959. *Steroids,* Reinhold, New York.

Fraenkel, G. S. 1959. The raison d'être of secondary plant substances. *Science* **129**: 1466–70.

Huang, I.-Y., and M. S. Bergdoll. 1970. The primary structure of staphylococcal enterotoxin B. III. The cyanogen bromide peptides of reduced and amino-ethylated enterotoxin B, and the complete amino acid sequence. *J. Biol. Chem.* **245**: 3518–25.

Kingsbury, J. M. 1964. *Poisonous Plants of the United States and Canada,* Prentice-Hall, Englewood Cliffs, NJ.

Kjaer, A. 1960. Naturally-derived isothiocyanates (mustard oils) and their parent glycosides. *Prog. Chem. Org. Nat. Products* **18**: 122–76.

Kleinkauf, H., and H. von Döhren. 1982. *Peptide Antibiotics: Biosynthesis and Function,* de Gruyter, Berlin, Germany.

Liener, I. E. 1980. *Toxic Constituents of Plant Foodstuffs,* 2nd Ed., Academic Press, New York.

Ohsaka, A., K. Hayashi, and Y. Sawai. 1976. *Animal, Plant, and Microbial Toxins. Vol. 1: Biochemistry,* Plenum Press, New York, NY.

Painter, E. P. 1941. The chemistry and toxicity of selenium compounds, with special reference to the selenium problem. *Chem. Rev.* **28**: 179–213.

Tanaka, K., K. Isselbacher, and V. Shih. 1972. Isovaleric and α-methylbutyric acidemias induced by hypoglycin A: Mechanism of Jamaican vomiting sickness. *Science* **175**: 69–71.

Tu, A.T. 1977. *Venoms: Chemistry and Molecular Biology,* John Wiley & Sons, New York.

VanEtten, C. H., M. E. Daxenbichler, and I. A. Wolff. 1969. Natural glucosinolates (thioglucosides) in foods and feeds. *J. Agr. Food. Chem.* **17**: 483–91.

Whittaker, R. H., and P. P. Feeny. 1971. Allelochemics: Chemical interactions between species. *Science* **171**: 757–70.

WHO. 1988. *Pyrrolizidine Alkaloids,* World Health Organization, Geneva, Switzerland Environ. Health Criteria 80.

Wieland, T. 1968. Poisonous principles of mushrooms of the genus *Amanita. Science* **159**: 946–52.

Wieland, T. 1986. *Peptides of Poisonous Amanita Mushrooms.* Springer Verlag, New York, NY.

Wieland, T., C. Götzendörfer, J. Dabrowski, W. N. Lipscomb, and G. Shoham. 1983. Unexpected similarity of the structures of the weakly toxic amanitin (S)-sulfoxide and the highly toxic (R)-sulfoxide and sulfone as revealed by proton nuclear magnetic resonance and x-ray analysis. *Biochemistry* **22**: 1264–71.

Yang, C. C. 1965. Crystallization and properties of cobrotoxin from Formosan cobra venom. *J. Biol. Chem.* **240**: 1616–18.

Yang, C. C., and C. C. Chang. 1976. Cationic groups and biological activity of cobrotoxin, in *Animal, Plant, and Microbial Toxins* (A. Ohsaka, K. Hyashi, and S. Sawai, eds.), Vol. 1: Biochemistry, Plenum Press, New York, NY.

Special Topic 12. Allelochemicals

Homo sapiens is not the only species that makes chemicals to control "pests." Many plants, animals, and microorganisms produce natural ***allelochemicals*** that affect their competition and interactions by altering the behavior, growth, health, or population dynamics of their neighbors. These substances may take the form of ***allomones*** which give adaptive advantage to the producer, or ***kairomones*** which give the advantage to the recipient. We do not include here the ***pheromones*** (*e.g.*, insect sex attractants), which transmit chemical signals within the same species and generally have little toxicological significance by themselves. Old but useful reviews of allelochemicals include those of Fraenkel (1959) and Whittaker and Feeney (1971). Chemical structures of a few key examples are shown in Fig. 12.7 if they have not appeared elsewhere in the book.

Microorganisms are continually at war with one another. Their principal weapons are a wide variety of antimicrobial chemicals, many of which have been adopted by humans for their own use as ***antibiotics***. For at least 3000 years, people applied moldy bread or spoiled bean curd to wounds to combat infections, but it remained for Sir Arthur Fleming to isolate the responsible agent, **penicillin,** from the bread mold *Penicillium notatum* in 1929 and initiate a worldwide search for new antibiotics. Penicillin suppresses the growth of a broad range of pathogenic bacteria that cause diseases in humans and domestic animals, generally without harming the host. It does so by preventing biosynthesis of unique bacterial cell wall components, causing exposed protoplasm literally to explode (Section 9.6.1). The besieged bacteria defend themselves with a hydrolytic enzyme, penicillinase, that rapidly degrades the antibiotic, the most efficient producers generating eventual genetic resistance in the intended victims.

Bacteria and fungi also attack higher plants with chemicals. They are responsible for the chestnut blight, Dutch elm disease, and pine blister rust that have destroyed so many of our nation's trees, as well as devastating losses in wheat, potatoes, and other food crops. Most homeowners who grow their own tomatoes have watched in dismay as their well-watered

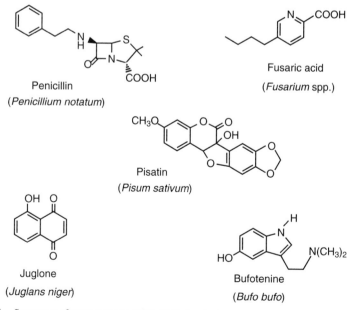

Penicillin
(*Penicillium notatum*)

Fusaric acid
(*Fusarium* spp.)

Pisatin
(*Pisum sativum*)

Juglone
(*Juglans niger*)

Bufotenine
(*Bufo bufo*)

Figure 12.7. Structures of some representative allelochemicals.

plants suddenly wilted and died due to a biotoxin called **fusaric acid** produced by the fungus *Fusarium.* Many commercial tomato plants are now bred to be "wilt-resistant," as most plants possess an immune system that can be genetically improved. However, unlike our own system of protein antibodies, that of plants is based on an ability rapidly to generate relatively simple antibiotics known as ***phytoalexins.*** Although the principal resistance factor of the tomato probably is the solanine relative, tomatine, many other plants produce phenolics. For example, upon an attack by fungi, pea plants quickly produce the fungicidal phytoalexin **pisatin.**

Plants use natural toxic chemicals to protect their interests against competition from other plants, a phenomenon called ***allelopathy.*** For example, farmers long ago observed that crops such as tomatoes and alfalfa would not grow near black walnut trees *(Juglans niger).* This proved to be due to the tree's production of a natural herbicide, **juglone** (5-hydroxy-1,4-naphthoquinone), which occurs in the roots, leaves, and fruit hulls as the naphthohydroquinone glucoside until released and air-oxidized to toxic form. Most healthy plants also resist insects by formation of natural insecticides such as nicotine and solanine, or more often by chemical repellents and feeding inhibitors. The cyanogenic glycosides found in seeds are an example.

Insects provide some of the most thoroughly studied examples of the use of chemicals for defense and aggression (Eisner and Meinwald, 1966). Some defense systems are simple, such as the highly distasteful formic acid produced by formicine ants. Others are complex: Bombardier beetles (*Brachinus* species) store a mixture of hydroquinone, methylhydroquinone, and hydrogen peroxide in a gland on the abdomen, and a mixture of peroxidase and catalase enzymes in an adjacent gland. When triggered, the gland secretions mix, and a caustic mixture of benzoquinones is ejected with explosive force at temperatures near 100°C directly onto the attacker. In another instance, the monarch butterflies (*Daneus* species) protect their young by laying eggs only on leaves of toxic milkweeds (*Asclepius* species). Although the caterpillars that hatch and feed on the leaves remain unaffected, birds learn to avoid them or risk poisoning by the cardenolides accumulated by the larvae (Brower, 1969).

Vertebrates, too, have evolved chemical defenses in addition to venoms. Toads (*Bufo* species) secrete the toxic and psychoactive indole, **bufotenine,** from glands on their skin, while New Guinea songbirds of the genus *Pitohui* accumulate a steroid alkaloid similar to the batrachotoxin of the Colombian arrow frog (Dumbacher et al., 1992). Of course, the well-known skunk (genus *Mephitis*), ordinarily a clean and odorless little animal, can warn and then douse an attacker with a horrendously smelly and persistent mixture of aliphatic sulfur compounds. Chemical warfare is common among the world's organisms, and it is not surprising that humans have adapted some allelochemicals for their own use, including antibiotics, pesticides, and even the arrow poisons from which the term *toxic* derives. Many other products useful to humans await development from natural allelochemicals and will become an intimate part of our future lives.

C H A P T E R

Industrial **13**
Chemicals

13.1 | INDUSTRIAL CHEMICALS

13.1.1. Minerals

Most pollutants are escaped industrial chemicals. They normally are either synthetic intermediates on their way to becoming consumer products or the natural raw materials from which the synthetics are made. Almost all articles of commerce, including plastics, pharmaceuticals, detergents, lubricants, paper products, fabrics, and paints are made today from synthetic chemicals or require synthetics or chemical processing for their production. Raw materials are all extracted from Nature: Animals provide soap, glue, wool, and leather; plants produce sugar, cellulose, and rubber; and important minerals include water, air, coal, oil, and mineral ores.

Minerals hold both toxicological and economic interest (Table 13.1). As the source of the world's metals, many ores (Table 13.2) are toxic and were mentioned in Chapter 11. Minerals are used directly or are converted to chemical intermediates that form the basis of industrial chemistry. For example, "phosphate rock," composed of the calcium phosphates phosphorite and apatite, is converted to phosphoric acid for synthetic fertilizer as well as to elemental phosphorus that is made into pesticides and many other products. Enough sulfur is mined each year in the United States to fill 400 football stadiums, 90% of it converted to sulfuric acid and the rest to agrochemicals, pulp and paper products, and specialty chemicals. Salt generates chlorine, sodium hydroxide, and sodium carbonate, and, while not "mined," air is liquefied to produce nitrogen, which in turn is made into ammonia. The importance of these raw materials is reflected in the list of the principal chemicals manufactured in this country (Appendix 13.1). Eleven of the top 15 are inorganic.

Coal, petroleum ("rock oil"), and natural gas are complex mixtures of organic compounds and are today's most important minerals for the manufacture of consumer products, although most of the world's production is burned as fuel rather than made into chemicals. At one time, coal tar, the distillate from making coke (carbon), was

TABLE 13.1
U.S. Production of Basic Minerals, 1993[a]

Mineral	Production (Million kg)		Major Uses
	1995	1992	
Bromine	212	177	Bromoalkanes
Calcium oxide[b]	18,720	15,866	Steel (flux)
Chlorine	11,390	10,675	Plastics, disinfection
Nitrogen	30,890	27,472	Ammonia
Phosphate rock	43,537	46,440	Phosphoric acid
Potash[c]	1,490	1,716	Fertilizer
Sodium carbonate[d]	10,113	9,024	Glass, cleaning
Sodium chloride[e]	39,694[f]	36,794	Chlorine, roads
Sulfur	12,011	11,585	Sulfuric acid

[a] Anon., 1996.
[b] From limestone and dolomite.
[c] KCl, K_2SO_4, and $K_2Mg(SO_4)_2$ (K_2O equivalent).
[d] Natural only.
[e] Kostick, 1994.
[f] Data for 1993.

the principal source of organic chemicals, but most of these now come from natural gas or petroleum.

13.1.2. Petroleum

Petroleum is a mixure of organic and organo-metallic compounds (Table 13.3). The *paraffins* consist primarily of C_1–C_{34} *n*-alkanes, the most volatile of which (methane to butane) constitute "natural gas." *Naphthenes* are alicyclic hydrocarbons, mostly cyclopentane and cyclohexane and their derivatives, while the *aromatics* represent unsaturated ring compounds ranging from benzene (b.p. 80°C) to polycyclic aromatic hydrocarbons (PAH) such as the seven-ringed coronene (b.p. 525°C). Most petroleum also contains nonhydrocarbon components, Prudhoe Bay crude oil typically over 3% of organosulfur compounds like benzothiophene and dibenzothiophene as well as phenols and heterocyclic nitrogenous bases like quinoline. Coal tar contains similar aromatic constituents but in different proportions.

TABLE 13.2
Some Metallic Ores

Metal	Mineral	Form
Be	Beryl	$Be_3Al_2(SiO_3)_6$
Cd/Zn	Zincblende	$(Cd,Zn)S$
Cu	Chalcopyrite	$Cu_2Fe_2S_4$
	Cuprite	Cu_2O
	Malachite	$Cu_2(OH)_2CO_3$
Cr	Chromite	$Fe(CrO_2)_2$
Hg	Cinnabar	HgS
Ni	Pentlandite	$(Ni,Cu,Fe)S$
Pb	Galena	PbS
Sn	Cassiterite	SnO_2

TABLE 13.3
Composition of Petroleum[a]

Component	Prudhoe Bay	Louisiana	Kuwait	#2 Fuel	Bunker C Oil
%S	0.94	0.25	2.44	0.32	1.46
%N	0.23	0.69	0.14	0.02	0.94
Ni, ppm	10.	2.2	7.7	0.5	89
V, ppm	20.	1.9	28.	1.5	73
Naphtha (20–205°C), %	**23.2**	**18.6**	**22.7**		
Paraffins $<C_{11}$	12.5	8.8	16.2		
Naphthenes	7.4	7.7	4.1		
Aromatics C_6–C_{11}	3.2	2.1	2.4		
High-boiling (> 205°C), %	**76.8**	**81.4**	**77.3**	**100**	**100**
Paraffins C_{11}–C_{32}	5.8	5.2	4.7	8.1	1.7
Isoparaffin	—	14.0	13.2	22.3	5.0
Naphthenes	28.5	37.7	16.2	31.4	15.2
Aromatics $>C_{11}$	25.0	16.5	21.9	38.2	34.2
1–2 Rings	16.9	10.5	9.7	28.1	8.6
PAH	8.1	6.0	12.2	10.1	25.6
N,S,O compounds, %	2.9	8.4	17.9	0.0	30
Pentane-insoluble, %	1.2	0.2	3.5	0.0	14.4

[a]NRC, 1985.

In order to be useful, crude oil must be converted to simpler mixtures and individual chemical constituents. This is accomplished by fractional distillation (refining) to give, at progressively higher temperatures, gasoline, jet fuel, kerosene, heating oil, diesel oil, lube oils, and bitumen (asphalt). Compounds of sulfur and nitrogen generally are removed somewhere in the refining process, although some still remain in the high-boiling fractions (Table 13.3).

Crude oil is black and sticky but not particularly toxic acutely to vertebrates (but see Section 13.4.2). Its well-publicized effects on oiled seabirds are due largely to loss of waterproofing (NRC, 1985). Likewise, humans are not normally affected, except in the unlikely event that they swallow a sizable quantity of the viscous, smelly stuff, although some skin phototoxicity is known. The low toxicity does not extend to aquatic invertebrates, which are poisoned by the water-soluble components that leach into their environment. The earlier life stages are especially susceptible, and aromatic and phenolic compounds appear to be the most toxic. Much of the toxicity can be related to the PAH content, as discussed in Section 13.4.

13.2 | PETROCHEMICALS

13.2.1. Sources

Most of the organic compounds listed in Appendix 13.1 are derived from petroleum fractions, and actually from only a few basic hydrocarbons including methane, ethane, propane, benzene, toluene, and xylene (Fig. 13.1). In order to maximize the amounts of these products, less profitable petroleum fractions are subjected to catalytic *reforming* to give gaseous alkanes and the more reactive olefins, which again are separated by fractional distillation. Staggering amounts of ethylene (39 billion pounds) and propylene (22 billion pounds) are produced in the United States each year (Appendix 13.1) and then converted to intermediates on the way to familiar plastics, synthetic fibers, solvents, fertilizers, and other useful products.

Despite the low mammalian toxicity of most petroleum fractions, many of the chemical intermediates derived from them show pronounced toxicity (indicated by ●

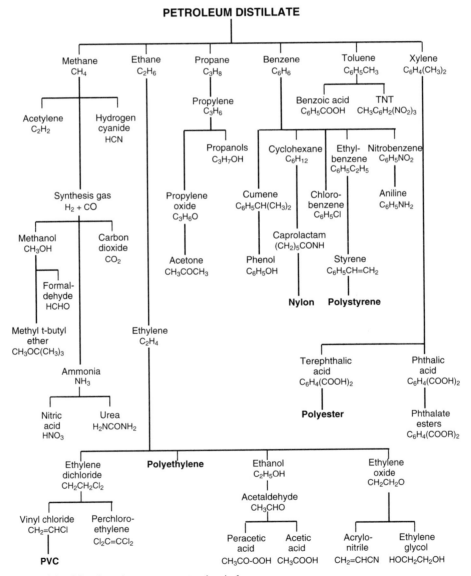

Figure 13.1. Manufacturing routes to petrochemicals.

in Appendix 13.1). Fortunately, people seldom come into contact with them except by accident, as they are immediately consumed in making other products. A few do receive extensive public use, sometimes with serious consequences, as the following examples show.

13.2.2. Major Petrochemicals

"We only stagger, we never fall; we sober up on wood alcohol." If these words of the college drinking song were to be taken literally, the results would probably be fatal. **Methanol,** the "wood alcohol" originally produced by destructive distillation of wood products, has been responsible for much human suffering and death, especially when access to grain alcohol was restricted. The toxic nature of methanol is hardly a secret: Methyl alcohol means "poison distillate" in ancient Arabic, and swallowing as little as one ounce can lead to blindness and death.

Ingestion of methanol by primates, or comparable respiratory or dermal exposure, leads progressively to inebriation, acidosis, and CNS damage, but only the first is observed in laboratory rodents. These effects are explained by the metabolism of methanol to formaldehyde, formic acid, and finally to carbon dioxide:

$$CH_3OH \xrightarrow[\text{or -2H}]{[O]} HCH{=}O \xrightarrow[\text{or -2H}]{[O]} HCOOH \xrightarrow{[O]} CO_2 \qquad (13.1)$$

Methanol is metabolized by alcohol dehydrogenase or by oxidation with catalase. Rodents rapidly carry out the oxidation to CO_2 and suffer only inebriation, but the slower dehydrogenation used by primates allows a buildup of the formic acid that results in the other effects. In methanol poisoning epidemics, about one-third of the victims recover, another one-third are blinded, and the remaining one-third die.

Methanol is manufactured by hydrogenation of carbon monoxide as well as by direct catalytic oxidation of methane. It has long seen use as a solvent, as fuel for portable stoves, and as a gasoline additive. However, its proposed use as an alternative motor fuel demands a lot of additional knowledge about possible exposures and effects in relation to dispensing, spills, and the generation of formaldehyde and formic acid in engine exhaust.

A major use for methanol is the manufacture of methyl *tert*-butyl ether (MTBE), a fuel additive that has become the organic chemical third largest in production volume (Appendix 13.1). MTBE initially was added to replace toxic tetraethyl lead as an octane booster, but discovery that it also reduced smog-producing engine emissions led it now to constitute about 11% of every gallon of gasoline sold in California. It is easily prepared by addition of methanol to isobutene, low-boiling (51°C), and soluble in water (48 g/L).

Although a vapor pressure of 245 torr at 25°C indicates rapid volatilization, its solubility tends to hold it in water (H' 0.024), and it has been found to accumulate in reservoirs and provide an unpleasant taste to the drinking water. The principal source now is thought to be from the exhaust gases of boats and jetskis, which emit as much as 60% of their fuel unburned. The additive aimed at reducing air pollution has turned out to be a major polluter of water! MTBE so far does not appear to be particularly toxic to people, but its effect on aquatic organisms remains unknown.

Formaldehyde is also a major industrial chemical and is produced by the catalytic air-oxidation of methanol. Most biology students recall the pungent odor of the formaldehyde in which biological specimens were preserved. Its principal uses are in plastics, wallboard, foam products, and building insulation, and most of the U.S. population today is continuously exposed to formaldehyde vapor from these sources. Sufficient exposure to the vapor is known to cause allergies in humans and nasal cancer in laboratory animals.

For thousands of years a product of the distillation of fermented grain, **ethanol** ("grain alcohol") is now made primarily by catalytic hydration of ethylene. It is available to the public in the form of "denatured alcohol" (solvents made unfit for drinking by unpleasant additives such as methanol or benzene), fortified alcoholic beverages such as brandy, a diluent in medicines, perfumes, and colognes, and often mixed with gasoline as automobile fuel. Thus, drinkers or not, most people are exposed to small amounts of ethanol every day. Its toxic effects were discussed in Section 7.4.2.

Ethylene oxide is a major industrial intermediate that is mostly converted to other chemicals such as ethylene glycol, ethanolamines, and detergents. Public contact at one time was through its use in the sterilization of medical and dental equipment, disinfection of packaged foods, and as a preservative in libraries and museums. It is especially reactive toward any nucleophiles it contacts, which has led to its withdrawal from most public access. For example, injection of ethylene oxide into packages of dried fruit to reduce spoilage was found to destroy the vitamin niacin (nicotinic acid) by reaction with the pyridine ring, and to lead to reaction with chloride ion in the fruit to form toxic 2-chloroethanol:

$$\underset{CH_2-CH_2}{\overset{O}{\frown}} + Cl^- \xrightarrow{\text{H}^+} HOCH_2CH_2Cl \tag{13.2}$$

Understandably, ethylene oxide is both toxic and carcinogenic.

Ethylene glycol is made by hydrolysis of ethylene oxide. It is used in polyester fibers and plastics, cosmetics, ink, and brake and hydraulic fluid, but the main source of human and animal exposure is antifreeze. "Glycol" is metabolized in mammals by the same route as are methanol and ethanol, and the metabolites again are more toxic than the parent. The end product before carbon dioxide is oxalic acid, which accumulates in the kidneys in the form of insoluble and needle-sharp crystals of calcium oxalate:

$$HOCH_2CH_2OH \longrightarrow HOCH_2CH=O \longrightarrow HOOC-COOH \tag{13.3}$$

Ethylene glycol   Oxalic acid

Ethylene glycol tastes sweet and so is attractive to animals and children. Dogs and cats are frequent victims, and a rare California condor released into the wild was found to have died by drinking waste antifreeze.

Chlorinated hydrocarbons. **Ethylene dichloride** (1,2-dichloroethane, EDC), ranked thirteenth in chemical production figures for 1995, was widely used at one time as a solvent and soil fumigant. Now, it serves primarily as the parent of a large family of useful chlorinated alkanes and alkenes (Fig. 13.2), which are low-boiling, highly reactive liquids that are hepato- and nephrotoxic and expected to be carcinogenic.

High-temperature chlorinolysis of EDC provides a chlorinated hydrocarbon mixture that is resolved by fractional distillation. One component, **chloroform** (trichloromethane), is familiar to readers of murder mysteries. It is indeed a narcotic at high concentrations and was used as an anesthetic for many decades until abandoned in the early 1900s because of its low margin of safety. Although now considered carcinogenic, it was widely used in cough medicine for years. Today, the principal route of exposure for most people is through drinking water, where it is formed by the chlorination of dissolved natural organic matter (Appendix 2.3). Another component, **carbon tetrachloride,** was used for decades as a home drycleaner and fire extinguisher, but now the main use is in manufacture of fluorinated monomers such as the precursor of Teflon®. The toxic damage from both compounds is due to formation of reactive metabolites (Eqs. 13.4–13.7), especially the highly reactive free radicals.

$$CHCl_3 \longrightarrow \bullet CCl_3 + \bullet H \tag{13.4}$$

$$CCl_4 \longrightarrow \bullet CCl_3 + \bullet Cl$$

$$\bullet CCl_3 + O_2 \longrightarrow Cl_3C-OO\bullet \tag{13.5}$$

$$\bullet CCl_3 + \bullet OH \longrightarrow Cl_3C-OH \longrightarrow O=CCl_2 \tag{13.6}$$

$$CCl_4 \longrightarrow Cl_2C: \quad \text{Dichlorocarbene} \tag{13.7}$$

Tetrachloroethylene (perchloroethylene, PCE) and **trichloroethylene** (TCE) are important cleaning and degreasing solvents. They become reactive in light and air (Eq. 13.8) but remain quite stable in the dark in ground water. This has led to extensive cleanup problems, for example, where PCE waste from drycleaning establishments or TCE from aircraft engine degreasing was allowed to seep into soil over many years. 1,1-Dichloroethylene (**vinylidene chloride**) and monochloroethylene (**vinyl chloride**) are polymerized to plastics used in packaging materials (*e.g.*, Saran) and PVC hard goods such as pipe and phonograph records. The damage done to liver and kidneys by these chloro-olefins, as well as the reported carcinogenicity, are due to reactive epoxide metabolites:

$$Cl_2C=CCl_2 \xrightarrow{[O]} Cl_2C\overset{O}{\overset{/\,\backslash}{-}}CCl_2 \longrightarrow Cl_3C\overset{O}{\overset{\parallel}{-}}CCl \xrightarrow{H_2O} Cl_3C-COOH \tag{13.8}$$

PCE   TCA

Most people's exposure to chloro-olefins is low but continuous, as the chemicals have become widespread contaminants of drinking water and air. "New car odor" is partly due to vinyl chloride released by plastic, vinyl chloride and vinylidene chloride leach out of PVC water pipes and plastic food wrap, and butter and other fatty foods are known to accumulate PCE from nearby drycleaning operations. Liver and kidney damage as well as cancer are reported in workers exposed to chloroethylene vapor

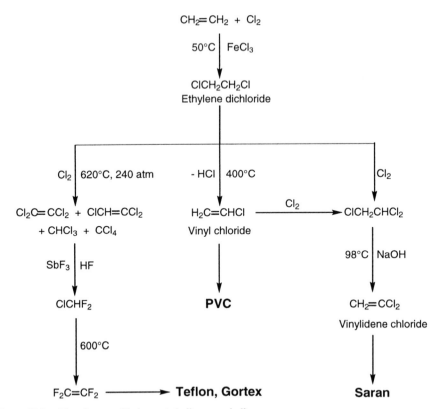

Figure 13.2. Manufacture of halogenated alkanes and alkenes.

on the job, but the consequences of the long-term, low-level inhalation exposure of the rest of us remains uncertain. Further, these substances are among the most serious threats to the earth's protective ozone layer (Special Topic 5).

Although **phenol** originally was manufactured by heating chlorobenzene with alkali or by extraction from coal tar, most of it now is made by oxidation of the cumene (isopropylbenzene) produced from benzene and propylene with an acid catalyst. For decades, the usual exposure to phenol was through antiseptics such as Listerine®, still in wide use. As implied by the old name, carbolic acid, it is caustic and causes painful burns on the skin. It also is poisonous, a lethal oral dose for most adult humans being about 15 grams, although as little as 1 gram has been known to be fatal. Chronic exposure results in damage to liver and kidneys. Most of the 4 billion pound annual production of phenol goes into the manufacture of plastics and resins or the disinfectant chlorophenols.

The three isomeric **xylenes** (dimethylbenzenes) also are separated by fractional distillation. p-Xylene is then oxidized to terephthalic acid (benzene-1,4-dicarboxylic acid), which is esterified with ethylene glycol to make a polyesters such as Dacron®. The o-xylene likewise is oxidized to phthalic acid (benzene-1,2-dicarboxylic acid), although this acid also is made by catalytic oxidation of naphthalene. While much of the phthalic acid is used for making alkyd resins, a significant proportion goes into the **phthalate esters** used as plasticizers (see Section 14.6).

Each industrial chemical has its own story, and each has become an integral part of our lives, whether we know it or not. They are all useful in some way, or they would not be made, but their toxicological implications for workers, wildlife, and often for everyone else are only now being explored. So far, people have just enjoyed the benefits and generally accepted the risks. But there is more to it; read on.

13.3 | TOXIC BYPRODUCTS AND CONVERSION PRODUCTS

13.3.1. Origins

Anyone who has studied chemistry knows that the "yield" of a chemical reaction seldom is 100% of that theoretically possible. There generally seem to be side reactions and byproducts. If you expected to get 10 grams of product but could only turn in 9 grams, you felt you were doing well with a 90% yield, the waste went into a bottle, and someone took it away. But what if you were expecting 100 million pounds and only got 90 million? People should be interested in the missing 10 million pounds and where it had gone. Efficient as it has to be to be profitable, chemical manufacture has the same problem as your lab preparation—a million pounds of waste here, another million there, and soon one is talking about real pollution potential. We have discussed the 700 billion pounds of beneficial chemical products made in the United States every year; now we will examine examples of the part not usually accounted for.

13.3.2. Polycyclic Aromatic Hydrocarbons (PAH)

Also called polynuclear aromatic hydrocarbons, this class of oil and coal constituents led to the first recognition in 1775 that a chemical substance (soot) could cause cancer. Many of these high-boiling, fat-soluble compounds are markedly car-

TABLE 13.4
Major Sources of PAH (ppm)

PAH	Fresh Oil[a]	Used Oil[a]	Road Asphalt[b]	Asphalt Paint[b]	Roof Tar[b]	Creosote[c]
Anthracene	—	—	0.32	0.07	3,400[d]	20,000
Phenanthrene	—	—	7.3	2.4		210,000
Fluoranthene	0.07	50.0	0.72	0.46	28,000	100,000
Pyrene	0.30	135.0	1.5	0.29	20,000	85,000
Benzo[a]pyrene	0.06	23.0	1.8	0.48	12,500	—
Chrysene	0.70	40.0	1.5	0.80	10,000	30,000

[a]IARC, 1984.
[b]IARC, 1985.
[c]Lorenz and Gjovik, 1972.
[d]Combined with phenanthrene.

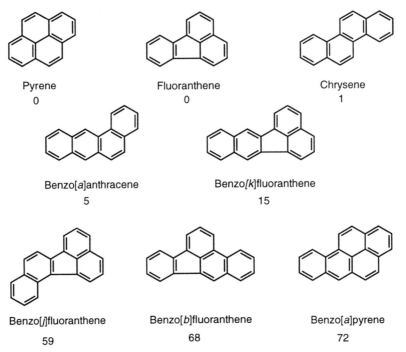

Figure 13.3. Structure and relative carcinogenicity of some PAH congeners. Numbers signify Iball Index, a ratio of skin tumor incidence (%) to the latency period (days) times 100 (Woo and Arcos, 1981).

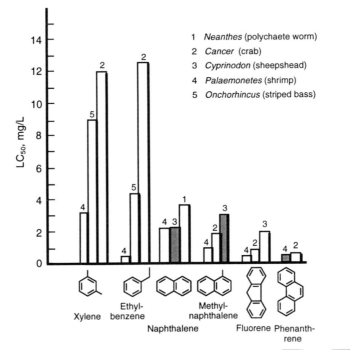

Figure 13.4. Toxicity of aromatic hydrocarbons to aquatic animals. LC$_{50}$, 96 h ▢, 24 h ▨. Adapted from NRC, 1985.

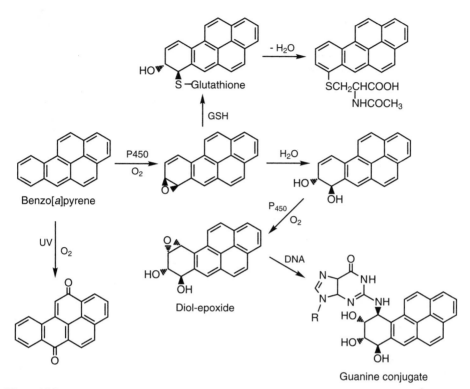

Figure 13.5. Metabolism and photo-oxidation of benzo[a]pyrene. See Dipple et al., 1984. GSH = glutathione, R = DNA or RNA.

cinogenic (Fig. 13.3), in humans as well as in other animals, and they are responsible for the high incidence of cancer among shale oil workers and roofers. They are widespread in the environment, including tobacco smoke, smoked foods, roasted coffee, charbroiled meat, and automobile exhaust, in fact anywhere organic matter is heated to a high temperature with restricted oxygen (Woo and Arcos, 1981).

These are the conditions under which asphalt, tar, and creosote are formed, and they, along with used motor oil, represent the most concentrated input of PAH to the environment (Table 13.4). The PAH congeners move into air, water, soil, and biota; for example, PAH found on the surface of orchard fruit was traced to a nearby asphalt road. U.S tap water contains up to 24 ng/L total PAH, urban runoff up to 10 μg/L of individual congeners, and 15–62 *mg/kg* of highly carcinogenic benzo[b]phenanthrene was found in urban soil (USPHS, 1990b).

Aquatic animals are very sensitive to some PAH congeners (Fig. 13.4), and over a wide range of aquatic species, toxicity increases as molecular weight and K_{ow} increase. These compounds are not acutely toxic to mammals, but both crude oil and coal tar can cause severe dermatitis in humans, especially if the contaminated skin subsequently is exposed to sunlight. This interesting phenomenon, called phototoxicity (Section 7.4.3), is readily observed in fish and aquatic invertebrates exposed to ppb levels of various PAHs in the water. The immediate effect is on respiration, as fish quickly rise gasping to the surface. Phototoxicity is thought to be due to the absorption of ultraviolet energy by the PAH followed by transfer of the energy to molecular O_2 with generation of singlet molecular oxygen, which then destroys gill or skin membranes (Larson, 1988). This discovery has important ecological implications, as oil spills into both fresh and salt water are frequent and widespread (Special Topic 6).

Fortunately, most PAH congeners are rapidly metabolized in mammals and other animals. Initial oxidation takes place on the A-ring to form an epoxide, which then can generate nontoxic diols and mercapturates (Fig. 13.5). Unfortunately, the oxidation may continue and produce a diol–epoxide, which reacts with nucleic acid purines such as guanine to initiate DNA damage and possible cancer (Dipple et al., 1984). Photooxidation also provides a route to the environmental destruction of PAH (Fig. 13.5), whereas microbial degradation often is slow (Special Topic 6).

13.3.3. Hex Waste

The production of chlorinated olefins (Fig. 13.2) always results in accumulation of a high-boiling residue composed of hexachlorobenzene, hexachlorobutadiene, octachlorostyrene, and other chlorinated compounds, termed "hex waste." Such very stable substances represent repeating polychloroethylene units (Eq. 13.9), condensed in the high-temperature chlorination of ethylene, although they also can be formed wherever carbon, chlorine, and heat come together, such as in the production of chlorine by the electrolysis of brine with carbon electrodes.

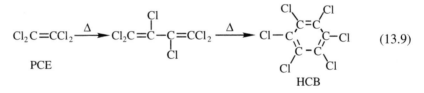

$$ (13.9) $$

Hexachlorobenzene now is detected everywhere in the environment. Chlorinated solvent production releases some 12,000 kg of it every year, pesticide manufacture has been responsible for another 13,000 kg, and releases from the production of chlorine and magnesium remain unaccounted (USPHS, 1990a). Introduced into the environment through waste disposal, HCB rapidly becomes redistributed. Its strong soil adsorption (log K_{oc} 6.08) assures that a large proportion will remain associated with soil and sediment, while its low water solubility and relatively high vapor pressure (Appendix 3.1) indicate rapid transfer from surface waters into the atmosphere ($H' = 0.028$).

HCB is very stable ($t_{1/2}$ 3–6 years in soil), and its log K_{ow} of 6.18 indicates extreme lipophilicity. In mammals, it produces porphyria (discolored skin) and damage to the liver and immune system. It is readily transferred into breast milk from food and so presents a particular hazard to the newborn (USPHS, 1990a). Almost everyone, no matter how remote, carries HCB in their fat, and control over its generation and disposal seems to be the only way to reduce the releases.

TABLE 13.5
Chlorinated Dioxins in Technical PCP (ppm)[a]

Sample	Source	Cl_4	Cl_5	Cl_6	Cl_7	Cl_8	PCDF[b]
PCP	[c]	<0.01	<0.03	<0.03	1.0	3.2	3.15
PCP	B	<0.01	<0.03	0.15	1.1	5.5	1.48
PCP	B	<0.02	<0.03	5.4	130	370	134.4
PCP	C	<0.02	<0.03	5.2	95	280	458.4
PCP	D	<0.02	<0.03	4.2	54	210	323.1
PCP	[d]	<0.02	<0.03	9.1	180	280	566.3
PCP-Na[e]	A	0.08	0.03	0.25	2.8	5.1	25.8
PCP-Na	C	0.06	0.03	0.40	4.2	11	8.80
PCP-Na	[d]	0.05	<0.03	3.40	40	115	85.1

[a] Adapted from Buser and Bosshardt, 1976.
[b] Total polychlorinated dibenzofurans, ppm.
[c] Analytical standard.
[d] Laboratory supplier.
[e] Sodium pentachlorophenate.

13.3.4. Chlorinated Dioxins

The world was shocked by the 1970s revelation of a highly toxic impurity, 2,3,7,8-tetrachlorodibenzo-*p*–dioxin **(TCDD)**, in a herbicide mixture called "Agent Orange" sprayed on jungle during the war in Vietnam. Actually, manufacturers had been aware of the toxicity of dioxins since 1950, as a persistent skin rash called chloracne was observed after industrial accidents involving 2,4,5-trichlorophenol, a common disinfectant and source of the widely used herbicide 2,4,5-T. However, it took a highly publicized 1976 explosion in a trichlorophenol disinfectant factory in Seveso, Italy, for the widespread occurrence and health implications of dioxins to be recognized.

Dioxins are generated when *o*-chlorophenols or their alkali metal salts are heated to a high temperature (Eq. 13.10). That was what happened when superheated steam instead of cooling water was accidentally delivered to a reactor at the ICMESA cosmetics factory in Seveso, resulting in an exothermic reaction that released a cloud of sodium trichlorophenate and TCDD from a vent pipe. Intermediate phenoxyphenols, called **predioxins**, are formed first and have been detected widely in water. They readily form the corresponding dioxins upon heating.

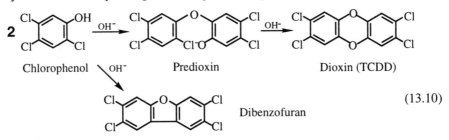

(13.10)

Manufacture of the widely used pesticide pentachlorophenol (PCP) likewise generates dioxins. As commercial PCP actually is a mixture of chlorophenols, a rather complex stew of penta-, hexa-, hepta-, and octachlorochlorodioxins is produced when PCP is "purified" by high-temperature distillation (Table 13.5). Toxic polychlorinated dibenzofurans (PCDF) also are generated, sometimes at levels higher than those of the dioxins. Asked what was done with the insoluble sludge of polychlorodioxins that always collected in barrels of aqueous sodium pentachlorophenate sold as an algicide and disinfectant, a company official acknowledged that it probably was just flushed into the nearest river.

Although commercial chlorophenols contain less than 0.1% dioxins and dibenzo-furans, such enormous quantities have been produced, and the impurities are so environmentally persistent (Section 14.4) that many schemes for dioxin disposal have been tried. So far, the most practical has been high-temperature incineration far out at sea, but adverse public response has caused this to be stopped. A less controversial photochemical method based on photochemical dechlorination (Crosby et al., 1971) also has been used successfully (Section 14.4).

As with the other industrial byproducts, no one *planned* for halogenated dioxins to be generated or was even particularly aware of the problems they might cause. However, it now seems essential that manufacturers and governments be aware of the chemical composition of industrial products and wastes, and that provisions be required for the destruction or removal of persistent or exceptionally toxic components before they are released into an unsuspecting environment (Special Topic 15).

13.4 | REFERENCES

Anonymous. 1996. Facts and figures for the chemical industry. *Chem. Eng. News* **74**(26): 38–79.

Buser, H.-R., and H. P. Bosshardt. 1976. Determination of polychlorinated dibenzo-*p*-dioxins and dibenzofurans in commercial pentachlorophenol by combined gas chromatography–mass spectrometry. *J. Assoc. Offic. Agr. Chemists* **59**: 562–69.

Crosby, D. G., A. S. Wong, J. R. Plimmer, and E. A. Woolson. 1971. Photodecomposition of chlorinated dibenzo-*p*-dioxins. *Science* **173**: 748–49.

Dipple, A., R. C. Moschel, and C. A. H. Biggar. 1984. Polynuclear aromatic carcinogens, in *Chemical Carcinogens* (C. E. Searle, ed.), American Chemical Society, Washington, DC. ACS Monograph 182, pp. 41–163.

Hileman, B. 1993. Concerns broaden over chlorine and chlorinated hydrocarbons. *Chem. Eng. News.* **71**(13): 11–20.

IARC. 1984. *IARC Monographs on the Evaluation of the Carcinogenic Risk of Chemicals to Humans: Polynuclear Aromatic Hydrocarbons, Part 2, Carbon Blacks, Mineral Oils (Lubricant Base Oils and Derived Products) and Some Nitroarenes,* International Agency for Research on Cancer, Lyon, France, **33**: 87–168.

IARC. 1985. *IARC Monographs on the Evaluation of the Carcinogenic Risk of Chemicals to Humans: Polynuclear Aromatic Compounds, Part 4, Bitumens, Coal-Tars and Derived Products, Shale-Oils and Soots,* International Agency for Research on Cancer, Lyon, France, **35**: 39–160.

Kostick, D. S. 1994. *Salt—1993,* Bureau of Mines, U.S. Department of the Interior, Washington, D.C.

Larson, R.A. 1988. Environmental phototoxicity. *Environ. Sci. Technol.* **22**: 354–60.

Lorenz, L. F., and L. R. Gjovik. 1972. Analyzing creosote by gas chromatography: Relation to creosote specifications. *Proc. Amer. Wood-Preserv. Assoc.* **68**: 32–42.

NRC. 1985. *Oil in the Sea,* National Academy Press, Washington, D.C.

USPHS. 1990a. *Toxicological Profile for Hexachlorobenzene.* U.S. Department of Health and Human Services, Washington, DC. TP-90-17.

USPHS. 1990a. *Toxicological Profile for Polycyclic Aromatic Hydrocarbons.* U.S. Department of Health and Human Services, Washington, DC. TP-90–20.

Woo, Y.-T., and J. C. Arcos. 1981. In *Carcinogens in Industry and the Environment* (J. M. Sontag, ed.), Marcel Dekker, New York, NY, pp. 167–281.

Special Topic 13. Why Chlorinate?

There is increasing public and scientific concern over the presence and effects of organochlorine compounds in the environment, including calls for the phaseout of chlorine and chlorinated products (Hileman, 1993). Starting with DDT, many heavily used chlorinated hydrocarbons were shown to be environmentally persistent and readily transported throughout the world via air and water. Eggshell thinning by DDT (actually by DDE) almost brought an end to several important bird species and introduced the public to the toxic potential of persistent chlorinated chemicals.

Now, such widely diverse diseases as cancer of the prostate, breast, and testicles, weakened immunity, birth defects, reduced fertility and hatching in birds, fish, and turtles, and behavioral abnormalities in wildlife are being associated with exposure to low levels of chlorinated organic contaminants such as TCDD, DDE, and BHC. For example, the significant reduction in sperm counts among human males during the past 50 years is being reviewed in relation to exposure of the male fetus to organochlorines known to cross the placental barrier. The chlorination of water as protection against microbial diseases generates chloroform and other chlorinated compounds, and dioxins are formed during the bleaching of paper and the incineration of chlorine-containing plastics such as PVC. Chlorinated chemicals are found everywhere, including the polar regions, the atmosphere, and the deep sea.

For perspective, *natural* chlorinated compounds now are detected widely in the environment. In addition to the simple chloroalkanes such as methyl chloride, chloroform, and carbon tetrachloride found in the atmosphere (Section 2.3.1), marine animals and plants produce a wide variety of natural organochlorine compounds for defense, and some of our most useful microbial antibiotics, such as the tetracyclines, contain chlorine. In fact, many organisms possess haloperoxidase enzymes which halogenate natural metabolites, arguably to increase toxicity and persistence. We could not avoid organochlorine compounds, even if we wanted to.

There is no question that the addition of chlorine atoms increases the stability and toxicity of many chemicals (Section 14.1). These were precisely the properties that led to the use of chlorinated compounds in the first place. In fact, some 15,000 such chemicals are in current use throughout the world as pharmaceuticals, crop protectants, disinfectants, and solvents as well as for the manufacture of important plastics (Hileman, 1993). Introduction of chlorine into an organic molecule can enhance or reduce reactivity, alter the solubility and polarity in predictable ways, and, not the least important economically, inexpensively add substantial weight. A pound of cheap ethylene easily becomes 6 pounds of the more valuable tetrachloroethylene.

Many manufacturers, such as makers of microprocessors, require chlorinated solvents, and large-scale chemicals such as caustic soda and magnesium are themselves "byproducts" of the chlorine industry. About 45% of U.S. industries are direct consumers of chlorine and its coproducts, and they account for over 1.4 million jobs in the United States and Canada (Kostick, 1994). Substitutes for chlorine-based products would cost an *additional* $91 billion annually in this country, and construction of factories to make them would cost another $67 billion. The choice might boil down to either extensive economic disruption or a possible health catastrophe.

Hopefully, both possibilities can be avoided. Although it probably is too late to do much about the environmental release and distribution of old persistent chlorinated hydrocarbons like DDE and PCB, discontinuance of their production and use in other countries and discovery of ways to ameliorate existing adverse effects still is possible. The use of chlorinated solvents could be minimized and their wastes tightly controlled. Emphasis could be placed on the removal from the market of the most persistent ones—already largely accomplished in this country—but because most chlorinated compounds actually have low environmental persistence, restriction must be on a case-by-case analysis which balances benefit against hazard.

Today's humans rely heavily on industrial chemicals, the production of which grew by 31% between 1983 and 1993. However, the time is approaching when the world's supply of petrochemicals will be limited by a simple lack of oil. The search for and development of essential replacements, especially for the organochlorines, will provide work for generations of new environmental chemists and toxicologists.

APPENDIX 13.1.
Top U.S. Chemical Products, 1995[a]

Rank	Chemical	10^9 kg	Rank	Chemical	10^9 kg
1•[b]	Sulfuric acid	43.29	26•	Ethylene oxide	3.46
2	Nitrogen	30.89	27•	Hydrochloric acid	3.40
3	Oxygen	24.28	28	Toluene	3.12
4	Ethylene	21.32	29	p-Xylene	2.84
5	Calcium oxide	18.72	30	Cumene	2.27
6•	Ammonia	16.16	31	Ammonium sulfate	2.38
7•	Phosphoric acid	11.89	32•	Ethylene glycol	2.37
8•	Sodium hydroxide	11.89	33•	Acetic acid	2.12
9	Propylene	11.66	34•	Phenol	1.89
10•	Chlorine	11.39	35•	Propylene oxide	1.82
11	Sodium carbonate	10.12	36	Butadiene	1.67
12	Methyl t-butyl ether	8.00	37	Carbon black	1.51
13•	Ethylene dichloride	7.84	38	Isobutene	1.47
14•	Nitric acid	7.83	39	Potash[d]	1.46
15	Ammonium nitrate	7.26	40•	Acrylonitrile	1.46
16•	Benzene	7.26	41	Vinyl acetate	1.31
17	Urea	7.08	42	Titanium dioxide	1.26
18•	Vinyl chloride	6.80	43	Acetone	1.25
19	Ethylbenzene	6.20	44•	Butyraldehyde	1.22
20	Styrene	5.17	45	Aluminum sulfate	1.09
21•	Methanol	5.13	46	Sodium silicate	1.02
22	Carbon dioxide[c]	4.94	47	Cyclohexane	0.97
23	Xylenes, mixed	4.25	48	Adipic acid	0.82
24•	Formaldehyde, 37%	3.68	49•	Nitrobenzene	0.75
25	Terephthalic acid	3.61	50•	Bisphenol A	0.74

Total organic chemicals	129.79
Total inorganic chemicals	210.70
Grand total 1995	340.49
Grand total 1994	324.41

[a] Anonymous, 1996.
[b] • Denotes highly toxic or corrosive.
[c] Solid and liquid.
[d] KCl, K_2SO_4, and $K_2Mg(SO_4)_2$ (K_2O basis).

Refractory Pollutants 14

14.1 | REFRACTORY CHEMICALS

The dictionary defines refractory as "difficult or impossible to manage." That certainly describes some chemicals that appear in the environment. Many common mineral ores, such as cinnabar (HgS), are very persistent but will react with air or acids or be decomposed by heat. However, some manmade organic chemicals can be even more difficult to deal with, DDE, PCB, and polychlorinated dioxins among them. Although they may be transported readily by environmental forces, their transformations often are slow—a matter of years. This chapter is about them.

One obvious characteristic of many refractory organic chemicals is a high degree of chlorination. Each of the electronegative chlorines withdraws electrons from the ring or chain, making it increasingly resistant to electrophilic reactions such as oxidation. Each added chlorine decreases the substance's volatility (Table 14.1) and increases its lipophilicity (K_{ow}) to result in slower dissipation and increased penetration through membranes.

Intoxication mechanisms also may be affected. A substance such as pentachlorophenol, whose acidity leads to uncoupling of oxidative phosphorylation (Section 9.3), has a pK_a of 4.86; the unchlorinated phenol, pK_a 10.0, is inactive by this mechanism. While another electron-withdrawing substituent, the nitro group, undergoes rapid biotransformation via reduction, organochlorine compounds, especially the aromatics, often are unreactive. The additive electron-withdrawing capacity (Hammett σ_p^+ 0.11, σ_m^+ 0.40, Section 16.3.2) signifies an increasingly refractory molecule, and the large size of the chlorine atom also provides steric hindrance toward an incoming reagent. The more chlorines a molecule has, the more stable it becomes.

The chlorines allow detection and measurement at very low levels. Chlorine's high electronegativity causes a strong response in the electron-capture detector of a

TABLE 14.1
Effects of Chlorination on Physical Properties

Chemical	Formula	log K_{ow}[a]	b.p.(°C)[b]
Benzene	C_6H_6	2.13	80
Chlorobenzene	C_6H_5Cl	2.84	132
1,2-Dichlorobenzene	$C_6H_4Cl_2$	3.38	181
1,4-Dichlorobenzene	$C_6H_4Cl_2$	3.52	174
1,2,4-Trichlorobenzene	$C_6H_3Cl_3$	3.97	214
1,2,3-Trichlorobenzene	$C_6H_3Cl_3$	4.05	218
1,2,4,5-Tetrachlorobenzene	$C_6H_2Cl_4$	4.56	243
Pentachlorobenzene	C_6HCl_5	5.17	277
Hexachlorobenzene	C_6Cl_6	5.31 (6.18[c])	322

[a] Watarai et al., 1982.
[b] Lide, 1992.
[c] USPHS, 1990.

gas chromatograph and permits reliable detection down to 10^{-11} grams, while the Hall conductivity detector is specific for Cl and detects it at 10^{-12} grams. A characteristic ^{35}Cl:^{37}Cl isotope ratio and the inherent sensitivity and specificity of a mass spectrometer allow measurement of organochlorine compounds down to about 10^{-14} grams (parts per quadrillion), and combining a gas chromatograph to resolve complex mixtures of chlorinated compounds with a mass spectrometer (GCMS) provides a quantitative fingerprint of each component of the mixture based on its isotopic composition. Considering the combination of relatively high H', chemical stability, and detectability, one should no longer be surprised to find refractory organochlorine compounds in water, soil, air, and biota anywhere in the world.

For example, analysis of California shellfish routinely detects ppb to ppm levels of as many as 35 different chlorinated hydrocarbons (Appendix 14.1). The tabulated data are from four sites: freshwater clams in an agricultural drain (Blanco Drain) near Salinas, marine mussels from a boat dock at Marina del Rey, San Francisco Bay at the mouth of Redwood Creek, and the harbor at Bodega Bay. The first represents a long-active agricultural area, the second runoff from a large city (Los Angeles), the third a creek through an urban residential area, and the fourth a small fishing village with no appreciable agricultural drainage. Agriculture consistently produces the highest levels, the city follows close behind, and the isolated bay almost always is lowest. Except for endosulfan and lindane, all application of these chemicals has long been banned in California.

The insecticide DDT, or one of its close relatives presented in the Appendix, appears in almost any environmental sample. It was the first organochlorine compound to focus public attention on refractory environmental chemicals, and a detailed discussion of it is instructive.

14.2 | DDT AND DDE

14.2.1. Background

DDT is probably the most controversial chemical of the twentieth century. On the one hand, it saved thousands of people all over the world from death and disease caused by insect-borne pathogens; on the other, its persistence and lipophilicity resulted in the near-extinction of some species of birds and other animals, and its long-term effects on humans and wildlife are still a matter of deep concern (Section 7.4.6).

DDT [1,1,1-trichloro-2,2-*bis*(*p*-chlorophenyl)ethane] was first synthesized by a Strasbourg University student, Othmar Zeidler, who published his work in 1874. As just one item in a dissertation, it was forgotten until it turned up in 1939 during routine screening for new insecticides conducted by Paul Mueller of the Swiss company, J.R. Geigy AG. DDT was not the first insecticidal chemical, nor even the first synthetic one, but it soon was recognized to have superior insecticidal properties. It controlled a broad range of insect pests including the flies, mosquitoes, and body lice that transmit malaria, typhus, and other terrible diseases. With it, the United States and many other nations became essentially malaria-free by 1953, and Mueller received the 1948 Nobel prize for his discovery. DDT had a long residual activity due to its stability to light and air, and it appeared safe for humans and other mammals. As Zeidler had shown, it also was easy and cheap to make, and annual U.S. production had reached 81.2 million kg by 1963.

The chemical is manufactured by reaction of chlorobenzene with chloral (trichloroacetaldehyde) in the presence of concentrated sulfuric acid:

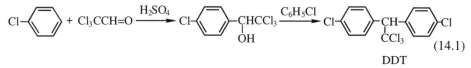

$$(14.1)$$

However, yields of the desired *p,p'*-isomer were not high (Table 14.2), and purification generally was required, usually with little regard for what happened to the wastes. The desired *p,p*-DDT was accompanied by significant amounts of insecticidally inactive *o,p'*- and *o,o'*-isomers, as well as by small proportions of *p,p'*- and *o,p'*-DDD and other byproducts. (DDD is DDT's 1,1-dichloro congener resulting from reaction of chlorobenzene with an impurity, dichloroacetaldehyde, in the chloral.) In due course, a number of other insecticidal 2,2-diphenylethane insecticides were found and marketed, including the methoxy analog of DDT [methoxychlor, 2,2-*bis*(*p*-methoxyphenyl)-] and the DDD analog perthane [2,2-*bis*(*p*-ethylphenyl)-]. Pest-control applications of *p,p'*-DDD also were developed.

14.2.2. Toxicity

DDT proved toxic to many species of harmful insects but also to beneficial insects, fish, and aquatic invertebrates. Acutely, it shows low toxicity in humans,

TABLE 14.2
The Composition of Technical DDT[a]

Compound	% Present
p,p'-DDT	65–73
o,p'-DDT	19–21
p,p'-DDD	0.17–4.0
o,o'-DDT	0.1–1.0
o,p'-DDD	0.04
Chlorobenzene	0.3
2-(p-Chlorophenyl)-1,1,1-trichloroethanol	0.2
1-(o-Chlorophenyl)-2,2,2-trichloroethyl p-chlorobenzenesulfonate	0.1–1.9
bis-(p-Chlorophenyl) sulfone	0.034–0.6
p-Dichlorobenzene	0.1
Na p-Chlorobenzoate	0.02
1-(p-Chlorophenyl)-1-chloroacetamide	0.01
1-(o-Chlorophenyl)-1-chloroacetamide	0.007
Inorganic	0.01–0.1

[a] Haller et al., 1945.

other mammals, and most birds, the only recorded human death being that of a 1-year-old child who drank a DDT formulation (USPHS, 1994). However, in the years following the 1962 publication of Rachel Carson's book, *Silent Spring,* the public came to associate nearly every form of illness and environmental damage with DDT (Beatty, 1973), and it was removed from the U.S. market in 1972 in what Beatty calls "the triumph of the amateurs." The concern led to literally thousands of animal toxicity studies and many conflicting results.

DDT, DDE, and DDD isomers, known collectively as DDTR, have received extensive recent review (USPHS, 1994), and there is general agreement that p,p'-DDT acts primarily on the central nervous system (Section 9.4.1). Typical median lethal doses include 1.7 mg/kg in honeybees *(Apis mellifera),* 113 mg/kg in laboratory rats, 205 mg/kg in Japanese beetles *(Popillia japonica),* 1296 mg/kg in pheasants *(Phasianus colchicus),* and 2000 mg/kg in bullfrogs *(Rana catesbeiana).* DDE and DDD do not produce neurotoxicity and show few acute effects in most species (but see Section 7.4.6).

In human tests, no effects were observed at an oral dose of 6 mg/kg of pure p,p'-DDT, but convulsions occurred above 16 mg/kg, and other reversible CNS symptoms appeared at about 22 mg/kg or 1.5 g total dose (USPHS, 1994). However, long-term occupational exposure to as much as 42 mg/day of technical DDT produced normal neurological exams and no adverse symptoms in workers. For perspective, the average person's 240 μg/day DDT exposure in the United States in 1970 had dropped to 22 μg/kg/day by 1981, and the World Health Organization now suggests 20 μg/kg/day (1.5 mg/person/day) as an acceptable daily intake of DDT. After reaching a high point before 1970, DDT residues in human fat declined steadily in this country (Fig. 14.1), although U.S. babies could have received as much as 300 μg/day of DDTR (mostly DDE) in breast milk in the late 1970s (Takei et al., 1983).

Figure 14.1. Loss of chlorinated hydrocarbons from human fat over time. Extrapolated . . . Adapted from Kutz et al., 1991.

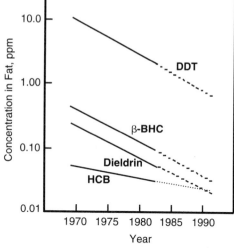

Although results differ among the various animal test species, no toxic effects of *p,p'*-DDT have been seen on human blood, liver, skin, immune system, reproductive system, or behavior (USPHS, 1994). With few exceptions, extensive research has produced no association between DDT, DDD, or DDE and human cancer, although the International Agency for Research on Cancer (IARC) considers DDT a possible human carcinogen, and for possible historic reasons, the U.S. EPA has designated it a probable carcinogen. One reason for the conflicting animal data is the unrealistically high doses often required for *any* effect. For example, liver hepatomas were observed in mice fed 32.5 mg/kg/day for 15–30 weeks, although no tumors were produced in rats or hamsters treated similarly. In general, acute effects are first seen in lab animals at between 30 and 100 mg/kg and chronic effects at 2–10 mg/kg (USPHS, 1994).

o,p'-DDT, which represents up to 20% of technical DDT, has low acute toxicity. However, it competes with natural estradiol for estrogen receptors, and estrogenic effects were originally seen in baby rats injected with large doses. However, more recent research raises grave concerns over the possible ability of this and other chlorinated hydrocarbons to affect estrogen balance at environmental levels of exposure (Section 7.4.6).

14.2.3. Transport and Transformation

Residues of DDT and, especially, DDE now are found everywhere in our environment, even in places where DDT was never applied. They are detected in antarctic ice, in the atmosphere over the Central Pacific Ocean, in dolphins off the coast of Japan, and in forest litter in Maine. DDTR are prime examples of predictably persistent and mobile environmental chemicals. Low aqueous solubilities (DDT 0.0012 mg/L, DDE 0.010 mg/L) and vapor pressures (DDT 1.88×10^{-7}, DDE 6.5

$\times$ 10^{-6} torr) result in Henry's law constants of 3.04 $\times$ 10^{-3} and 1.11 $\times$ 10^{-2}, respectively, meaning that they are easily volatile and the atmosphere will provide the main route of dispersal. Their very high log K_{oc} values (DDT 5.77, DDE 6.00) show that they bind tightly to soil and can be transported in air and water on particles, and the log K_{ow} (DDT 6.91, DDE 6.96) explains why they have bioconcentrated in biota worldwide.

DDT actually is quite reactive (Fig. 14.2). It is dehydrochlorinated to DDE by heat or alkaline soil, it is readily reduced to DDD by the iron porphyrins in stored blood, the CCl_3 can be hydrolyzed stepwise to COOH (DDA), and oxidation on the benzylic carbon gives a tertiary alcohol known as dicofol (Kelthane®). The reactions occur both environmentally and metabolically, and DDE is the degradation product most commonly observed. Several schemes have been proposed to account for the observed metabolites, as many as 13 of which have been isolated from DDT-treated birds alone. Metabolism is the principal source of the resistance to DDT intoxication seen in insects and other animals (Special Topic 7), and genes have been identified for DDT dehydrochlorination, its oxidation by MFO, and alteration of Ca^{2+}-ATPase that reduces neuron sensitivity to DDT by 98% (Matsumura, 1985).

In addition to the usual products of metabolic oxidation, reduction, and hydrolysis, a particularly interesing DDT metabolite is the methyl sulfone analog of DDE (Jensen and Jansson, 1976). First detected in Baltic seal blubber, this type of com-

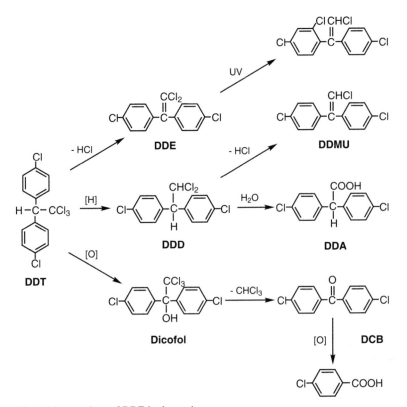

Figure 14.2. Major reactions of DDT in the environment.

pound is now found widely in animals exposed to chlorinated aromatic hydrocarbons such as DDT, HCB, and PCBs. They are formed by initial epoxidation of the ring, followed by reaction with glutathione, subsequent conversion to thiol, *S*-methylation, and final oxidation of the sulfide to the methyl sulfone (Section 6.3.2).

14.2.4. DDE

DDE is a problem. It is stable to oxidation, reduction, and hydrolysis, so it is still circulating in our environment a quarter-century after the use of DDT was abandoned by most countries. Not surprisingly, DDE is the main DDT relative found in the environment; typical 1984 human fat samples contained 0.70 ppm DDT but 3.42 ppm DDE. DDE also was responsible for the much-publicized egg-shell thinning in several bird species in the 1960s. Carnivorous birds at the top of their foodchains developed high levels of the chemical in fat, 150 ppm in Western Grebes, for example, while most seed-eating species concentrated much less. At such high concentrations, egg shells were as much as 20% thinner than normal, thin enough to break when settled on by the parents. The thinning mechanism remains unknown, although it apparently is associated with Ca metabolism and perhaps with the Ca-fixing enzyme, carbonic anhydrase. DDE may well be around after most other chlorinated hydrocarbons have disappeared.

14.3 | CHLORINATED ALICYCLICS

14.3.1. Benzene Hexachloride (BHC)

BHC (hexachlorocyclohexane) is the oldest chlorinated hydrocarbon insecticide. Its preparation was first described by English chemist Michael Faraday in 1825, at the very beginnings of organic chemistry. In an interesting history, Brooks (1974) describes the discovery of its insecticidal activity in France and England in the late 1930s and its scientific and practical development during World War II. BHC is easily prepared by reaction of benzene with chlorine in the presence of ultraviolet radiation:

$$\bigcirc + 3Cl_2 \xrightarrow{UV} \text{BHC} \tag{14.2}$$

BHC

According to Brooks, small quantities were made by the ICI laboratories in Widnes, England, ". . . by chlorinating benzene on the roof of a convenient building in the fitful Widnes sunlight and filtering the solid in a large Büchner funnel. When the sun went in [behind a cloud], chlorine would dissolve in the benzene, only to react violently when the sun came out." By 1955, annual production in the United States alone, in more sophisticated equipment, reached almost 50 million kg per year.

Sixteen stereoisomeric forms of hexachlorocyclohexane are possible. However, just five, designated α, β, γ, δ, and ϵ, are found in the chlorination mixture (Table 14.3), together with more highly chlorinated products. Only the γ-isomer, called **lindane** after its discoverer, L. van der Linden, shows meaningful insecticidal activity, and it constitutes less than 20% of technical BHC. Commercial lindane is almost 100% pure γ-isomer obtained by crystallization from methanol; it has been used extensively against agricultural pests, for wood treatment against termites, to control nuisance insects around the home, and medically to combat ectoparasites of humans and domestic animals. The noninsecticidal isomers are of no direct use but can be converted by base to trichlorobenzenes and then to many other useful chemical intermediates.

Lindane is neurotoxic to a broad range of insects, acting at the same receptor as picrotoxinin (Section 9.4.1). The topical (external) LD_{50} is 3.8 mg/kg in the German cockroach *(Blattella germanica)*, 1.0 mg/kg in the housefly *(Musca domestica)*, and 3.4 mg/kg in the grasshopper *(Melanoplus differentialis)*, while the oral LD_{50} in the latter is 6.7 mg/kg. Unfortunately, lindane also is toxic to fish, and a typical 96 h LC_{50} is only 27 μg/L in rainbow trout. Median lethal doses lie in the mg/kg range in molluscs and tadpoles, in the hundreds of mg/kg in birds, and in the thousands for most other laboratory animals (Brooks, 1974; IPCS, 1991). Symptoms of intoxication are seen in humans from acute doses of 15–50 mg/kg, but human dietary intake of lindane in the United States is estimated at <0.003 μg/kg/day (IPCS, 1991), mostly from food and especially from fish.

Extensive testing has shown that BHC isomers have no particular effect on liver

TABLE 14.3
Properties of Principal BHC Isomers[a]

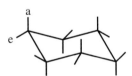

Property	α-BHC	β-BHC	γ-BHC	δ-BHC
Configuration	aaeeee[b]	eeeee	aaaee	aeeee
% in BHC	65–70	7–10	14–16	~7
m.p., °C	159.2	311.2	112.9	140.8
Vapor pressure, torr[c]	2.5×10^{-5}	2.8×10^{-7}	3.3×10^{-5}	1.7×10^{-5}
Solubility, mg/L[c]	1.63	0.70	7.90	21.3
Xylene Solubility, g/L[c]	85	33	247	421
H'	2.5×10^{-4}	6.4×10^{-6}	6.7×10^{-5}	1.3×10^{-5}
log K_{ow}	3.82	3.80	3.72	4.14
Dehydrochlorination[d]	0.169	3×10^{-6}	0.045	"fast"
LD_{50}, rat oral, mg/kg[e]	177	>6000	76	1000

[a] IPCS, 1991.
[b] a = axial, e = equatorial chlorine.
[c] 20°C.
[d] Initial rate, L/s/mole (Cristol, 1947).
[e] Sweet, 1987.

or kidney function, reproduction, immune response, or behavior, and they are not mutagenic, teratogenic, or carcinogenic at normal exposure levels. As with DDT, reports of their toxic effects often are clouded by excessive test levels. For example, in mice receiving 100, 300, and 600 mg/kg of lindane in their diet for 32 weeks, 30% of female and 75% of male survivors showed liver tumors at the highest exposure level but none at the other levels (IPCS, 1991). Note that the reported acute oral LD_{50} of lindane in the mouse is listed at between *55 and 250 mg/kg.*

Lindane is the most volatile of the BHC isomers (Table 14.3), and it long was used in electric vaporizers to ward off flying insects. However, the small white tablets were easily mistaken for candy by children, and poisoning became so frequent that this use was stopped. The compound is lipophilic (log K_{ow} 3.2) but also quite reactive. With two of its six chlorines *trans* to the adjacent hydrogens, it is rapidly dehydrochlorinated to γ-pentachlorocyclohexene (PCCH), which goes on to provide a variety of metabolic products including 2,4,5-trichlorophenol (Fig. 6.6).

Not so the β-isomer, which also is quickly bioconcentrated into fat but eliminated slowly (Fig. 14.2). This isomer possesses only equatorial chlorines, not prone to dehydrochlorination. Human breast milk from the United States that contained detectable γ-BHC in 3.9% of the samples (44 ppb maximum, lipid basis), and α-BHC in 22.6% of samples (850 ppb maximum), contained β-BHC in 83.3% at a maximum concentration of 470 ppb (Takei et al., 1983). Extensive application of technical BHC in Japan in the decade after World War II resulted in human fat levels of the persistent β-isomer that still averaged 6.55 **ppm** over 25 years later. However, even β-BHC eventually is degraded to 2,4,6-trichlorophenol.

14.3.2. Cyclodienes

This group of related organochlorine compounds, dating back to 1945, is prepared from hexachlorocyclopentadiene by the Diels–Alder reaction followed by further reactions to provide insecticides such as **dieldrin, endrin, chlordane,** and **heptachlor.**

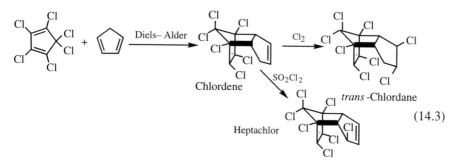

$$(14.3)$$

The cyclodienes first arose from efforts to find uses for cyclopentadiene, a large-scale byproduct from the manufacture of butadiene for synthetic rubber. Their composition and stereochemistry can be complex, arising initially from the steric course of the Diels–Alder reaction. For example, chlorination of chlordene (Eq. 14.3) results in technical chlordane, which is a mixture of at least 14 components including unchanged chlordene (20%), *cis*-chlordane (20%), *trans*-chlordane (25%), and the substitution products heptachlor (10%) and *cis-* and *trans*-nonachlor (7%) (Brooks, 1974). These all are insecticidal and act at the picrotoxinin binding site on the GABA receptor (Fig. 9.4; Tanaka et al., 1984).

The cyclodiene insecticides are highly toxic to insects, fish, and birds, but, unlike DDT and the BHC isomers, they are toxic to mammals as well. For example, dieldrin's acute oral LD_{50} in the rat is 46 mg/kg, and endrin's is 7.5 mg/kg. Like other organochlorine insecticides, cyclodienes are very lipophilic (log K_{ow} of heptachlor is 5.44), so they have occurred almost universally in body fat and human breast milk (Fig. 14.2). With the possible exception of mirex, they are not carcinogenic in humans. The mechanism of toxic action is that represented by chlordane (Section 9.4).

Most cyclodienes share the low aqueous solubilities and vapor pressures of DDTR and BHC. With Henry's law constants often less than that of water, they actually become more concentrated as a water body evaporates. They are exceptionally persistent, the half-life of aldrin in soil being 1–4 years, dieldrin 1–7, chlordane 2–4, and heptachlor 7–12 years. Except for a photoaddition reaction, the hexachlorosubstituted ring is very stable, although environmental and metabolic oxidation, reduction, and dehydrochlorination can occur on other parts of the molecule.

For example, the unguarded double bonds of aldrin and heptachlor are epoxidized to dieldrin and heptachlor epoxide, respectively, products that are lipophilic, stable, and toxic:

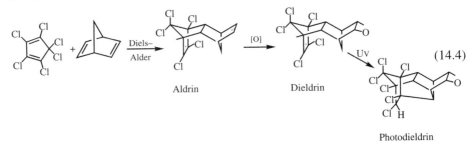

$$(14.4)$$

Aldrin   Dieldrin

Photodieldrin

Sunlight or UV radiation convert dieldrin to the cage-like photodieldrin, and other cyclodienes behave similarly. As cyclodienes do not absorb sunlight UV, the reaction mechanism remains obscure, although some bacteria are able to perform the same conversion via metabolism. The photoproducts often are more toxic than the corresponding cyclodienes.

For decades, Hawaiian pineapple growers found heptachlor to be effective in controlling the ants that encouraged a deadly wilt disease. Cattle growers started using the waste pineapple leaves, known as "green chop," as a cheap and nutritious feed, and it soon made its way into dairies. In early 1982, the heptachlor and its epoxide were discovered at elevated levels in both commercial milk and breast milk, and a public furor ensued. Attempts were made to depurate the cows with clean feed, which the insecticides' bioconcentration and persistence rendered futile. The Hawaiian dairy industry collapsed, and to this day, Hawaii's milk is flown in from the mainland. A similar mishap in Arkansas, Missouri, and Oklahoma in 1986 again forced the destruction of thousands of gallons of milk and the bankruptcy of hundreds of dairymen (Farley, 1987).

14.3.3. Other Organochlorine Insecticides

Hexachlorocyclopentadiene dimerizes in the presence of acid catalysts to form the cage structures **chlordecone** (Kepone®) and **mirex**. Chlordecone also can be chlorinated to mirex:

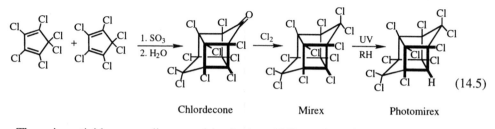

Chlordecone   Mirex   Photomirex   (14.5)

These insecticides were discovered in the late 1950s and used primarily in ant baits, as mirex proved to be the best control available against imported fire ants in the South. Predictably, they are stable and very persistent. They are not appreciably metabolized, although sunlight slowly dechlorinates them reductively in the presence of a hydrogen-donating solvent (Alley et al., 1974). The mechanism of toxic action is unclear and in fact may not involve the picrotoxinin binding site used by other cyclodienes (Fig. 9.4).

Kepone achieved notoriety in 1975, when the manufacturing plant in Hopewell, Virginia, was discovered to be severely contaminating its employees, the James River, and large stretches of Chesapeake Bay with the insecticide. Where the acute oral LD_{50} of Kepone in rats had been reported to be a relatively safe 95 mg/kg, a 6-month chronic exposure was found to reduce the figure to 1.5 mg/kg, placing chlordecone with mirex among the most cumulative poisons known. Many workers were severely injured, the river was closed to fishing, and seafood as far away as the Atlantic coast was found to contain elevated levels of kepone (Huggett and Bender, 1980).

Still another class of organochlorine insecticides, the chlorinated terpenes, has been produced in many countries and sold as Toxaphene (campheclor) in the United States, Chlorothene in Russia, and Kanechlor in Japan. At one point, campheclor represented 40% of the weight of all chlorinated insecticides sold in this country. Made by chlorinating the natural hydrocarbon mixture extracted from pine stumps to 55–69% Cl content by weight, these insecticides represent extremely complex mixtures of chlorinated alicyclic compounds. The structures of several of the principal toxic components have been determined (Casida et al., 1974; Turner et al., 1975), one of which is shown in Fig. 9.4. In general, toxaphene exhibits about the same mammalian toxicity as lindane, and it appears to have a similar mechanism of toxic action. In fish, it causes a characteristic "broken-back" effect, in which the spine is deformed or ruptured even at low levels of exposure. Toxaphene is very persistent in soil ($t_{1/2}$ 10 years) but dissipates rapidly from leaves and water by volatility, photolysis, and dehydrochlorination (Brooks, 1974). Although still used in other countries, it is no longer applied in the United States.

14.4 | CHLORINATED DIOXINS

The chlorinated dioxins were discussed in Section 13.3.4 as byproducts from the manufacture of chlorophenols. However, they also are formed in the environment and occur in tobacco smoke, engine exhaust, bleached paper products, and burning wood and plastics (Table 14.4). Although pesticides long were thought to be their main source, the most important generators have proved to be municipal and hospital incinerators that burn chlorinated plastics such as PVC and Saran. Dioxins even occur naturally as the products of forest fires. All together, there are 75 possible polychlorinated dibenzodioxin (PCDD) congeners and 135 chlorinated dibenzofurans (PCDF), but only 6 PCDDs and 11 PCDFs, with chlorines in the 2-, 3-, 7-, and 8-positions, are unusually toxic. 2,3,7,8-Tetrachlorodibenzodioxin (TCCD) is the most toxic of all.

The dioxins usually exist as mixtures, each congener with its own chemical and toxic character, so the confusion over dioxin effects is not surprising. The lowest acute oral LD_{50} of TCCD, 2.5 μg/kg in the guinea pig, is widely cited, but oral LD_{50} values for other toxic chlorinated dioxins in the guinea pig are higher (*e.g.*, 120 μg/kg for 2,3,4,6,7,8-hexachloro) and are higher still for congeners lacking the complete 2,3,7,8-pattern (*e.g.*, 1125 μg/kg for the 1,2,4,7,8-pentachloro- and 1.5 $\times$ 10^7 μg/kg for the 1,3,6,8-tetrachlorodioxins). In order to make practical sense of all this, toxic equivalency factors (TEF) are widely used; they represent the fraction of TCCD toxicity exhibited by a congener or group of congeners. The TEF for 1,2,3,7,8-pentaCCD is 0.5, 0.1 for each of the three hexaCDDs, 0.01 for the heptaCDD, and 0.001 for octaCDD, so a mixture containing 50% each of TCDD and a hexaCDD would have a calculated toxicity 0.505 times that of pure TCDD (1 $\times$ 50%, + 0.1 $\times$ 50%). Factors for the PCDFs are calculated in the same way, but are based on TCDD.

TABLE 14.4
Environmental Sources of TCCD[a]

Source	Date	TCCD (pg/kg)[b]
Whole milk, U.S.	1992	1.8
Chicken broth	1992	1.1–1.5
Ground beef	1992	17–62
Yellow perch	1984	26,000
Facial tissue	1988	1.1
Auto muffler	1980	nd[c]–4,000
Urban soil, Chicago, IL	1986	4200–9200
Incinerator flue gas, U.S.	1986	3.5[d]
Industrial wastewater, MI	1986	15[d]
Waste disposal site, AR	1986	14,000[d]
Love Canal sediment, NY	1985	nd–9.6 $\times$ 10^9

[a] USPHS, 1989.
[b] pg/kg = parts per quadrillion.
[c] None detected.
[d] pg/L.

Besides the extreme animal sensitivity, TCCD toxicity has two unusual features. After exposure to lethal doses, animals do not die quickly but instead stop eating, lose weight, and eventually just waste away. The weight loss but not lethality can be prevented by intravenous feeding. Also, in some species, including humans, an early symptom of dioxin exposure is a persistent but nonfatal skin rash called chloracne (Section 7.4.3). Other effects in experimental animals can include immunotoxicity, embryotoxicity, and carcinogenicity (actually, tumor promotion), although none of these has been confirmed yet in people. Half of human exposure is via meat, with another 20% each from fish and dairy products. Total *predicted* adult TCDD intake in the United States is about 35 pg/day, or about half the estimated no-effect level when exposed daily over a lifetime (USPHS, 1989). Humans do not appear to be particularly sensitive to dioxins.

The very low vapor pressures and aqueous solubilities of dioxins, and their exceptionally high log K_{ow} and log K_{oc} values (in the range of 6–8), result in high environmental persistence, bioconcentration, and binding to soil. Because of their electron-withdrawing chlorines and coplanar structure, dioxins are largely unreactive, although they are biotransformed slowly by hydroxylation in an unsubstituted position. However, in the presence of a source of abstractable hydrogen atoms, they are converted by UV radiation or sunlight into a series of less toxic congeners ending in unchlorinated dibenzo-*p*-dioxin (Crosby et al., 1971).

This photoreduction is probably a free radical exchange (Eq. 14.6). The necessary H donor can be a pesticide formulating agent, solvent, waste oil, or even plant leaf waxes, and so dioxin photodegradation is thought to be common in the environment (Crosby and Wong, 1977). Of course, absorption of the UV energy is required for any reaction to occur, and TCDD does absorb sunlight UV at about 300 nm.

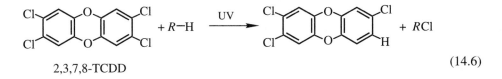

2,3,7,8-TCDD

$$(14.6)$$

However, a dark waste oil contaminated with TCDD was used to lay summer road dust in the small Missouri town of Times Beach in the early 1970s, and the town subsequently had to be abandoned because of the persistence of the dioxin. The remainder of the waste oil eventually was diluted with an organic solvent and circulated past UV lamps in order to destroy the dioxins, but the dirt was scraped up and incinerated. The expenssive process was finally completed in January, 1997.

Following the TCDD accident in Seveso, application of organic solvents such as cyclohexanone or olive oil to contaminated grass, leaves, and tile roofs was shown to destroy the dioxin in the summer sunlight (Crosby, 1978). However, official uncertainty and political demands delayed action until leaf fall and winter rain made decontamination impossible. The populace was relocated and the contaminated soil and plant materials covered by a thick dome of concrete, assuring that the TCCD would remain for their grandchildren to worry over.

14.5 | POLYCHLORINATED BIPHENYLS

Polychlorinated biphenyls (PCBs; Table 14.5) illustrate how a molecule's low electronegativity can predestine its unreactivity and environmental persistence. PCBs actually predate the chlorinated hydrocarbon insecticides by more than a decade. Prepared easily and cheaply by catalyzed chlorination of the biphenyl byproduct from the manufacture of phenol, they were applied widely as insulation fluids in electrical equipment and as plasticizers and pesticide carriers. They were added to printing ink, adhesives, duplicating paper, and anywhere that a stable, viscous liquid was needed. PCB mixtures were manufactured commercially in the United States from 1929 until 1977, with a peak 1970 production of almost 40 million kg.

The crude chlorination product was separated by distillation into fractions according to chlorine content and sold in this country primarily under the name Aroclor. Aroclor 1248 contained 48% Cl, Aroclor 1254 contained 54% Cl, and so on. Each fraction represented a complex mixture of many chlorinated biphenyls, although no Aroclor contained all 209 possible congeners (Table 14.5). For example, the principal constituents of Aroclor 1254 were the 2,3,6,3',4'-, 2,4,5,3'4'-, 2,4,5,2',5'-, 2,4,5,2',4'-, and 2,3,6,2',4'-pentachlorobiphenyls (IPCS, 1993). The physical, chemical, and toxic properties listed for any Aroclor represent the average of those of its individual constituents.

The toxicity pattern of the PCBs is similar to that of the chlorinated dioxins, except that they are not nearly as toxic. The acute oral LD_{50} of Aroclor 1254 in the rat is 1010 mg/kg, that of Aroclor 1242 is 4250 mg/kg, and Aroclor 1260 is not toxic to rats at 1.25 mg/kg/day over a period of 105 weeks (USPHS, 1993). Higher doses cause liver and kidney damage and chloracne, partly because the PCB mixtures also contain chlorinated dibenzofurans. The mechanism of toxic action is the same as that

TABLE 14.5
Aroclor Composition (%)[a]

$$Cl_x \underset{4}{\bigcirc}\!\!-\!\!\underset{4'}{\bigcirc} Cl_y$$

PCB	Isomers	1016	1221	1242	1248	1254	1260
No Cl	1	nd[b]	10	nd	nd	nd	nd
Cl_1	3	2	50	26	nd	nd	nd
Cl_2	12	19	35	13	1	nd	nd
Cl_3	24	57	4	45	2	1	nd
Cl_4	42	22	1	31	49	15	nd
Cl_5	46	nd	nd	10	27	53	12
Cl_6	42	nd	nd	nd	2	26	42
Cl_7	24	nd	nd	nd	nd	4	38
Cl_8	12	nd	nd	nd	nd	nd	7
Cl_9	3	nd	nd	nd	nd	nd	1

[a] USPHS, 1993.
[b] None detected.

of chlorodioxins if the PCB structure conforms to the right dimensions (Scheme 9.5): It must be coplanar, with chlorines in the 3- and/or 4-positions. The estimated total intake of PCB by human adults currently is <1 ng/kg/day, mostly from eating fish.

Low aqueous solubilities combined with low vapor pressures provide relatively high volatilities, and so PCBs are widely detected in air and rainwater. As expected, they are unreactive and only slowly metabolized via oxidation. However, upon incineration in the presence of air, they are converted to dibenzofurans and dioxins. A February, 1981, fire in electrical equipment in a new State Office Building in Binghamton, New York, caused smoke containing PCB, PCDD, and PCDF to be distributed throughout the structure via air shafts. The soot, which contained 3–10 ppm of PCB and ~3 ppm TCCD, coated everything—walls, ceilings, desks, appliances—but especially the areas above the ceilings and inside the walls. The cleanup, which cost about as much as dismantling the building and starting over, finally was completed and the offices reoccupied late in 1994. Fortunately, the fire occurred at 5:30 A.M., and no one was injured.

14.6 | PHTHALATE ESTERS

Although the organochlorines immediately come to mind when discussing environmentally persistent chemicals, any substance that has low electronegativity, and so resists oxidation or hydrolysis, is a candidate. Polycarboxylic acids and esters such as phthalate plasticizers and terephthalates provide examples.

Plasticizers soften otherwise hard and brittle plastics and make them pliable. PVC bags would be as stiff as refrigerator containers were it not for large proportions of phthalate plasticizers. The phthalates, which are lipophilic liquids, are not chemically bonded to their plastics and so can migrate, for example into blood plasma when it is stored in plasticized bags or contacts plasticized tubing. While the phthalates are not acutely toxic, chronic exposure to high levels causes damage to liver and testicles in mammals and death to some aquatic species (IPCS, 1992). They also are suspected of being endocrine disrupters (Section 7.4.6).

Phthalates now are ubiquitous in air, water, and food, as they are used in packaging, clothing, floor coverings, air conditioners, cosmetics, and printing ink. The **di(2-ethylhexyl)** ester (DEHP) represents about half of all phthalate plasticizers in use, although the dibutyl and butyl benzyl esters also are important. The low vapor pressure of DEHP (6.5×10^{-6} torr) combines with low aqueous solubility (45 μg/L) to provide easy volatility from water (H' 0.003), and a K_{ow} of 10^5 ensures high bioconcentration. While atmospheric photolysis has been reported, phthalates are otherwise surprisingly unreactive, and their biotransformation is slow and almost entirely on the side chain. All this adds up to a high degree of persistence and worldwide detection. Phthalates are easily mistaken for chlorinated hydrocarbons by gas chromatography, and more careful analysis has shown that much of the DDT reported to occur in seawater is plasticizer instead (Section 2.2).

Terephthalates (benzene-1,4-dicarboxylates) are represented by the widely used herbicide **chlorthal-dimethyl** (Dacthal®), dimethyl tetrachloroterephthalate, also called DCPA. Like phthalate esters, DCPA is essentially nontoxic in both vertebrates and invertebrates, environmentally stable, moderately volatile, and very persistent. In plants, it affects growth of developing roots by inhibiting cell division after prophase. It is microbially hydrolyzed to even more persistent tetrachlorophthalic acid, the pesticide derivative most frequently detected in U.S. drinking water. DCPA contains byproduct TCDD and HCB and so can exceed EPA's $1:10^6$ cancer limit (Cox, 1991).

14.7 | PERSPECTIVE

Probably no refractory organic chemical is completely stable, but the reactivity of DDE, HCB, β-BHC, and a few others is so low that they will remain in us and our environment for decades. The levels do decline eventually, as reflected in a slow loss from human fat (Fig. 14.2), and the 1971 average DDT residue of 7.95 ppm has fallen to almost undetectable levels today (Kutz et al., 1991). Only the body levels of HCB, a manufacturing byproduct, have remained steady. Adverse effects of these chemicals on the health of the general human population remain to be demonstrated, and even pesticide factory workers whose fat levels exceeded 500 ppm over extended time periods showed no obvious harmful effects. However, see Section 7.4.6 for new concerns. The present daily intake of DDT and its relatives DDD and DDE (DDTR) among U.S. adults is only 30 ng/kg, almost all as DDE from eating contaminated fish or from vegetables grown in contaminated soil.

Most refractory chemicals were manufactured for good reason. The chlorinated insecticides were inexpensive, easy to make, and stayed around long enough to be effective. PCBs had long-lasting stability under harsh conditions, and phthalates kept plastics flexible for long periods. In many applications, the replacements necessarily are just as persistent. For example, environmentally stable triphenyl phosphate is now used in place of PCBs in transformers, but it, too, will likely be found undesirable. Society will have to decide just how much chemical pollution it is willing to tolerate in order to enjoy the benefits of refractory chemicals.

DDT manufacture and use ended in the United States by 1974, and most other persistent chlorinated hydrocarbons have now been removed from our trade. However, production has continued in many other countries. Average lipid levels of DDTR were 15.4 ppm in residents of Delhi in 1984 and 59.3 ppm in Costa Ricans in 1982 (Kutz et al., 1991). As persistent residues obviously circulate freely in air, food, and other media, worldwide exposure could again increase (Coulston, 1985). Without irrefutable evidence that present exposures are harming people or wildlife, the great unplanned experiment on the biological effects of chronic exposure to refractory organic chemicals seems destined to continue.

14.8 | REFERENCES

Alley, E. G., B. R. Layton, and J. P. Minyard, Jr. 1974. Identification of the photoproducts of the insecticides mirex and kepone. *J. Agr. Food Chem.* **22:** 442–45.

Amato, J. R., D. I. Mount, E. J. Durhan, M. T. Lukasewycz, G. T. Ankley, and E. D. Robert. 1992. An example of the identification of diazinon as a primary toxicant in an effluent. *Environ. Toxicol. Chem.* **11:** 209–16.

Beatty, R. G. 1973. *The DDT Myth: Triumph of the Amateurs,* John Day, New York.

Brooks, G. T. 1974. *Chlorinated Insecticides,* CRC press, Cleveland, OH, 2 Vols.

Casida, J. E., R. L. Holmstead, S. Khalifa, J. R. Knox, T. Ohsawa, K. J. Palmer, and R. Y. Wong. 1974. Toxaphene insecticide: A complex biodegradable mixture. *Science* **183:** 520–21.

Coulston, F. 1985. Reconsideration of the dilemma of DDT for the establishment of an acceptable daily intake. *Regul. Toxicol. Pharmacol.* **5:** 332–83.

Cox, C. 1991. DCPA (Dacthal). *J. Pesticide Reform.* **11**(3): 17–20.

Cristol, S. J. 1947. The kinetics of the alkaline dehydrochlorination of the benzene hexachloride isomers. The mechanism of second-order elimination reactions. *J. Amer. Chem. Soc.* **69:** 338–42.

Crosby, D. G. 1978. Conquering the monster: The photochemical destruction of chlorinated dioxins, in *Disposal and Decontamination of Pesticides* (M. V. Kennedy, ed.), American Chemical Society, Washington, DC. ACS Sympos. Ser., pp. 1–12.

Crosby, D. G., and A. S. Wong. 1977. Environmental degradation of 2,3,7,8-tetrachlorodibenzo-*p*-dioxin (TCDD). *Science* **195:** 1337–38.

Crosby, D. G., A. S. Wong, J. R. Plimmer, and E. A. Woolson. 1971. Photodecomposition of chlorinated dibenzo-*p*-dioxins. *Science* **173:** 748–49.

Farley, D. 1987. From tainted feed to mothers' milk. *FDA Consumer,* March, 38–40.

Haller, H. L., P. D. Bartlett, N. C. Drake, M. S. Newman, S. J. Cristol, C. M. Eaker, R. A. Hayes, G. W. Kilmer, B. Magerlein, G. Mueller, A. Schneider, and W. Wheatley. 1945. Chemical composition of technical DDT. *J. Amer. Chem. Soc.* **67:** 1591–602.

Huggett, R. J., and M. E. Bender. 1980. Kepone in the James River. *Environ. Sci. Technol.* **14:** 918–23.

IPCS. 1991. *Lindane,* WHO, Geneva, Switzerland. Environmental Health Criteria 124.

IPCS. 1992. *Diethylhexyl Phthalate,* WHO, Geneva, Switzerland, Environmental Health Criteria 131.

IPCS. 1993. *Polychlorinated Biphenyls and Terphenyls,* Second Edition, WHO, Geneva, Switzerland. Environmental Health Criteria 140.

Jensen, S., and B. Jansson. 1976. Methyl sulfone metabolites of PCB and DDE. *Ambio* **5:** 257–60.

Kutz, F. W., P. H. Wood, and D. P. Bottimore. 1991. Organochlorine pesticides and polychlorinated biphenyls in human adipose tissue, *Rev. Environ. Contam. Toxicol.* **120:** 1–82.

Lide, D. R. 1992. *CRC Handbook of Chemistry and Physics,* CRC Press, Boca Raton, FL.

Matsumura, F. 1985. *Toxicology of Insecticides,* 2nd Ed., Plenum Press, New York, NY.

Sweet, D. V. 1987. *Registry of Toxic Effects of Chemical Substances, 1985–86,* NIOSH, USDHHS, Washington, DC. Publ. 87–114, Vol. 2, pp. 1748–49.

SWRCB. 1988. *California State Mussel Watch, Ten Year Summary, 1977–1987,* Calif. State Water Resources Control Board, Sacramento, CA. Water Quality Monitoring Report 87–3.

Takei, G. H., S. M. Kauahikaua, and G. H. Leong. 1983. Analysis of human milk samples collected in Hawaii for residues of organochlorine pesticides and polychlorobiphenyls. *Bull. Environ. Contam. Toxicol.* **30:** 606–13.

Tanaka, K., J. G. Scott, and F. Matsumura. 1984. Picrotoxinin receptor in the central nervous system of the American cockroach: Its role in the action of cyclodiene-type insecticides. *Pestic. Biochem. Physiol.* **22:** 117–27.

Turner, W. V., S. Khalifa, and J. E. Casida. 1975. Toxaphene toxicant A. Mixture of 2,2,5-*endo,6-exo*,8,8,9,10-octachlorobornane and 2,2,5-*endo,6-exo*,8,9,9,10-octachlorobornane. *J. Agr. Food Chem.* **23:** 991–94.

USEPA. 1986. *Pesticides in Ground Water: Background Document,* U.S. Environmental Protection Agency, Washington, DC.

USPHS. 1989. *Toxicological Profile for 2,3,7,8-Tetrachlorodibenzo-p-dioxin,* Public Health Service, USDHHS, Washington, DC. ATSDR/TP-88-23.

USPHS. 1990. *Toxicological Profile for Hexachlorobenzene,* Public Health Service, USDHHS, Washington, DC. TP-90-17.

USPHS. 1993. *Toxicological Profile for Selected PCBs (Aroclor-1260, -1254, -1248, -1242, -1232, -1221, and -1016),* Public Health Service, USDHHS, Washington, DC. TP-92/16.

USPHS. 1994. *Toxicological Profile for 4,4'-DDT, 4,4'-DDE, 4,4'-DDD (Update),* Public Health Service, USDHHS, Washington, DC. TP-93/05.

Watarai, H., M. Tanaka, and N. Suzuki. 1982. Determination of partition coefficients of halobenzenes in heptane/water and 1–octanol/water systems and comparison with scaled particle calculation. *Anal. Chem.* **54:** 702–05.

Special Topic 14. Environmental Persistence

The term *persistence* has appeared often in this and other chapters, but what does it really mean? A dictionary says that persistence is "when something exists for a longer than usual time," but what is "usual"? A farmer battling insect infestations thinks 10 days is an unusually short persistence for a pesticide, while the worried environmentalist thinks it is unusually long. Such a subjective definition is unsatisfactory.

The chemical definition of persistence relates to dissipation rate, which generally is pseudo-first order (Section 5.1). If a chemical has a half-life of 4 days, its aqueous concentration of 4 mg/L falls to 2 mg/L in just 4 days, 90% is gone within two weeks, and the substance is said to be nonpersistent. However, even after *three months,* a concentration of 1 ng/L is predicted to remain, well within the capability of modern analytical detection. Theoretically, the rate law applies until the last molecule is gone, but physical and chemical dissipation become chronologically so slow at low concentrations that the level may seem almost unchanging.

High fugacity (vapor pressure and aqueous solubility) leads to rapid dissipation. Likewise, low persistence is associated with high reactivity toward environmental reagents, most importantly oxidation and hydrolysis. If vapor pressure, solubility, and reactivity are low, persistence will be high. Thus the microenvironment that surrounds a substance becomes very important. Dissipation is faster under warm rather than cool temperatures, under moist compared to dry conditions, and in sunlight and air rather than under shaded and less aerobic conditions. No wonder published figures on persistence are so variable.

However, environmental dissipation has limits. The familiar household insecticide diazinon provides an example. Although its vapor pressure of close to 10^{-4} torr suggests moderate volatility, the relatively high aqueous solubility (60 mg/L) leads to a Henry's law constant of 2.5×10^{-5}, meaning only slow volatilization from water. As an organophosphate, environmental oxidation and hydrolysis might also be expected to limit persistence, but TIE analysis

TABLE 14.6
Potential for Groundwater Contamination[a]

Property	Permissible Limit
Aqueous solubility	<30 ppm
K_d	>5
K_{oc}	$>300–500$
Henry's law constant (H')	>0.4
Hydrolysis half-life	<25 weeks
Photolysis half-life	<1 week
Field half-life	<3 weeks
Speciation	No negative charge at ambient pH

[a] USEPA, 1986.

(Section 8.4.4) of municipal wastewater from across the country consistently revealed diazinon levels of up to about 1 $\mu g/L$ (Amato et al., 1992), enough to kill aquatic invertebrates if not fish. In this case, dilution may have made breakdown so slow that the *biodegradation threshold* (Section 6.2.6) assured this "nonpersistent" chemical would remain for a long time at about 1 ppb. As many pollutants behave similarly, the chemical definition of persistence also proves unsatisfactory.

The persistence and toxicity of degradation products also can be significant. When the carbamate insecticide aldicarb (Temik®) first appeared, its very high toxicity was thought to be offset by a lack of persistence. However, it soon became evident that the two principal degradation products, aldicarb sulfoxide and sulfone, were not only as toxic as their parent but were more persistent. Similarly, the organophosphorus insecticide parathion was found to be degraded on cotton leaves by sunlight, but the products were still about as toxic as the original. Are the parent compounds to be considered persistent if they produce long-lived offspring?

Perhaps no there is acceptable scientific definition of persistence. Perhaps the best we can do is just to give it an operational definition in terms of some threshold physical and chemical properties that will trigger action, as has been done for groundwater pollution (Table 14.6). Although such an approach certainly does not consider all the environmental variables that can influence how long a particular chemical will remain at toxicologically pertinent levels, it seems like a good way to start.

APPENDIX 14.1.
Chlorinated Hydrocarbons in California Shellfish (ppb)[a]

Hydrocarbon	Salinas	Marina del Rey	SF Bay	Bodega Bay
Total DDTR	**3769.0**	**295.7**	**20.7**	**1.8**
p,p'-DDT	467.2	114.4	2.5	nd[b]
o,p'-DDT	146.0	nd	nd	nd
p,p'-DDE	1971.0	114.4	10.2	0.4
o,p'-DDE	42.3	10.9	1.2	nd
p,p'-DDD	876.0	114.4	3.6	1.4
o,p'-DDD	233.6	10.9	2.1	nd

APPENDIX 14.1. (*continued*)

Hydrocarbon	Salinas	Marina del Rey	SF Bay	Bodega Bay
Benzene hexachloride	**nd**	**3.7**	**0.3**	**1.8**
α-Isomer	nd	1.3	0.3	1.8
β-Isomer	nd	0.7	nd	nd
γ-Isomer (lindane)	nd	1.7	nd	nd
Cyclodienes	**293.8**	**158.4**	**27.2**	**2.4**
cis-Chlordane	19.0	54.6	6.1	1.1
trans-Chlordane	15.3	37.0	4.6	nd
Heptachlor[c]	nd	1.0	nd	nd
Nonachlors	35.0	45.8	3.4	nd
Aldrin	15.3	nd	nd	nd
Dieldrin	182.5	16.0	12.9	1.2
Endosulfan	17.5	nd	0.3	nd
Toxaphene	**1679.0**	**nd**	**nd**	**nd**
HCB	**1.2**	**nd**	**nd**	**nd**
Total PCB	**292.0**	**316.8**	**103.5**	**2.0**

[a]SWRCB, 1988.
[b]None detected.
[c]Heptachlor and epoxide.

CHAPTER **15**

Reactive Pollutants

15.1 | REACTIVITY

Although some toxic chemicals may seem quite inert (Chapter 14), toxicity, the "biochemical lesion," normally requires reactivity. Unreactive DDE and β-BHC are virtually nontoxic, while their more reactive relatives DDT and lindane are insecticidal. Reactive metabolites often account for the toxicity of other poorly reactive chemicals, as in the case of benzo[*a*]pyrene. Reactivity also controls both metabolic and environmental transformation: The higher the reactivity, the greater the toxicity and the lower the environmental persistence. Toxicants that act by physical rather than biochemical means, such as halocarbon anesthetics, are exceptions.

What is "reactivity"? Reactivity is a measure of the breaking and remaking of chemical bonds and so implies certain bond strengths, activation energies, and electronegativities in the atoms and molecules involved. Energy is required to break a bond, and the bond dissociation energies listed in Table 5.1 are typical (but apply only to the bonds indicated). D values generally are measured or calculated for gas-phase free-radical reactions, but those for reactions in solution also depend on the influence of neighboring atoms or groups, molecular size and shape, solvent, and attacking reagent if any, and especially on the ionic character of the molecule.

The principal types of environmental reactions were shown in Chapters 5 and 6 to include hydrolysis and other nucleophilic processes, oxidation (in this context, oxygenation), reduction (meaning hydrogenation), elimination, and a few others. The present chapter discusses a selection of common types of environmental chemicals that readily participate in these reactions, and some consequences of that reactivity.

15.2 | ALKYL HALIDES

The most important of the environmental reactions of alkyl halides is the alkylation of nucleophiles, generally by an SN2 mechanism. Already introduced in Section 6.3.2

and Special Topic 9, the formation of powerfully electrophilic carbonium ions, which can methylate ("alkylate") electronegative sites such as the amino groups in DNA or the SH in glutathione, cause alkylating agents to be acutely poisonous and often genotoxic. **Methyl bromide,** a gaseous fumigant very common in agriculture and previously in structural termite control, is very toxic but apparently not carcinogenic (USPHS, 1991). It reacts so rapidly with glutathione and other thiols that it apparently does not persist long enough to reach DNA in the nucleus.

$$CH_3Br + {}^-SCH_2\overset{\overset{\displaystyle NH^-}{|}}{C}HCO^- \longrightarrow Br^- + CH_3SCH_2\overset{\overset{\displaystyle NH^-}{|}}{C}HCO^- \qquad (15.1)$$

Its most harmful effect in mammals is neurotoxicity, and in California, at least, methyl bromide has been responsible for more occupational deaths than any other pesticide. In soil, alkylation reactions of methyl bromide, including hydrolysis, liberate bromide ion; taken up by pasture grass, the bromide has caused illness and death among grazing horses. Although intended to kill plant-destroying nematodes, methyl bromide undoubtedly is lethal to most other soil dwellers as well.

Being volatile, both natural and manmade methyl bromide makes its way into the troposphere, where it survives about a year. A small proportion reaches the stratospheric ozone layer and is photolyzed there to destructive bromine atoms (Special Topic 5). It is readily attacked by hydroxyl radical to form $BrCH_2O_2\bullet$ and the related $BrCH_2O\bullet$ and formyl bromide, $HCOBr$ (Orlando and Tyndall, 1996).

Another fumigant, **ethylene dibromide** (EDB, 1,2-dibromoethane), also reacts readily with –SH groups but generates another alkylating carcinogen by doing so (Eq. 6.23). There was a public clamor in the early 1980s when EDB residues from grain fumigation were detected in packaged bakery mixes and again when people learned that EDB had been used for years as an additive in leaded gasoline. In one particularly acrimonious instance in Hawaii, high EDB residues in soil, blamed on nematicidal fumigation by the pineapple industry, were later discovered to be due to a massive underground gasoline leak from a military fuel line.

The alkyl bromides, as well as methyl iodide produced biochemically by marine plants (Section 2.4), alkylate water and so undergo slow hydrolytic detoxication in soil. The nucleophilic reactivity of water is poor, and if its reaction rate (nucleophilicity) is taken as 1, that of Cl^- is 1000, OH^- is 1.6×10^4, and SH^- is 1.3×10^5 (Hine, 1962). Alkyl halides also dissociate photochemically in the stratosphere to alkyl and bromine radicals, placing them among the worst threats against Earth's protective ozone layer (Special Topic 5). By the time the size and degree of halogenation reach that of DDT, steric shielding of the α carbon, oxidation, and elimination at the β carbon almost preclude alkylation.

15.3 | CHLOROPHENOLS

Chlorophenols have been common items of commerce and serious environmental pollutants for decades. Like many chemicals, the phenol ring becomes electronically activated upon absorption of UV radiation (Section 5.2), a good example being the widely used wood preservative **pentachlorophenol** (PCP). PCP is stable indefinitely in aqueous solution in the dark but is degraded rapidly in sunlight (Wong and Crosby, 1981). Although photoreduction and oxidation take place, the most important abiotic transformation is photonucleophilic displacement of chloride from the UV-activated ring by an hydroxide ion from water (Section 5.5.2). For each chlorine, replacement by hydroxyl leads to a dihydroxybenzene and is followed by oxidations to cleave the aromatic ring to acids such as DCMA and finally give small organic fragments, chloride ions, and water (Fig 15.1). In sunlight at pH 7.3, the PCP half-life is about 1 day.

PCP is a general biocide (Section 9.3), affecting plants, animals, and microorganisms. Consequently, it has been used for decades as a wood preservative against insects and fungi, as a herbicide, an algicide in cooling tower water, and a disinfectant, and it had become one of the most widely used pesticides by the late 1970s (Rao, 1978). In the summer of 1980, a barge loaded with solid PCP collided with another vessel in the New Orleans Ship Canal, the brackish waterway that connects the Port of New Orleans with the Gulf of Mexico. A large storage container filled with PCP was knocked overboard, and the channel was closed to shipping because of a concern that spilled chemical would be stirred up by the ships' wakes and distributed to nearby oyster and shrimp beds. Actually, there seems never to have been much danger from the PCP, as residues declined rapidly in the warm alkaline water under a summer sun. The PCP was gone within a short time, and the port could reopen. Of more concern were the persistent dioxin impurities (Table 13.5), which remained in the bottom mud for weeks.

Tri- and tetrachlorophenols (called Dowicides®), circulated through open towers as algicides in industrial cooling water, soon lost their effectiveness and required

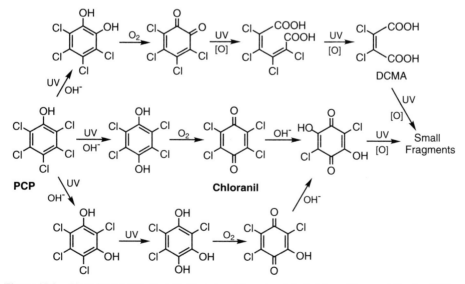

Figure 15.1. Photodegradation of pentachlorophenol in water. Adapted from Wong and Crosby, 1981.

frequent replacement. Although most organisms degrade chlorophenols along much the same lines as shown in Fig. 15.1, photonucleophilic degradation in sunlight was shown to have been primarily responsible in this case.

15.4 | DIVALENT SULFUR COMPOUNDS

Thiols, thioethers, and phosphorothioates are environmental sulfur compounds of particular interest, and oxidation is the principal route by which they react. This process is rapid and usually results in detoxication. Thiols react with even mild oxidizing agents to form an S–S (disulfide) bond, as in the case of DHL (Fig. 9.2). Relatively few thiols persist long enough in the environment to present a threat, although the environmental impact of widely used rubber antioxidants (accelerators), such as the skin sensitizer 2-mercaptobenzothiazole, has received little attention.

The environmental oxidations on sulfur often involve free-radical oxidants such as hydroxyl radicals or molecular oxygen, but ionic reagents like hydrogen peroxide and hypochlorite also serve.

$$R\!-\!\overset{..}{\underset{R'}{S}}: +\ \underset{H}{O}\!-\!OH \quad\xrightarrow{-OH^-}\quad R\!-\!\overset{+}{\underset{R'}{S}}\!-\!OH \quad\xrightarrow{-H^+}\quad R\!-\!\overset{+}{\underset{R'}{S}}\!-\!O^- \qquad (15.2)$$

This reaction is especially important for organophosphorus insecticides. The World War II search for chemical weapons by both sides resulted in the development of an important class of pesticides, the organophosphorus esters (Scheme 15.3). Esters of phosphoric acid normally are not especially toxic, as illustrated by the triphenyl phosphate used for insulation in electrical transformers. However, incorporation of a high-energy bond, as for paraoxon (Section 6.1), provided toxic agents for use against many of man's most serious pests. As expected from their chemical warfare origins, the first organophosphate insecticides were extremely poisonous, but later versions were tamed somewhat by introduction of the less electronegative sulfur in place of the coordinate-covalent oxygen to form phosphorothionates and phosphorodithioates (Scheme 15.3), collectively called phosphorothioates.

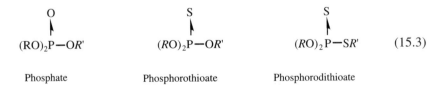

$$\underset{\text{Phosphate}}{\overset{\displaystyle O}{\overset{\uparrow}{(RO)_2P\!-\!OR'}}} \qquad\qquad \underset{\text{Phosphorothioate}}{\overset{\displaystyle S}{\overset{\uparrow}{(RO)_2P\!-\!OR'}}} \qquad\qquad \underset{\text{Phosphorodithioate}}{\overset{\displaystyle S}{\overset{\uparrow}{(RO)_2P\!-\!SR'}}} \qquad (15.3)$$

In order for phosphorothioates to become toxic, they first must be oxidized to the corresponding phosphate, termed an "oxon":

$$(CH_3O)_2\overset{\overset{\displaystyle S}{\|}}{P}-SCH-COOC_2H_5 \quad\xrightarrow{\text{[O]}}\quad (CH_3O)_2\overset{\overset{\displaystyle O}{\|}}{P}-SCH-COOC_2H_5$$
$$\underset{CH_2COOC_2H_5}{|} \qquad\qquad\qquad \underset{CH_2COOC_2H_5}{|}$$

$$\qquad\qquad\qquad\qquad\qquad\qquad\qquad\qquad\qquad\qquad (15.4)$$

Malathion   Malaoxon

Oxons are much more reactive toward nucleophiles than are the thions and so act more rapidly on enzymes such as acetylcholinesterases (Section 9.4.1). Pure **malathion** [S-(1,2-dicarbethoxyethyl) O,O-dimethylphosphorodithioate] is itself almost nontoxic, and what toxicity is shown by the commercial product is due to traces of oxon impurities such as byproduct $(CH_3O)_2P(O)SCH_3$ (Table 15.1). Many kinds of environmental oxidizing agents convert P–S bonds to P–O, including air, ozone, aqueous Cl_2, hydrogen peroxide, and cytochrome P450. The mechanism involves initial oxygenation of the sulfur, formation of a three-membered P–O–S ring, and extrusion of elemental sulfur to generate the oxon (Eq. 15.5). The chemistry, mode of action, and toxicity of organophosphorus insecticides have been reviewed recently by Chambers and Levi (1992), although the 1961 book by Heath still contains a lot of useful information.

TABLE 15.1
Impurities in Technical Malathion[a]

Formula	%	LD$_{50}$ (mg/kg)[b]
$(CH_3O)_2\,P(S)SCHCOOC_2H_5$[c] $\quad\quad\quad\quad\vert$ $\quad\quad\quad CH_2COO\,C_2H_5$	96+	12,500
$(CH_3O)_2\,P(S)SCH_3$	1.1	660
$(CH_3O)(CH_3S)P(O)SCHCOOC_2H_5$ $\quad\quad\quad\quad\quad\quad\vert$ $\quad\quad\quad\quad\quad CH_2\,COO\,C_2H_5$	0.2	120
$(CH_3O)_2\,P(O)SCH_3$	0.04	260
$(CH_3S)_2\,P(O)OCH_3$	0.003	110
$(CH_3O)_2P(S)SCHCOOC_2H_5$ $\quad\quad\quad\quad\vert$ $\quad\quad\quad CH_2COOH$	0.6–0.8	—[d]
$(CH_3O)_2\,P(S)S\text{-}P(S)(OCH_3)_2$	0.5	—[d]

[a] Umetsu et al., 1977.
[b] Rat oral.
[c] Malathion.
[d] Not reported.

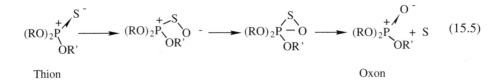

Thion   Oxon

(15.5)

Organic sulfides (thioethers) similarly are oxidized to sulfoxides and sulfones, which can have properties very different from those of the parent. An example is the highly toxic but nonpersistent carbamate insecticide, **aldicarb** (Temik®), $CH_3SC(CH_3)_2CH = NO\text{-}CONHCH_3$, which is environmentally oxidized to its persistent sulfoxide and sulfone, which are even more toxic.

15.5 | DITHIOCARBAMATES

Dithiocarbamates are widely used as fungicides (Section 9.6.2) and as antioxidants in rubber. Like other divalent sulfur compounds, they are easily air oxidized to disulfides, although an environmentally more significant reaction may be the classic β-elimination to form isothiocyanates (Scheme 9.11).

In July, 1991, a railroad tankcar carrying 19,000 gallons of a 37% aqueous solution of **metam-sodium** (Vapam®, sodium N-methyldithiocarbamate) derailed near the mountain town of Dunsmuir, California, spilling over half its contents into the upper Sacramento River. Vapam is a soil sterilant that generates toxic and volatile **methyl isothiocyanate** (MITC, Reaction **2**) from the dithiocarbamic acid formed by hydrolysis of the salt (Reaction **1**).

$$\underset{\text{Metam-sodium}}{CH_3NH-\overset{\overset{\text{S}}{\|}}{C}-SNa} \xrightarrow[\mathbf{1}]{H_2O} CH_3NH-\overset{\overset{\text{S}}{\|}}{C}-SH \xrightarrow[\mathbf{2}]{-H_2S} \underset{\text{MITC}}{CH_3N=C=S} \quad (15.6)$$

Almost all plant and animal life in and along the river was damaged or destroyed, and methyl isothiocyanate vapor filled the narrow valley and choked the Dunsmuir residents. A front of chemical moved 40 miles downstream to Lake Shasta and dissipated there. Fortunately, no one seems to have been killed or seriously injured, although massive lawsuits were filed. Within three years, the water and biota had largely returned to normal, but it is worth considering what might have happened if the spilled chemical had been one that did not decompose and dissipate so readily.

15.6 | NITROARENES

Environmental and metabolic reduction (hydrogenation) of carbonyls, organochlorines, and nitro compounds is common. The environmental reduction of a nitro group to an amine takes place stepwise via intermediate nitroso and hydroxylamino compounds and may employ either sequential uptake of electrons and hydrogen ions (Section 5.4.2), hydride ion transfer (say, from glucose), or addition of H atoms:

$$RN\overset{+}{\underset{\backslash\backslash O}{\diagup}}\!O^- \xrightarrow{2\text{H}\cdot} RN\overset{\diagup OH}{\underset{\backslash OH}{}} \xrightarrow{-\text{H}_2\text{O}} RN{=}O \xrightarrow{\text{H}\cdot} R\overset{\bullet}{N}{-}OH \xrightarrow{\text{H}\cdot} RNH{-}OH \xrightarrow[-\text{H}_2\text{O}]{2\text{H}\cdot} RNH_2$$

Nitro $\qquad\qquad$ Nitroso $\qquad\qquad$ Hydroxylamine $\quad$ Amine

$$(15.7)$$

The result usually is detoxication, as when parathion is reduced to aminoparathion in soil (Section 6.1), but there are exceptions. The highly reactive hydroxylamine intermediates can become metabolically activated in a manner analogous to the N-hydroxyamides (see Eq. 15.9), and some nitro and nitroso compounds thus become carcinogenic. Although carcinogenicity and mutagenicity long have been observed in simple nitroaromatic compounds (nitroarenes), a wide range of more complex nitroarenes, such as **2-nitropyrene,** are found in diesel exhaust, emissions from portable gas heaters, the carbon black used in inks, and atmospheric particulate matter from all over the world (Appendix 15.1). They are formed by either thermal or photochemical nitration of PAH by atmospheric nitrogen oxides (IARC, 1989). Recognition of the significant environmental levels of nitropyrenes and nitrobenzopyrenes now has contributed to improvements in both engine design and combustion characteristics and to better quality in ink pigments.

One of the most widely used nitro compounds is the military explosive, **TNT** (2,4,6-trinitrotoluene). TNT and related compounds enter the environment as wastes from the production and discard of munitions, and to whatever extent they escape destruction during an explosion. TNT is manufactured by stepwise nitration of toluene, and processing wastewater—known as "pink water"—contains mono-, di-, and trinitrotoluenes, aminodinitrotoluenes, and at least 17 other related chemicals (Marinkas, 1996).

"Pink water" gets its name primarily from the color of TNT photodegradation products formed when the water is exposed to sunlight. TNT and the dinitrotoluenes are rapidly photolyzed (Fig. 15.2), and the red color is due in part to azoxybenzenes. In sunlight, one major product is trinitrobenzene, presumably derived from trinitrobenzaldehyde via trinitrobenzoic acid, and another is dinitroanthranil (Yinon, 1990).

Considering the power and heat generated in a TNT explosion, detonation products have been understandably difficult to capture and identify. As the explosion can take place in an inert atmosphere, this is not an ordinary combustion but rather a series of internal reactions (Fig. 15.2). As with any process involving the action of heat and pressure on an organic material, some PAH is produced, but much of the

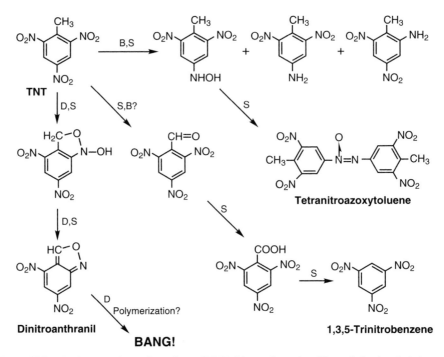

Figure 15.2.   Environmental transformations of TNT: Biotransformation (B), sunlight photolysis in water (S), and detonation (D).

original carbon can be accounted for as microscopic diamond dust, a discovery that was quickly commercialized.

TNT also is readily metabolized by animals, plants, and microorganisms, primarily by reduction. About half of the total metabolites in animals and plants appear to be sugar conjugates, but as they do not yield trinitrobenzyl alcohol upon hydrolysis, one may suspect that they represent the novel conjugates of nitrophenylhydroxylamines already known as metabolites. However, the tetranitroazoxybenzenes reported as metabolites almost certainly arise from the isolation procedure instead.

TNT is readily absorbed through the skin and lungs, but although TNT poisoning presents an occupational hazard that has injured or killed thousands of workers, the compound still is not considered highly toxic to mammals (acute oral LD_{50} in rats and mice is about 1000 mg/kg). Chronic exposure in man and other mammals results in skin rash, aplastic anemia, methemoglobinemia, and severe liver damage and jaundice, and a maximum safe exposure level of 4 μg/kg/day has been proposed. TNT is toxic to most other life forms, including bacteria, fungi, algae, higher plants, aquatic invertebrates, and fish, and it is mutagenic in several assays. The urine of workers exposed regularly to TNT also is mutagenic, possibly due to the phenylhydroxylamine metabolites. The carcinogenicity of pure TNT is questionable and could be due to the carcinogenic 2,6-dinitrotoluene impurity found in technical TNT.

15.7 | AMINES AND THEIR DERIVATIVES

Both the darkening of aromatic amines upon air oxidation and their occasional toxicity and carcinogenicity have been recognized for many decades, but the possible connection has not. For example, the common headache remedy, **acetaminophen** (Tylenol, paracetamol), normally is excreted as conjugates in the urine, but a small proportion is oxidized metabolically to a highly reactive and hepatotoxic acetimidoquinone:

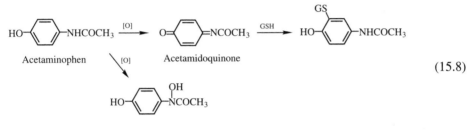

Acetaminophen   Acetamidoquinone

N-Hydroxyacetaminophen

(15.8)

This highly conjugated system rapidly depletes the protective hepatic glutathione via Michael addition to the quinoid ring, and a similar reaction can then take place with protein SH groups.

Like another pain reliever, phenacetin (*p*-ethoxyacetanilide, APC), acetaminophen is carcinogenic at high doses due to hydroxylation on the nitrogen (Searle, 1984). Conjugation of the resulting hydroxylamine (Eq. 15.9) followed by loss of sulfate leaves behind powerfully electrophilic nitrenium ions which react with purines and pyrimidines of DNA. Many carcinogenic amines, such as *o*-toluidine and benzidine, utilize this mechanism.

$$ R-\overset{\overset{\displaystyle OH}{|}}{N}COCH_3 \longrightarrow R-\overset{\overset{\displaystyle OSO_3^-}{|}}{N}COCH_3 \xrightarrow{-SO_4^{2-}} R-\overset{+}{N}COCH_3 \xrightarrow{DNA-NH_2} R-\overset{\overset{\displaystyle NH-DNA}{|}}{N}COCH_3 \qquad (15.9) $$

Nitreniumion   DNA Conjugate

N-Hydroxylation also must occur in the environment, but the resulting hydroxylamines are so unstable that any toxic consequences remain unknown. While one might not think of acetaminophen as a particularly "environmental" toxicant, 96% of the millions of pounds consumed each year to relieve the world's aches and pains is excreted into sewers either unchanged or as hydrolyzable sulfate and glucuronide conjugates. We can well ask what ever becomes of it all.

Aliphatic and aromatic secondary amines easily form **nitrosamines** (*N*-nitroso amines) in the presence of nitrous acid or nitrogen oxides, or by bacterial action. Nitrosamines are found widely in the environment and have raised considerable concern because of their unexpected toxicity. Imagine the consternation in the mid-1950s when a common industrial intermediate and solvent, *N,N*-dimethylnitrosamine (DMN), also used as rocket fuel, was found to be hepatotoxic and highly carcinogenic in all species tested. In sheep, for example, a single dose of 5 mg/kg or 12 daily doses of 0.5 mg/kg proved lethal. DMN and other nitrosamines subsequently were found in food, beer, and tobacco products (Appendix 15.2), and particularly in popular foods such as bacon that are cured or preserved with nitrites or nitrates.

Alkylnitrosamines, especially, are powerful carcinogens. They become metabolically activated by oxidation and subsequent generation of carbonium ions which can alkylate DNA (Searle, 1984):

$$CH_3\overset{\overset{\displaystyle N=O}{|}}{N}CH_3 \xrightarrow{[O]} CH_3\overset{\overset{\displaystyle N=O}{|}}{N}CH_2OH \xrightarrow{-HCHO} [CH_3NH-N=O \rightleftharpoons CH_3N=N-OH] \xrightarrow[-OH^-]{-N_2} \overset{+}{C}H_3 \quad (15.10)$$

DMN   Methylcarbonium

Tobacco-induced cancer is thought to be due in part to nitrosamines such as nitrosonornicotine. While liver cancer is the most common result, the type and location of tumors are both species and organ specific. Considering the common occurrence of nitrite in the environment, nitrosamines, too, must occur frequently, but little is known about their impact on nonmammalian species.

15.8 | ESTERS

15.8.1. Phosphate Esters

The widespread use of phosphate esters as pesticides, heat transfer media, electrical insulation, fireproofing agents, and solvents ensures their entry into the environment. Simple phosphates are unreactive, as their uses indicate, and their hydrolysis is very slow (Table 15.2). Consequently, with the exception of tri-*o*-cresyl phosphate (Section 7.4.2), they are not particularly toxic.

As discussed in Section 9.4.1, organophosphate insecticides depend on a high-energy ester link for toxic action. The rate of their reaction with, and inhibition of, AChE is directly correlated with the rate of alkaline hydrolysis (Table 15.2), as both serine 203 and hydroxide anion attack the backside of the tetrahedral P (Eq. 15.11) to expel a weaker nucleophile, X. Consequently, the less nucleophilic X is, the the better leaving group it becomes.

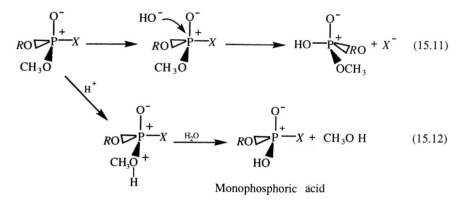

Monophosphoric acid

TABLE 15.2
Some Important Organophosphorus Esters

$$
\begin{array}{c}
X^{-} \\
| \\
R-\overset{+}{P}-R' \\
| \\
R''
\end{array}
$$

Common name	X	R	R'	R"	LD$_{50}$ (mg/kg)[a]	$t_{1/2}$ (hrs)[b]
TEP[c]	O	OEt[d]	OEt	$-$OEt	>1600	>10^6
Acephate	O	OMe	SMe	$-$NHCOCH$_3$	945	60
Dichlorvos	O	OMe	OMe	$-$OCH$=$CCl$_2$	80	48
Disulfoton	S	OEt	OEt	$-$SCH$_2$CH$_2$SEt	3	3144
Paraoxon	O	OEt	OEt	$-$O$-\langle\bigcirc\rangle-$NO$_2$	2	370[e]
Parathion	S	OEt	OEt	$-$O$-\langle\bigcirc\rangle-$NO$_2$	14	3120
Diazinon	S	OEt	OEt	$-$O$\overset{\text{CH}_3}{\diagdown}N\diagup$CH(CH$_3$)$_2$	108	4440
Malathion	S	OMe	OMe	$-$SCHCOOEt$\mid$CH$_2$COOEt	1375	f

[a] Acute oral toxicity in male rats (Gaines, 1960).
[b] Hydrolysis halflife at pH 9 and 25°C.
[c] Triethyl phosphate.
[d] Et = ethyl, Me = methyl.
[e] pH 8.
[f] Carboxylate ester hydrolyzed preferentially.

Also, anything that causes the phosphorus to have more positive charge (such as coordinate-covalent O rather then S) increases the reaction rate. This explains why paraoxon is more toxic than either triethyl phosphate or parathion. As alkyl phosphates are unreactive toward nucleophiles, a phosphorylated serine is resistant to hydrolysis, but the hydrolysis of both the alkyl phosphates and phosphorothionates is greatly accelerated by Cu^{2+} (Heath, 1961).

Acid and neutral hydrolysis are slower than alkaline hydrolysis, act at an alkoxy substituent, and follow a different mechanism (Eq. 15.12). Phosphorylated enzymes undergo this type of transformation to give unreactive monophosphoric acids, so treatment of a poisoning victim with an antidote such as 2-PAM (Section 9.4.1) must be done quickly. The half-life of this "aging" is 58 hours for diethylphosphorylated

AChE from human red blood cells but only 50 minutes for the dimethylphosphory-lated enzyme (Chambers and Levi, 1992).

However, in small enough doses, the phosphate ester may never arrive at the target. Esters like paraoxon are hydrolyzed by esterases in the blood, although the rate varies between animal species. Rabbit serum, for example, is about 20 times more active in phosphate hydrolysis than that of a rat and 40 times that of a mouse (Aldridge, 1953), the reaction taking place at the high-energy bond. Selective toxicity can also be due to species differences in the rates of oxidative biotransformation, and toxicity may be increased by interfering with or diverting oxidative metabolism in a target organism with a readily oxidized chemical synergist. Piperonyl butoxide has been the synergist used most widely.

Organophosphorus insecticides long were considered to have low environmental persistence, but this is not universally true (Special Topic 14). Their residues are well known to be toxic to wildlife, especially in nontarget animals such as wild birds. Extensive losses among redtailed hawks in California's Central Valley were traced to organophosphorus insecticides accumulated in fieldmice and insects on which they fed and to absorption through the the the bird's feet as they perched in sprayed trees.

15.8.2. Carboxylate Esters

Esters of carboxylic acids are common in the environment, both as natural products such as amyl acetate and as pollutants including phthalate ester plasticizers and persistent herbicides (*e.g.,* chlorthal dimethyl, Section 14.6). As with phosphates, their most important reaction generally is hydrolysis to the corresponding acid and alcohol, a rapid S$_N$2 reaction under alkaline conditions. The hydrolysis of esters such as malaoxon (Section 6.3.1) often results in detoxication, but herbicidal esters like those of 2,4-D (Table 5.6) simply form their toxic acids.

The ester mixture that is the active component of **pyrethrum** contains six related insecticidal **pyrethrin** esters and is extracted from chrysanthemum flowers, especially from *C. cinerariaefolium*. Used originally by tribesmen of Asia Minor to protect themselves against body lice, "Persian insect powder" was introduced into Europe in the early 1800s, and its production shifted later to Yugoslavia, then Kenya and Japan, and most recently to Australia. The harvested flowers are dried, powdered, extracted with petroleum solvent, and the filtered extract concentrated to provide the "pyrethrum" oleoresin of commerce (30–35% pyrethrins). The highly variable annual production of refined pyrethrins reached 200,000 kg in recent years, and the filter cake of powdered flowers still containing some pyrethrins is processed into the familiar "mosquito coils."

The individual insecticides, called pyrethrins I and II, cinerins I and II, and jasmolins I and II, are esters of two series of cyclopropanecarboxylic acids (Table 15.3) with keto-alcohols, called rethrolones, containing olefinic sidechains of various lengths. Pyrethrin chemistry has been investigated for more than a century by scientists from many countries (Crombie, 1995). The pyrethrins' outstanding characteristics of rapid knockdown of flying insects, very low mammalian toxicity (acute oral LD$_{50}$ 2370 and 1030 mg/kg of the ester mixture in male and female rats, respectively), and rapid deactivation have led to widespread home and garden use but only

TABLE 15.3
Natural Pyrethrins

Name	R	R'
Cinerin I	$-CH_3$	$-CH_2-CH=CH-CH_3$
Pyrethrin I	$-CH_3$	$-CH_2-CH=CH-CH=CH_2$
Jasmolin I	$-CH_3$	$-CH_2-CH=CH-CH_2CH_3$
Cinerin I	$-COOCH_3$	$-CH_2-CH=CH-CH_3$
Pyrethrin I	$-COOCH_3$	$-CH_2-CH=CH-CH=CH_2$
Jasmolin I	$-COOCH_3$	$-CH_2-CH=CH-CH_2CH_3$

limited use in agriculture. The mechanism of pyrethrin toxicity is the same as that of DDT, blockage of neuronal Na^+ channels.

Hydrolysis of the pyrethrins occurs readily and is the basis of its metabolic detoxication and selectivity. However, natural pyrethrins break down even more rapidly under UV irradiation, and the half-life of pyrethrin I on a sunlit surface is only a few minutes (Chen and Casida, 1969). Isomerization and polymerization of the rethrolone side chain and extensive free-radical oxidation on the cyclopropane ring predominate (Crosby, 1995). A pyrethrin bug bomb sprayed into a large sunlit room dissipated from the air and horizontal glass plates within 2 days, leaving behind only what were identified as ozonolysis products (Class and Kintrup, 1991). This instability to light has llmited agricultural uses and encouraged development of more stable synthetic analogs.

The synthetic *"pyrethroid"* insecticides provide a wide range of environmental stabilities, but their structural relation to pyrethrins sometimes strains the imagination (Fig. 15.3). The unstable rethrolone part was replaced by a photochemically unreactive group (as in tetramethrin), or the reactive side-chain on the cyclopropanecarboxylate replaced by a group resistant to oxidation (permethrin). Eventually, the cyclopropane ring was dispensed with entirely (fenvalerate), then the ester link (flufenprox), and finally an insecticidal molecule emerged in which no remnant of the pyrethrin structure remained (silafluofen). The mode of toxic action remains the same, but environmental half-life can now exceed 30 days.

15.8.3. Carbamate Esters

Carbamate esters, *R*NH-COO*R'*, represent another important class of insecticides. Like the organophosphates, certain **N-methylcarbamates** inhibit acetylcholinesterase (Section 9.4.1), carbamoylating the Ser 203 rather than phosphorylating it. However, formation of the enzyme–substrate complex with carbamates is reversible, and hydrolytic regeneration in the carbamoylated enzyme is rapid. The carbamoyl group fits into the enzyme's "acyl pocket" (Fig. 9.3), metabolic stabilization of the

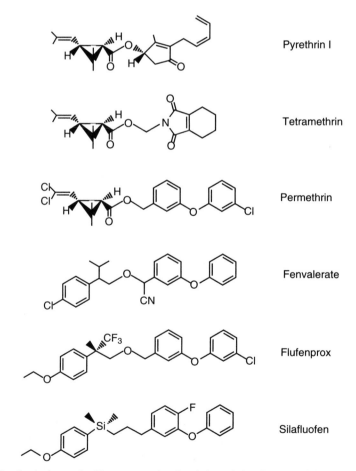

Pyrethrin I

Tetramethrin

Permethrin

Fenvalerate

Flufenprox

Silafluofen

Figure 15.3. Synthetic pyrethroid structures, showing their evolution from pyrethrin I.

complex ("aging") does not occur, and 2-PAM is not an effective antidote. Unlike organophosphates, reactivity as measured by hydrolysis rate does not seem to be the primary factor in phenylcarbamate toxicity, and affinity for the active site may be more important (Kuhr and Dorough, 1976).

It was originally assumed that carbamates would be less toxic than phosphates, and the first commercially successful carbamate insecticide, **carbaryl** (Sevin®, 1-naphthyl N-methylcarbamate), indeed showed very favorable mammalian toxicity. Unfortunately, this was not necessarily true of later arrivals (Table 15.4), and aldicarb (Temik®) proved to be one of the most toxic of all insecticides to mammals. Due to low metabolic detoxication rates, even carbaryl proved very toxic to bees and to many aquatic invertebrates, a hazard moderated only by rapid environmental hydrolysis. However, there is evidence that carbaryl's toxicity to molluscs, at least, may actually be due largely by its hydrolysis product, 1-naphthol (Stewart et al., 1967).

Hydrolysis indeed is the most important biotransformation route of carbamates in humans and bacteria (Eq. 15.13). As with phosphate insecticides, the toxic carbamates all possess a high-energy O–CO ester bond (Table 15.4) that is unusually sus-

TABLE 15.4
Major N-Methylcarbamate Insecticides

$$R-O\overset{\overset{\displaystyle O}{\displaystyle \|}}{C}-NHCH_3$$

Common name	Trade name	R	LD_{50} (mg/kg) [a]
Aldicarb	Temik	$CH_3S-\overset{\overset{\displaystyle CH_3}{\displaystyle |}}{\underset{\underset{\displaystyle CH_3}{\displaystyle |}}{C}}CH=N-$	0.8
Carbaryl	Sevin		850
Carbofuran	Furadan		37
Methomyl	Lannate	$CH_3S-\overset{}{\underset{\underset{\displaystyle CH_3}{\displaystyle |}}{C}HCH=N-$	21
Propoxur	Baygon	$OCH(CH_3)_2$	83
Mexacarbate	Zectran	$CH_3\diagdown\diagup CH_3$ $N(CH_3)_2$	37

[a] Acute oral toxicity in male rats (Gaines, 1969).

ceptible to both alkaline and enzymatic hydrolysis. Oxidation occurs initially on the *N*-methyl group (Eq. 15.14), although aromatic ring oxidation also can be extensive.

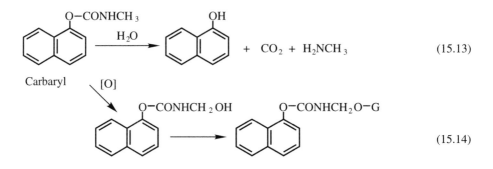

$$\text{Carbaryl} \quad + \quad CO_2 + H_2NCH_3 \tag{15.13}$$

$$\tag{15.14}$$

Most vertebrates rely on oxidation for 60–80% of their carbamate detoxication, insects 80–90%, and plants 100% (Matsumura, 1985), and hydroxylated metabolites are excreted as sugar conjugates (Eq. 15.14, G = glucose or glucuronic acid). Environmental residues of most carbamates are negligible due to hydrolysis and photooxidation, although aldicarb sulfoxide and sulfone remain in soil for long periods (Section 15.4).

15.9 | CARBON MONOXIDE AND CYANIDES

Chelation of metal ions with organic ligands is a common chemical reaction. Two important environmental toxicants, **carbon monoxide** and **cyanide,** coordinate to iron but with different results. Carbon monoxide was introduced in Section 1.6 as a major toxic hazard responsible for over 20% of U.S. poisoning deaths (Table 1.2). The principal sources of exposure are smoking, defective heaters and furnaces, and automobile exhaust, although CO also is a normal product of metabolism and a variable component of the natural atmosphere.

With each breath, hemoglobin in blood combines with atmospheric oxygen to form oxyhemoglobin, which is transported through the body to allow cells to respire. Hemoglobin is composed of a globular protein and an iron complex in which hexacoordinate Fe is bonded to four porphyrin nitrogens ("heme") and to a single imidazole nitrogen from the protein's histidine. This leaves one of the Fe orbitals open to combine with molecular oxygen and provide oxyhemoglobin (Fig. 15.4). However, carbon monoxide binds to the iron about 250 times more tightly than does oxygen and so competes strongly to form carboxyhemoglobin, COHb, which no longer transports oxygen.

About 0.5% of the hemoglobin in normal human blood is present as COHb, but the COHb level in smokers is 5–12%. Impairment of the nervous system starts at about 2.5%, headache at 10–20%, dizziness and vomiting at 30–40%, collapse at 40–

Figure 15.4. Structures of hemoglobins: Oxyhemoglobin, $X = O_2$; carboxyhemoglobin, $X = CO$; cyanohemoglobin, $X = CN$. His = protein-bound histidine. The Fe atom ● is bonded to all four porphyrin nitrogens.

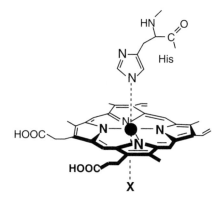

50%, with death when COHb exceeds 60%. See Section 7.4.4. Although breathing air that contains even 100 mg/m^3 of CO results in only about 5% COHb in the blood, the effect is cumulative. The *total* exposure to CO is what counts.

Being isoelectronic with CO, cyanide combines similarly with hexacoordinate iron. In this case, the iron is a key part of the cytochromes that transfer electrons from the electron transport system to molecular oxygen (Evered and Harnett, 1988). Cells become unable to utilize O_2, even though it may be present in abundance, and death is due to respiratory failure. Cytochrome P450 also is inhibited, so most detoxication stops. The actual quantity of available cytochromes always is small and the cyanide binding strong; very little cyanide is required to shut things down. A level of 1 μg/mL in blood is lethal and corresponds to a total dose of about 50 mg of HCN in an adult human. Antidotal treatment starts with conversion of the ferrous iron to ferric by administration of amyl nitrite or sodium nitrite, as Fe^{III} does not coordinate strongly with cyanide. This is followed by sodium thiosulfate to react with excess cyanide and give less toxic thiocyanate and sulfite:

$$CN^- + S_2O_3^{2-} \longrightarrow SCN^- + SO_3^{2-}$$

(15.15)

Cyanide occurs naturally in plants and in human food (Section 12.3), sometimes at dangerous levels. Cyanide salts, especially **sodium cyanide,** are widely used in the electroplating of metals, in steelmaking, as sources of HCN for insecticidal fumigation, and for extraction of gold and silver from their ores. In the state of Nevada alone, over 100 million pounds of cyanide is used each year to recover the 6 million ounces of gold for which the state is famous. Despite a good human health record— no deaths in 15 years—the cyanide has had a substantial impact on migratory birds. Once the gold is extracted from the crushed ore with aqueous cyanide, cyanide-containing tailings and spent solution are deposited in huge impoundments, some exceeding 500 acres in area, that offer inviting ponds for migrating waterfowl. In the 1980s and early 1990s, a series of major bird kills forced changes in mining operations that resulted in a tenfold reduction in cyanide concentrations, to <50 mg/L. Although sublethal effects still are seen in birds at 10 mg/L, no major kills have been reported since 1991.

Hydrogen cyanide (hydrocyanic acid) is a major industrial chemical, although most of it is processed into other products such as acrylonitrile. HCN is a very poisonous, volatile liquid that smells like bitter almonds, but 20% of the U.S. population is genetically incapable of detecting the odor. After losing several workers for this reason, at least one major manufacturer forbade anyone to go into storage areas alone, and the production unit was equipped with a burnoff tower where any waste gas could be ignited automatically by a gas flame or an electric spark. As a last resort, each shift supervisor had to be able to fire a flaming arrow across the top of the 30-foot tower.

15.10 PERSPECTIVE

The practical utility of a chemical requires a balance between persistence and reactivity. Pesticides are a good example. Natural pyrethrins are too reactive to be used in field applications, while the environment is still suffering the effects of the overly persistent DDT. Although reactivity admittedly is responsible for toxicity, it also permits environmental self-cleaning, assures the biogeochemical recycling of matter, and prevents the accumulation of the large majority of the world's toxic chemicals. With only a few exceptions, such as the fossilized pitch called amber, there are no environmentally "permanent" organic chemicals, natural or synthetic.

15.11 REFERENCES

Aldridge, W. N. 1953. Serum esterases. 2. An enzyme hydrolyzing diethyl-*p*-nitrophenyl phosphate (E600) and its identity with the A-esterase of mammalian sera. *Biochem. J.* **53:** 117–24.

Barles, R. W., C. G. Daughton, and D. P. H. Hsieh. 1979. Accelerated parathion degradation in soil inoculated with acclimated bacteria under field conditions. *Arch. Environ. Contam. Toxicol.* **8:** 647–60.

Barr, D. P., and S. D. Aust. 1994. Pollutant degradation by white rot fungi. *Rev. Environ. Contam. Toxicol.* **138:** 49–72.

Chambers, J. E., and P. E. Levi. 1992. *Organophosphates: Chemistry, Fate, and Effects,* Academic Press, San Diego, CA.

Chen, Y.-L., and J. E. Casida. 1969. Photodecomposition of pyrethrin I, allethrin, phthalthrin, and dimethrin. *J. Agr. Food. Chem.* **17:** 208–15.

Class, T. J., and J. Kintrup. 1991. Pyrethroids as household insecticides: analysis, indoor exposure, and persistence. *Fresenius J. Anal. Chem.* **340:** 446–53.

Crombie, L. 1995. Chemistry of pyrethrins, in *Pyrethrum Flowers* (J. E. Casida and G. B. Quistad, eds.), Oxford University Press, New York, NY pp. 123–93.

Crosby, D. G. 1995. Environmental fate of pyrethrins. In *Pyrethrum Flowers* (J. E. Casida and G. B. Quistad, eds.), Oxford University Press, New York, pp. 194–213.

Evered, D., and S. Harnett. 1988. *Cyanide Compounds in Biology,* Wiley-Interscience, New York, NY.

Gaines, T. B. 1960. The acute toxicity of pesticides to rats. *Toxicol. Appl. Pharmacol.* **2:** 88–99.

Gaines, T. B. 1969. Acute toxicity of pesticides. *Toxicol. Appl. Pharmacol.* **14:** 515–34.

Heath, D. F. 1961. *Organophosphorus Poisons,* Pergamon Press, Oxford, U.K.

Helz, G. R., R. G. Zepp, and D. G. Crosby. 1994. *Aquatic and Surface Photochemistry,* Lewis Publishers, Boca Raton, FL.

Hine, J. 1962. *Physical Organic Chemistry,* McGraw-Hill, New York, NY.

IARC. 1978. *IARC Monographs on the Evaluation of the Carcinogenic Risk of Chemicals to Humans. Some N-Nitroso Compounds.* WHO, Geneva, Vol. 17.

IARC. 1989. *IARC Monographs on the Evaluation of the Carcinogenic Risk of Chemicals to Humans. Diesel and Gasoline Engine Exhausts and Some Nitroarenes.* WHO, Geneva, Vol. 46.

Kuhr, R. J., and H. W. Dorough. 1976. *Carbamate Insecticides: Chemistry, Biochemistry, and Toxicology,* CRC Press, Cleveland, OH.

Marinkas, P. L. 1996. *Organic Energetic Compounds,* Nova Science Publishers, Commack, NY.

Matsumura, F. 1985. *Toxicology of Insecticides,* Plenum Press, New York, NY.

Orlando, J. J., and G. S. Tyndall. 1996. Atmospheric oxidation of CH_3Br: Chemistry of the CH_2BrO radical. *J. Phys. Chem.* **100:** 7026–33.

Rao, K. R. 1978. *Pentachlorophenol: Chemistry, Pharmcology, and Environmental Toxicology,* Plenum Press, New York, NY.

Searle, C. E. 1984. *Chemical Carcinogens,* 2nd Edition, Vol. 2. Amer. Chem. Soc., Washington, DC. ACS Monograph 182.

Stewart, N. E., R. E. Milleran, and W. P. Breese. 1967. Acute toxicity of the insecticide Sevin and its hydrolytic product 1-naphthol to some marine organisms. *Trans. Amer. Fish. Soc.* **96**: 25–30.

Umetsu, N., F. H. Grose, R. Allahyari, S. Abu-El-Haj, and T. R. Fukuto. 1977. Effect of impurities on the mammalian toxicity of technical malathion and acephate. *J. Agr. Food Chem.* **25**: 946–54.

USEPA. 1990. Hazardous waste management system: Identification and listing of hazardous waste. Final rule. *Fed. Register* **55**(61): 11798–877.

USPHS. 1991. *Toxicological Profile for Bromomethane.* U.S. Public Health Service, Washington, DC. TP-91/06.

Wong, A. S., and D. G. Crosby. 1981. Photodecomposition of pentachlorophenol in water. *J. Agr. Food Chem.* **29:** 125–30.

Yinon, J. 1990. *Toxicity and Metabolism of Explosives,* CRC Press, Boca Raton, FL.

Special Topic 15. Hazardous Waste

The United States generates over *260 million tons* of hazardous waste every year. As "hazardous" denotes a probability that contact with a substance will produce injury, *any* harmful material discarded or inadvertently released into the environment must be considered to be hazardous waste.

The legal definition of hazardous waste is complex and is found in RCRA (Appendix 10.3). It concerns "substantial present or potential threat to human health or living organisms" due to a material's reactivity, corrosiveness, ignitability, or toxicity. We focus here on toxic chemicals. While the responsibility of RCRA is solid waste, the law actually includes liquids, semisolids, and gases. "Solid waste" apparently is not necessarily solid.

Another legislative attempt to control hazardous waste is CERCLA, the Comprehensive

Environmental Response, Compensation, and Liability Act of 1980. By a tax on petroleum and 42 chemicals as well as by government subsidy, it raised over $1 billion for remediation of toxic chemical storage and disposal sites. Despite severe criticism of its implementation—only about 150 of 538 priority sites had received any attention—it was eventually reauthorized in 1986 as SARA, or Superfund Amendment and Reauthorization Act, but still has not fully lived up to its intent or promise billions of dollars and a decade later.

Over 400 chemical substances are listed by the USEPA as hazardous constituents of waste, although as of 1990, only about 40 were assigned regulatory levels (Appendix 15.3). These include 8 inorganics, 16 chlorinated hydrocarbons, and 9 old pesticides, most of which, except for the inorganics, are reactive and nonpersistent. Curiously, the toxicologically fashionable TCDD, DDT, PAHs, and PCBs do not appear on this list (although they *are* regulated).

Most industrial operations result in some kind of waste (Appendix 2.1). For example, every organic chemistry student knows that a reaction seldom results in only a single product, and the byproduct mixture often is unusable and becomes "waste" (Section 13.4). There have been many notorious attempts to dispose of large-scale wastes: Love Canal, the New York disposal site for industrial chlorination byproducts; Times Beach, Missouri, where TCDD-containing liquid waste was applied to roads to reduce dust; and various strategies for storing spent nuclear fuel rods. However, toxic wastes come from local businesses, too, including waste TCE from dry cleaning and cyanide from electroplating. Domestic wastes include left-over pesticides, used motor oil, and used lead and Nicad batteries. Almost everyone has hazardous waste to dispose of.

Many disposal methods have been tried. Solids such as batteries and some pesticides usually have been dumped in landfills, liquids such as unused pesticide spray and waste solvents often go into surface impounds, deep holes called injection wells were used by our government to get rid of chemical warfare agents, and munitions such as TNT and the waste oil from tanker ships long were dumped into the sea. However, most Class 1 (chemical) landfills in the United States now are full, impounds are strictly regulated, most injection wells have become illegal, and the ocean dumping of toxic chemicals is widely prohibited. Still, industrial and municipal wastes continue to be generated in ever-increasing quantities. What to do?

Table 15.5 lists the present disposal options. Following the old adage that "the solution to pollution is dilution," direct disposal into rivers, lakes, and the ocean still remains the most common method, although it is better regulated now than in the past. Adsorption of pollutants onto solid sorbents such as steam-activated carbon pellets is one of the most effective ways to clean up contaminated water and can be especially efficient if treatment is repeated. For example, chlorobenzene and *p*-dichlorobenzene were 98% removed by a single pass through a column of granular activated carbon.

Volatile solvents such as TCE can be removed from contaminated water by passing it downward through a tall tower against a rapid stream of air bubbles from below (air stripping), although simply allowing evaporation from open impounds is more common. Algae have been used to bioconcentrate lipophilic pollutants out of water, usually followed by return to the land surface as mulch or soil amendment. The obvious difficulty with all these physical methods is that they amount to little more than shuffling the toxic chemicals from one place to another. Detoxication would be much better.

A common method of oxidative detoxication is incineration, although the possible formation of byproduct HCl, dioxins, and HCB must be considered. A variation is the "molten salt process," in which waste is fed continuously into a liquid mixture of sodium carbonate and sodium sulfate at >1000°C, which converts everything but metals to nontoxic forms. Wet oxidation at 200–300°C and 300–3000 psi pressure is effective in removing 99.8% of phenols and >98% of pesticides from dilute aqueous waste. Several processes utilize ultraviolet radia-

TABLE 15.5
Methods of Hazardous Waste Remediation

Physical Methods	Examples
Dissolution (in water)	Traditional disposal; leaching of salts from soil
Adsorption	Activated carbon; ion exchange (zeolites); landfills
Volatilization	Air stripping; steam stripping; open ponds and ditches (impoundments)
Precipitation	Flocculation (alum); sulfide precipitation; electromotive displacement
Solvent partitioning	Continuous solvent extraction
Bioconcentration	Uptake into algae; uptake of salts and metals into plants

Chemical Methods	
Abiotic degradation	Thermal oxidation (incineration); molten salt; wet oxidation; chemical oxidation; acid or base hydrolysis
Photodegradation	UV-ozone; UV-peroxide; UV-solvent; solar photolysis; catalyzed photooxidation
Biodegradation	Mixture with soil (landfill); composting; standard sewage treatment (trickling filter); genetically enhanced biodegradation; chemically enhanced biodegradation

tion, usually in the presence of oxidants such as hydrogen peroxide or ozone, and success with the semiconductor catalysts TiO_2 and ZnO has been reported (Helz et al., 1994) in which the oxides catalytically generate superoxide and hydroxyl radicals from oxygen and water (Section 5.4).

Various forms of microbial biodegradation, including landfill and standard sewage treatment, are probably the oldest of the destructive methods. Recently, bioremediation of spills through genetically selected (acclimated) or engineered microorganisms has shown some success. For example, *Pseudomonas stutzeriii* hydrolyzes spilled parathion to *p*-nitrophenol, which *P. aeruginosa* in the same inoculum then oxidizes (Barles et al., 1979). However, exotic species have difficulty competing with native microorganisms, and more success has been achieved by simply fertilizing a spill to assist indigenous species in their normal biodegradation.

A promising variation is waste degradation by isolated enzymes. For example, the white-rot fungus *Phanerochaete chrysosporium,* has been demonstrated to oxidize many types of organic compounds (Barr and Aust, 1994). The responsible enzyme, an iron-containing lignin peroxidase, has been isolated and shown to oxidize refractory chemicals such as DDT, TCDD, and PAH *in vitro* in the presence of H_2O_2. However, the best and essential future strategy is to reduce drastically the volume of hazardous waste generated, although regulation and remediation also must continue to be pursued vigorously. Even under the best of circumstances, hazardous waste is likely to be with us for a long time to come.

APPENDIX 15.1.
Nitroarenes in the Environment[a]

Nitroarene	Source	Conc. (ppb)[b]
1,3-Dinitropyrene	Diesel particles	600
	Heavy duty diesel	1600
	Gas heater emission	600
	Carbon black[c]	6300
1,6-Dinitropyrene	Diesel particles	810
	Heavy duty diesel	1200
	Atmospheric particles, Michigan	310
	Home gas heater	1880
1,8-Dinitropyrene	Heavy duty diesel	3400
	Atmospheric particles, Michigan	460
	Photocopier toner[c]	300
6-Nitrobenzo[a]pyrene	Diesel car exhaust	50,000
	Gasoline-powered car, leaded fuel	32,800
	Gasoline-powered car, unleaded fuel	17,300
	Home fireplace smoke	120[d]
2-Nitrofluorene	Atmospheric particles, Tokyo	21,800[d]
1-Nitronaphthalene	Diesel exhaust	700[d]
	Atmospheric particles, Philadelphia	3,500
1-Nitropyrene	Diesel car, 1979 model	2,030,000
	Gasoline-powered car	2500
	Used diesel oil	500[d]
	Home fireplace smoke	110[d]
	Atmospheric particles, Bermuda	720[d]

[a] IARC, 1989.
[b] Concentration in solvent extract unless otherwise noted.
[c] Made prior to 1980.
[d] Concentration in substrate rather than in extracts.

APPENDIX 15.2.
Nitrosamines in the Environment[a]

Nitrosamine	Source	Maximum Conc. (ppb)
Nitrosodimethylamine	Fried bacon (Germany)	12
	Ham (Germany)	11
	Dried fish (Japan)	34
	Pilsner beer (Germany)	7
	Pale ale (England)	8
	Cured tobacco	16
	Cigarette smoke, U.S., mainstream	43
	Cigarette smoke, U.S., sidestream	1770
	Rubber toys (Germany)	25

Nitrosamine	Source	Maximum Conc. (ppb)
Nitrosodimethylamine	Toy balloons (Germany)	150
	Air in a bar	0.24[b]
Nitrosopyrrolidine	Fried bacon (U.S.)	139
	Ham (Germany)	36
	Sausage (Canada)	105
Nitrosonornicotine	Cigarettes (U.S.)	1.7
	Small cigar (U.S.)	45
	Snuff (U.S.)	39
Nitrosodiethanolamine	Cosmetics (U.S.)	48,000
	Cigarettes (U.S.)	194
	Shampoo (U.S.)	260
Nitrosopiperidine	Nursing bottle nipples (Germany)	100
	Rubber gloves (Germany)	100

[a] Searle, 1984; IARC, 1978.
[b] Concentration in $\mu g/m^3$.

APPENDIX 15.3.
Some Toxic Constituents of Hazardous Waste[a]

Constituent	Regulatory level (mg/L)	Constituent	Regulatory level (mg/L)
Arsenic	5.0	Hexachloro-1,3-butadiene	0.5
Barium	100.0	Hexachloroethane	3.0
Benzene	0.5	Lead	5.0
Cadmium	1.0	Lindane	0.4
Carbon tetrachloride	0.5	Mercury	0.2
Chlordane	0.03	Methoxychlor	10.0
Chlorobenzene	100.0	Methyl ethyl ketone	200.0
Chloroform	6.0	Nitrobenzene	2.0
Chromium	5.0	Pentachlorophenol	100.0
o-Cresol	200.0	Pyridine	5.0
m-Cresol	200.0	Selenium	1.0
p-Cresol	200.0	Silver	5.0
2,4-D	10.0	Tetrachloroethylene	0.7
1,4-Dichlorobenzene	7.5	Toxaphene	0.5
1,2-Dichloroethane	0.5	Trichloroethylene	0.5
1,1-Dichloroethylene	0.7	2,4,5-Trichlorophenol	400.0
2,4-Dinitrotoluene	0.13	2,4,6-Trichlorophenol	2.0
Endrin	0.02	2,4,5-TP (Silvex)	1.0
Heptachlor and epoxide	0.008	Vinyl chloride	0.2
Hexachlorobenzene	0.13		

[a] USEPA, 1990

Predicting Environmental Fate and Effects

16

16.1 | QUANTITATIVE PREDICTION

From the preceding chapters, one sees that the environmental fate and effects of many xenobiotics are qualitatively predictable. Substances with high vapor pressures often move readily from water into the atmosphere, and organophosphorus esters containing high-energy bonds probably will inhibit acetylcholinesterases. However, as the number of chemicals of toxicological interest increases, together with the cost and complexity of environmental fate and toxicity measurements, the need for *quantitative* predictions becomes more urgent.

Several forms of prediction are available. They include (1) simple mathematical relationships between a compound's structural features and its physical and chemical properties, (2) mathematical models of a chemical's fate in the environment, (3) quantitative relations between structure and biological activity, and (4) microcosms and "model ecosystems." The following sections discuss each type.

16.2 | PREDICTING ENVIRONMENTAL TRANSPORT

16.2.1. Linear Free Energy Relations

Every physical or chemical system contains energy, generally expressed as heat *(H)*. Not all that energy is available for work, as some is lost to the disorder, or entropy *(S)*, of the system. What remains is called *free energy,* signified by *G* ("Gibbs free energy"). Free energy makes things happen. As the system undergoes a change Δ at temperature *T,* so do *G, H,* and *S:*

$$\Delta G = \Delta H - T \Delta S \qquad (16.1)$$

Transport of a chemical from one phase to another involves an energy change, the free energy of transfer. If the phases are, say, 1-octanol and water, this change can be denoted as ΔG_{ow}. ΔG_{ow} and its practical measure, log K_{ow}, are observed to increase in a uniform and linear way as an increasing number of the same substituent are substituted for hydrogens in a chemical structure (Table 14.1). In the example, plotting log K_{ow} against the number of chlorines on the benzene ring produces a straight line, an example of a *linear free energy relation* (LFER) between chemical structure and physical properties. The equilibrium and rate constants for any reaction also are directly and linearly related to the free energy change under a standard set of conditions, $\Delta G°$, also an LFER:

$$\log K = -\Delta G°/2.303 \, RT = -1.75 \times 10^{-4} \, \Delta G° \text{ J/mole} \qquad (16.2)$$

A good thermodynamic description of LFER is provided by Schwarzenbach et al. (1993).

16.2.2. Octanol–Water Partition Coefficients

K_{ow} provides a key to the prediction of many other physical properties. Any chemical structure, such as that of p-dichlorobenzene, can be pictured as a combination of various parts (fragments). In this case, these include the aromatic ring (ϕ), four hydrogen atoms, the two Cl atoms, and the bonds connecting them. The H atoms often are combined with the ring for simplicity. Hansch and Leo (1979) showed that each fragment could be represented by a *fragment constant* (f) and each bond by *bond factors* (F), empirically derived from the chemical structure and representing ΔG, which can be added together to provide an estimate of log K_{ow}. In other words, for each part of the molecule, a ΔG exists which is separable and additive (Eq. 16.3). For a chlorobenzene containing n chlorines, signified by $\phi(Cl)_n$,

$$\Delta G_{\phi(Cl)n} = \Delta G_\phi + n\Delta G_{Cl} \qquad (16.3)$$

The f and F values shown in Table 16.1 represent only a few of the many available, and calculations can become complicated as structural complexity increases. Hansch and Leo (1979) or Lyman et al. (1990) should be consulted for more detail.

Log K_{ow} is calculated as the sum of the appropriate fragment constants and bond factors, modified by a few additions (*e.g.*, other factors, shown here as F_x) and some rules. Where n is the number of bonds not involving hydrogen or within an independent group such as COOH:

$$\log K_{ow} = \Sigma (f) + \Sigma F_b(n-1) \pm \Sigma F_x \qquad (16.4)$$

For example, to calculate the K_{ow} of toluene, $C_6H_5CH_3$ (Table 16.1), f for C_6H_5 (1.90) is added to those for C (0.20) and for 3H (0.69); so $\Sigma(f)$ is 2.79. There is only one applicable bond, between the C_6H_5 and CH_3; so F_b is -0.12, but $(n-1) = (1-1) = 0$. Thus the sum, log K_{ow}, is calculated to be **2.79**. The measured log K_{ow} of toluene is **2.69**.

In the example of 1,1,1-trichloroethane (Cl_3C-CH_3), $2f_C = 0.40$, $3f_H = 0.69$, and $3f_{Cl} = 0.18$, so $\Sigma(f) = 1.27$. Adding F_b for 4 bonds (one C–C and 3 C–Cl)

TABLE 16.1
Representative Fragment and Bond Constants[a]

Fragment Constants (f)					
Fragment	f	f^{ϕ}	Fragment	f	f^{ϕ}
>C<	0.20	0.20	–H	0.23	0.23
–CN	–1.27	–0.34	–Cl	0.06	0.94
>C=O	–1.90	–1.09	–Br	0.20	1.09
–C(O)O–	–1.49	–0.56	–O–[b]	–1.82	–0.61
–C(O)OH	–1.11	–0.03	–S–	–0.79	–0.03
–CH=O	–1.10	–0.42	–NO$_2$	–1.16	–0.03
–C(O)NH$_2$	–2.18	–1.26	–NH–	–2.15	–1.03
–C(O)N<	–3.04	–2.80	–NH$_2$	–1.54	–1.00
C$_6$H$_5$–	1.90		–OH	–1.64	–0.44

Bond Constants (F)		
F_b	Single bond	–0.12
F_{br}	Single bond, alicyclic ring	–0.09
$F_=$	Double bond	–0.55
$F_=$	Double bond, conjugated in chain	–0.38
$F_=$	Double bond, conjugated to ϕ	–0.42
F_t	Triple bond	–1.42
F_{GX}	Geminal halogens $(X–C–X)^c$, $n=2$	0.30
	$n=3$	0.53
F_{VX}	Vicinal halogens $(X–C–C–X)^c$	$0.28 \times (n-1)$

[a] Hansch and Leo (1979). ϕ denotes attachment to a single aromatic ring.
[b] Methyl ethers and ethylene oxide -1.54.
[c] $X =$ halogen, $n =$ number of X.

gives $(n-1) = 3$, and F_{GX} for 3 geminal halogens ($3 \times 0.53 = 1.59$) gives a total of **2.50** for log K_{ow}. The measured value is **2.49**. For the related DDT, the benzene rings each represent f_ϕ 1.90, less 0.23 for each H atom replaced by an aromatic chlorine:

$$\log K_{ow} = \mathbf{6.78} = (2f_\phi\, 3.80 - 2f_H\, 0.46 + 2f_{\phi Cl}\, 1.88) + 2f_C\, 0.40 \\ + f_H\, 0.23 + 3f_{Cl}\, 0.18 + 7F_b - 0.84 + 3F_{GX}\, 1.59. \quad (16.5)$$

The best literature value is **6.91**. It must be noted that these are *log* values; the predicted value of K_{ow} in this case is 6.03×10^6 and the measured value 8.13×10^6. This degree of error is considered acceptable, especially with large numbers.

Some error actually is to be expected. For a series of 76 chemicals representing a variety of structural types, the average difference between the calculated and observed log K_{ow} values was 0.14 (Chou and Jurs, 1979). The maximum deviation was -0.94, or about an order of magnitude, but two-thirds of the values were within <0.1 log unit of each other. The uncertainty arising from the calculation of each f and F constant is estimated to be 0.02–0.05 log K_{ow} units (Lyman et al., 1990), but uncertainty also exists in the measured K_{ow}. Published values for the log K_{ow} of diphenylamine, for example, are 3.22, 3.34, 3.50, and 3.72 (calculated to be 3.51). More extreme examples of deviation include benzotrifluoride (observed log K_{ow} 2.90, calculated 3.60) and 2-hydroxy-1,4-naphthoquinone (observed 1.46, calculated 2.40).

If a structure is complicated, or if uncertainty must be reduced, an estimate can be made by adding or subtracting fragments and bonds in the known K_{ow} of a similar compound. For example, DDD is the analog of DDT with only two chlorines on the ethane. To calculate log K_{ow} for DDD, the known value for DDT (6.91) is adjusted for loss of one Cl (-0.06), addition of one H ($+0.23$), and change from three geminal Cl to two (difference of 0.99) to give a log K_{ow} of 6.09. Reported values are 6.02 and 6.22.

16.2.3. Bioconcentration Factors

In Section 3.2, the bioconcentration factor (BCF) was defined as the ratio of the concentration of a chemical in an organism to that in its environment, a partition coefficient in which one solvent is water and the other is body lipid. The additivity illustrated by Eq. 16.3 can be applied to the free energy change as a solute moves between *any* pair of immiscible solvents, say, chlorobenzene between water and lipid (lw) as well as between water and octanol:

$$\Delta G_{ow} = \Delta G^{\phi}_{ow} + n\,\Delta G^{x}_{ow} \tag{16.6}$$
$$\Delta G_{lw} = \Delta G^{\phi}_{lw} + n\,\Delta G^{x}_{lw}$$

ΔG_{lw} corresponds to log BCF just as ΔG_{ow} is related to log K_{ow}, confirming that log BCF and log K_{ow} are linearly related to each other (Section 3.3). The constant **a** represents the slope of the regression line and **b** the x-axis intercept:

$$\log \text{BCF} = \mathbf{a}(\log K_{ow}) + \mathbf{b} \tag{16.7}$$

Indeed, a plot of log BCF against log K_{ow} for a series of compounds shows this to be true (Fig. 3.4). Depending on the kind of organism in which the BCF is measured, its lipid content, and other factors, various regression equations of this same form can be generated (Lyman et al., 1990), such as those of Veith et al. (1980) and Mackay (1984) applied to bioconcentration into fish:

$$\log \text{BCF} = 0.76 \log K_{ow} - 0.23, \qquad r^2\ 0.82 \tag{16.8}$$

$$\log \text{BCF} = \log K_{ow} - 1.32, \qquad r^2\ 0.95 \tag{16.9}$$

An r^2 (correlation coefficient) of 1.00 represents 100% correlation. As the partition principle remains the same, BCF can likewise be predicted for terrestrial mammals (Kenaga, 1980), although with less certainty:

$$\log \text{BCF} = 0.50 \log K_{ow} - 3.46, \quad r^2\ 0.79 \tag{16.10}$$

A concept similar to that of BCF is the *biotransfer factor,* in which the daily intake of a chemical by an animal is translated into residues in food. For fresh meat (B_b) of 25% fat content or in whole milk (B_m) containing the normal 3.68% fat (Travis and Arms, 1988):

$$B_{b \text{ or } m} = \frac{\text{concentration in beef or milk (ppm)}}{\text{intake of chemical (mg/day)}}$$

$$\log B_b = \log K_{ow} - 7.60, \qquad r\ 0.81 \tag{16.11}$$

$$\log B_m = \log K_{ow} - 8.10, \qquad r\ 0.74 \tag{16.12}$$

16.2.4. Aqueous Solubility

If a reliable K_{ow} is available for a compound, the aqueous solubility can be estimated. As with BCF, a plot of log molar solubility against log K_{ow} for a series of chemicals shows a linear relationship from which a regression equation can be derived. The best fit is obtained with nonpolar or weakly polar compounds such as PAH or phthalate esters, as polar interactions interfere (Lyman et al., 1990).

The useful equation of Chiou et al. (1977) for calculation of aqueous solubility is derived from values for 34 chemicals of mixed polar and nonpolar types (Eq. 16.13), and that of Hansch et al. (1968) from 140 chemicals (Eq. 16.14).

$$\log S = -1.49 \log K_{ow} + 7.46 \ \mu moles/L, \qquad r^2 \ 0.97 \qquad (16.13)$$

$$\log S = -1.214 \log K_{ow} + 0.85 \ moles/L, \qquad r^2 \ 0.91 \qquad (16.14)$$

If the substance is a solid at room temperature, crystal lattice interactions must be accounted for and are represented conveniently as a function of melting point, T_m:

$$\log S = -1.12 \log K_{ow} + 7.30 - 0.017 T_m \ \mu moles/L, \qquad r \ 0.96 \quad (16.15)$$

Deviations from measured values in these calculations are due primarily to errors in the measurement or calculation of K_{ow}.

Aqueous solubilities can also be estimated by addition of fragment constants developed by Irmann (1965), but these constants not the same as those in Table 16.1 (see Lyman et al., 1990). The aqueous solubility of toluene at 25°C is calculated to be 0.26 g/L by Eq. 16.13 and 0.35 g/L by Eq. 16.14 (log K_{ow} 2.69), Irmann's method predicts 0.56 g/L, and the measured value is 0.52 g/L. Unfortunately, the most common source of error lies again in the measured value (Section 3.2), especially for slightly soluble substances.

16.2.5. Volatilization

The tendency to volatilize is most often represented by a Henry's law constant (Section 3.4), where the nondimensional H' represents a partition coefficient. Log H', too, can be calculated from fragment constants (see Schwarzenbach et al., 1993), although again not the ones shown in Table 16.1. Like aqueous solubility, log H (and so log H') may be roughly approximated for compounds of moderate vapor pressure from K_{ow} values:

$$\log H = 1.52 \log K_{ow} - 3.90 \qquad (16.16)$$

Volatilization rates from water may be estimated by the simple expression $t_{1/2} = 0.69Z / K_L$, where Z is the water depth in cm and K_L is the mass transfer coefficient in cm/h. However, the half-lives are affected strongly by air and water motion, and appropriate nomograms to estimate these are provided by Lyman et al. (1990). For practical purposes, even a rough idea of the volatilization rate often may suffice, so Table 16.2 qualitatively relates H and H' to volatility, as does Fig. 4.4.

TABLE 16.2
Relation of Henry's Law Constants to Volatility[a]

H^b	H'	Volatility (Examples)[c]
$<10^{-7}$	$<4\times10^{-6}$	Less than water, chemical will concentrate (glycerine)
$10^{-7}-10^{-6}$	$4\times10^{-6}-4\times10^{-5}$	Slow, rate controlled by diffusion through air (lindane)
$10^{-6}-10^{-5}$	$4\times10^{-5}-4\times10^{-4}$	Moderate in shallow rivers (pentachlorophenol)
$10^{-5}-10^{-3}$	$4\times10^{-4}-0.04$	Significant in all water bodies (DDT, nitrobenzene)
$10^{-3}-10^{-2}$	$0.04-0.4$	Rapid in all waters (naphthalene, chloroform, Aroclor 1254)
$>10^{-2}$	>0.4	Very rapid (methyl bromide, gasoline)

[a] Adapted from Lyman et al. (1990).
[b] Henry's law constant in atm m^3/mole.
[c] Based on $t_{1/2}$ in water.

16.2.6. Soil Sorption Coefficients

Provided K_{ow}, soil sorption coefficients based on organic carbon, K_{oc}, can be estimated (Section 3.5). From data on 19 chemicals of various types, including some rather polar pesticides, Brown and Flagg (1981) arrived at Eq. 16.17, but later observation shows that K_{oc} and K_{ow} often are more nearly equal (Eq. 16.18), as demonstrated by Di Toro (1985).

$$\log K_{oc} = 0.937 \log K_{ow} - 0.006 \qquad r^2\ 0.95 \qquad (16.17)$$
$$\log K_{oc} = 0.983 \log K_{ow} - 0.0002 \qquad\qquad (16.18)$$

If the types and polarities of the subject compounds is narrowed, the correlation is even better, with 10 nonpolar aromatic hydrocarbons conforming to the following equation (Karickoff et al., 1979).

$$\log K_{oc} = 1.00 \log K_{ow} - 0.21, \qquad r^2\ 1.00 \qquad (16.19)$$

Sorption coefficients also can be related to the proportion of *soil organic matter* (om) rather than organic carbon (Chiou et al., 1983):

$$\log K_{om} = 0.904 \log K_{ow} - 0.779, \qquad r^2\ 0.93 \qquad (16.20)$$

However, organic matter content is more variable and difficult to measure than is organic carbon and often is estimated from the latter via the relationship $K_{om} = 0.580\ K_{oc}$. Both K_{om} and K_{oc} are dependent on soil type, temperature, and other experimental variables, and their values vary considerably from source to source. For example, Eqs. 16.17, 16.18, and 16.19 predict a log K_{oc} of 3.48, 3.66, and 3.51, respectively, for the chlorinated hydrocarbon lindane (log K_{ow} 3.72), while measured values of 3.03 and 3.11 have been reported. One always wonders whether it is the calculated or the measured K_{oc} that is in error.

16.3 PREDICTING ENVIRONMENTAL TRANSFORMATIONS

16.3.1. Reactions and Products

The number of major types of environmental organic reactions is surprisingly limited (Appendix 5.4). Oxidations, reductions, hydrolyses including dehydrochlorinations, and some photodegradations account for most. Biodegradation reactions (Appendix 6.1) follow the same pattern, even down to some of the photodegradation products, with the addition of conjugation (synthesis). While the ease with which these reactions proceed will vary greatly with molecular structure, the general impression is that nitro reduction is the most facile, followed by oxidation and then hydrolysis. Conjugation also appears to occur easily but often must await the outcome of primary (Phase I) metabolism. Consequently, to predict the probable environmental transformation products of some particular chemical, it often would only be necessary to consult Appendices 5.3 and 6.1.

16.3.2. Equilibria and Rates

LFER extend to the prediction of reaction rates and equilibria. In the 1930s, Louis Hammett showed that the dissociation constants of substituted benzoic acids could be dissected into a fragment constant (k_H), representing the parent benzoic acid, and another (k_x) representing a substituted acid. Another factor, ρ, indicating the susceptibility of the equilibrium or reaction to electronic effects, was arbitrarily set at 1.00 for benzoic acid dissociation in water at 25°C. The result was the **Hammett equation,** which has its theoretical basis in Eq. 16.2:

$$\log k_x - \log k_H = \sigma\rho = \log (k_x/k_H) \tag{16.21}$$

The Greek letter σ denotes the **Hammett substituent constant,** a number representing the electronic influence of a substituent m or p to the carboxyl or other reacting group on the benzene ring; *ortho* substituents generate steric effects and seldom produce satisfactory results. The σ for hydrogen is arbitrarily set at zero, with electron-releasing atoms or groups such as –OH assigned negative values and electron-withdrawing ones such as Cl assigned positive values. Hansch et al. (1995) and Exner (1978) provide σ values for hundreds of substituents, and a few are illustrated in Table 16.3.

Among the many uses of Hammett σ constants is the calculation of pK_a, that is, the -log of acid dissociation constants. For example, to calculate the pK_a of p-chlorobenzoic acid and given the benzoic acid pK_a of 4.20 and $\rho = 1.00$, $\log k_x = \sigma + \log k_H = 0.23 + (-4.20) = -3.97$. The calculated pK_a is 3.97, the measured value 3.98. For the dissociation of substituted phenols, $\rho = 2.008$; for hydrolysis of ethyl benzoates in 85% ethanol at 25°C, $\rho = 2.55$; and for the reaction of anilines with benzoyl chloride in benzene at 25°C, $\rho = -2.694$. Although numerous ρ values have been determined (Jaffé, 1953), few bear any relation to environmental processes.

TABLE 16.3
Representative Hammett Sigma Constants[a]

Substituent	σ_p	σ_m	σ_p^+	σ_m^{+} [b]	σ_p^-
N(CH$_3$)$_2$	−0.83	−0.16	−1.70		−0.12
NH$_2$	−0.66	−0.16	−1.30	−0.16	−0.15
OH	−0.37	0.12	−0.92		−0.37
OCH$_3$	−0.27	0.12	−0.78	0.05	−0.26
CH$_3$	−0.17	−0.07	−0.31	−0.17	−0.17
C$_6$H$_5$	−0.01	0.06	−0.18	0.11	0.02
H	0	0	0	0	0
SCH$_3$	0.00	0.15	−0.60		0.06
F	0.06	0.34	−0.07	0.35	−0.03
Cl	0.23	0.37	0.11	0.40	0.19
Br	0.23	0.39	0.15	0.41	0.25
CHO	0.42	0.35	0.73		1.03
COOH	0.45	0.36	0.42	0.32	0.77
COOCH$_3$	0.45	0.36	0.49	0.37	0.64
CF$_3$	0.54	0.43	0.61		0.65
CN	0.66	0.56	0.66	0.56	1.00
NO$_2$	0.78	0.71	0.79	0.67	1.27

[a] Hansch et al. (1995).
[b] March (1992).

The Hammett equation applies to any reaction on an aromatic ring or its substitu-ents, whether electrophilic, nucleophilic, or free radical. However, if a listed substitu-ent has a resonance interaction with the reaction center in the transition state, a differ-ent value (σ^+) applies where an electron-withdrawing group interacts with the developing positive charge, as in electrophilic aromatic substitution reactions like ring hydroxylation. Where an electron-donating group interacts with the developing negative charge, σ^- is used. For more background, consult Hansch and Leo (1995) or March (1992).

An important feature of σ is that it indicates the relative electron density on the aromatic ring. For example, the ring in anisole ($X = $ –OCH$_3$, σ_p −0.27) is more electron rich than that in nitrobenzene ($X = $ –NO$_2$, σ_p 0.78) and thus more reactive toward electrophilic reagents such as environmental oxidants. As Hammett constants are additive, the ring of p-chlorophenol ($\sigma = -0.37 + 0.23 = -0.14$) is more electronegative than that of p-nitrophenol ($\sigma = -0.37 + 0.78 = 0.41$). Just as Hammett constants apply to aromatic rings, a similar set of **Taft constants** relates to aliphatic compounds by isolating only inductive electronic effects (Hansch and Leo, 1995; March, 1993).

16.4 | MODELING ENVIRONMENTAL FATE

Mathematical modeling of environmental distribution, transport, and transformations can be combined and extended to give a picture of the fate of a chemical in the

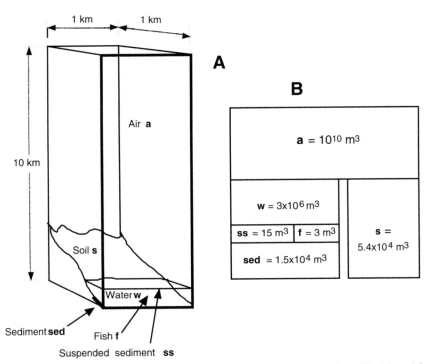

Figure 16.1. A unit world, stylized in three dimensions (A) and in two dimensions (B). Adapted from McCall et al., 1983.

whole environment (McCall et al., 1983). One can imagine a "unit world" (Fig. 16.1A), perhaps with a total volume **V** of 1 km $\times$ 1 km $\times$ 10 km, or 10^{10} m^3, composed of an air compartment, soil, water, suspended sediment, and living biota. This representation can be redrawn in two dimensions (Fig. 16.1B) to illustrate the important equilibria already discussed in Chapter 3: BCF between water and aquatic life, K_d and K_{oc} for soil, and $1/H'$, the reciprocal of the Henry's law constant, termed K_w, for transfer from water to air.

An amount, **M** moles, of a chemical introduced into this unit world would quickly distribute itself among the air (represented by a), water (w), fish (f) representing all biota, sediment (sed), soil (s), and soil water (sw). The sediment and soil density is assigned a value of 2.5. As **M/V** represents concentration, the distribution becomes:

$$(\%\mathbf{M}_{sed}/2.5\%\mathbf{V}_{sed})/\%\mathbf{M}_w/\%\mathbf{V}_w = Kd_{sed} \tag{16.22}$$

$$(\%\mathbf{M}_s/2.5\ \%\mathbf{V}_s)/(\%\mathbf{M}_{sw}/\%\mathbf{V}_{sw}) = Kd_s \tag{16.23}$$

$$(\%\mathbf{M}_w/\%\mathbf{V}_w)/(\%\mathbf{M}_a/(\%\mathbf{M}_w/\%\mathbf{V}_a) = K_w = 1/H' \tag{16.24}$$

$$(\%\mathbf{M}_f/\%\mathbf{V}_f)/(\%\mathbf{M}_w/\%\mathbf{V}_w) = BCF \tag{16.25}$$

From these expressions, the proportion of chemical to be found in any compartment, such as that in water, $\%\mathbf{M}_w$, can be calculated:

$$\%M_w = 100/[2.5Kd_{sed}(\%V_{sed}/\%V_w) + BCF(\%V_f/\%V_w) +$$
$$(1/K_w)(\%V_a/\%V_w) + 2.5Kd_s(\%V_s/\%V_w) + \%V_{sw}/\%V_w] \quad (16.26)$$

If one plausibly assumes that all environmental transformations—chemical, photo-chemical, and biological—conform to pseudo-first-order kinetics, the environmental degradation rate will be the sum of the rates k_1, k_2, etc. in the various compartments in relation to their $\%M$:

$$t_{1/2} = 0.693/(k_1\%M_s + k_2\%M_{sed} + k_3\%M_w + k_4\%M_a)/M \quad (16.27)$$

From these relationships, the percentage, concentration, and degradation rate of a chemical can be estimated for each compartment and for a designated feature of the environment, such as a pond, if the appropriate physical properties and environmental description are provided. Table 16.4 shows the resulting distribution of several common pesticides in a hypothetical pond, based on the properties shown. It is not necessary to specify the input of chemical to arrive at a % distribution, but the concentrations here are based on a total 200 kg of each chemical. As expected, those having a high K_{oc} concentrate in sediment, the more soluble ones are found in water, and concentrations in fish depend on the BCF.

The same principles apply to these chemicals in the whole environment (Table 16.5). Soil generally plays a greater concentrating role than might be expected, and fish a lesser role. Understand that these are all maximum initial concentrations, but that chemicals react. Although the volatile soil fumigant, 1,3-D (1,3-dichloropropene), is the most water soluble of the examples, it soon dissipates into the atmosphere where it is rapidly degraded; the actual equilibria will shift constantly. The chlorinated hydrocarbon lindane has a fairly high BCF of 325, but the actual proportion in fish is small due to the chemical's metabolism and dissipation into the atmosphere.

TABLE 16.4
Calculated Chemical Distribution in a Pond[a]

Physical Properties Chemical	M	S(mg/L)	P(torr)	K_w	log K_{oc}	log BCF
DDT	354.5	0.0017	1.9×10^{-7}	470	5.18	4.79
Chlorpyrifos	350.6	1.2	1.9×10^{-5}	3340	3.79	2.67
Lindane	290.8	0.15	3.2×10^{-5}	300	3.11	2.51
1,3-D	111.0	2700	25	18	1.84	0.48
2,4-D	221.0	900	6.0×10^{-7}	$<10^8$	1.78	0.48

Distribution Chemical	$\%w$[b]	$\%$sed	$\%$sus[c]	$\%f$	C_w[d]	C_{sed}	C_{sus}[c]	C_f
DDT	1.31	98.6	0.010	0.08	0.26	1580	0.020	16,200
Chlorpyrifos	24.7	75.3	0.075	0.01	4.94	1200	0.015	2,320
Lindane	60.6	39.4	0.039	0.02	12.1	630	0.008	3,940
1,3-D	96.7	3.22	0.003	<0.01	19.3	52	<0.001	63
2,4-D	96.8	3.16	0.003	<0.01	19.2	51	<0.001	61

[a] Data adapted from McCall et al. (1983).
[b] Compartment volumes are shown in Fig. 10.2.
[c] Sus means suspended sediment.
[d] C means concentration in ppb (μg/L); total input is 200 kg of each chemical.

TABLE 16.5
Calculated Chemical Distribution in the Environment[a]

Chemical	%a	%w	%s	%sed	%f	C_a[b]	C_w	C_s	C_{sed}	C_f
DDT	4.72	0.65	44.8	49.8	0.040	0.009	0.44	1160	2650	27,300
Chlorpyrifos	12.8	12.8	35.3	39.1	0.006	0.026	8.55	870	2090	4,020
Lindane	83.2	7.5	4.4	4.9	0.002	0.166	4.99	110	260	1,620
1,3-D	99.4	0.5	0.02	0.02	<0.001	0.199	0.36	0.47	1	1
2,4-D	0.003	93.8	3.1	3.1	<0.001	<0.001	62.5	78	160	200

[a]Data adapted from McCall et al. (1983).
[b]C means concentration in ppb (μg/L); total input is 200 kg of each chemical.

Such calculations would be tedious except for computers. There are a number of software programs that will perform the computations. The best known comes from the USEPA and is called EXAMS (*Ex*posure *A*nalysis *M*odeling *S*ystem). When fed the appropriate physical properties and environmental parameters, EXAMS provides proportions, concentrations, and dissipation rates for each compartment. Other programs include ALOHA (an atmospheric plume model) developed by NOAA, and PRZM, a soil mobility model.

The most serious pitfall in using such computer models lies in unquestioning belief of the results. The introduction to the ALOHA manual provides an important reminder: "A model is simply a tool, and should never be endowed with wisdom or responsibility beyond its power."

16.5 | QUANTITATIVE STRUCTURE–ACTIVITY RELATIONS (QSAR)

If one concedes that (1) absorption of a toxicant through cell membranes is related to the octanol–water partition coefficient K_{ow}, (2) metabolic degradation largely governs intoxication and represents the sum of chemical reactivities as expressed by Hammett σ constants, and (3) toxicity results from a toxicant–receptor interaction also related to σ, it is not surprising that toxicity often can be predicted quantitatively. Modern ideas of quantitative structure–activity relations (QSAR) are traced to the work of Corwin Hansch and others in the 1960s, but practical applications in the design of new drugs, pesticides, and other bioactive chemicals has stimulated widespread interest and rapid advances in this field (Hansch and Leo, 1995).

QSAR has three components: hydrophobicity, chemical reactivity, and three-dimensional structure. Hydrophobicity is the most important. Although K_{ow} (called P by Hansch) provides the most frequent measure, Fujita's fragment constant π also is used. Like Hammett's σ, π is additive and derived from Eq. 16.2, but it represents the hydrophobicity of a functional group or fragment.

$$\pi_x = \log K_{ow}^{RX} - \log K_{ow}^{RH} \quad \pi_H = 0$$

$$(16.28)$$

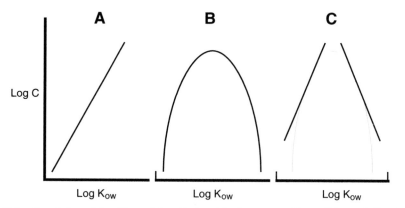

Figure 16.2. Correlation of log K_{ow} and log C plotted in three ways: Linear (A), parabolic (B), and bilinear (C). C is the concentration required to produce some specific effect.

Often, K_{ow} is the only variable specified, and the regression is linear (Fig. 16.2A). For example, enzyme inhibition can be predicted:

50% inhibition of brain ATPase: $\log 1/C = 0.77 \log K_{ow} + 0.53$, n 14, r^2 0.98

50% inhibition of AChE by ROH: $\log 1/C = 0.72 \log K_{ow} + 0.30$, n 8, r^2 0.92

as can the absorption of chemicals through human skin.

Rate of penetration of alcohols: $\log k = 0.57 \log K_{ow} + 0.01$, n 8, r^2 0.99

Rate of penetration of phosphate esters: $\log k = -0.31 \log K_{ow} - 0.02$, n 5, r^2 0.98

In these equations, n is the number of examples, r^2 the correlation coefficient, and C the molar concentration. The use of $\log 1/C$ avoids the sometimes inconvenient -log C. Remember that no mechanism is implied by these relationships; they are empirical.

These simple regressions set the stage for the prediction of toxic effects in whole animals (Hansch et al., 1989), the Special Topic for this chapter. All together, QSAR has a tantalizing future. While rough approximations may still be made by simple linear relations, better computers will allow the development of much more sophisticated and accurate models and even the eventual prediction of toxic effects in very complex environmental systems (Special Topic 16).

16.6 | MICROCOSMS (MODEL ECOSYSTEMS)

Despite the advances in QSAR, the approach seldom has been applied to environmental situations in which a variety of types of living organisms are exposed to

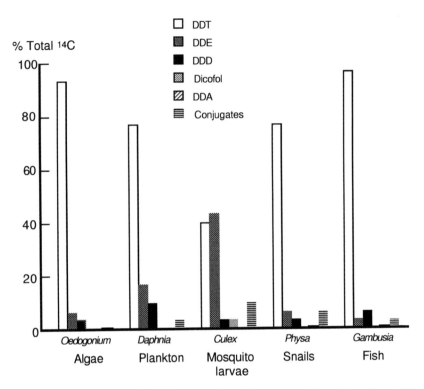

Figure 16.3. Biotransformations of DDT in an aquatic model ecosystem. Adapted from Lu and Metcalf, 1975. Bars represent the relative proportion of each metabolite as measured against total ^{14}C.

xenobiotics while interacting with each other and with their surroundings—that is, in real life. To gain better control over environmental experiments, the laboratory microcosm, which literally means "tiny universe," has come into increasing use. Among the expectations for a microcosm is that it will adequately represent a specific community if not an entire ecosystem (Special Topic 1), that it will allow comparison of the fate or effects of a variety of chemicals, that its results will be reproducible, and that it will mimic the distribution of chemicals in the real world.

In the simplest forms of microcosm, a group of aquatic organisms that constitute a very simple food chain is held under controlled conditions in aerated water in a glass flask, an isotopically labeled chemical is introduced, and samples of water and organisms are analyzed for parent compound and metabolites at the end of the experiment (Lu and Metcalf, 1975). When DDT was tested in such a system composed of algae *(Oedogonium cardiacum),* crustaceans *(Daphnia magna),* mosquito larvae *(Culex pipiens quinquifasciatus),* snails *(Physa* sp.), and fish *(Gambusia affinis),* it was found to be metabolized only slightly (<10%) by the algae and fish within 96 hours, while the snails and Daphnia metabolized about 20% of it (Fig. 16.3). However, mosquitoes, the very animals DDT was intended to control, were able to biodegrade more than 60% of the insecticide!

A somewhat more elaborate "model ecosystem," introduced by R. L. Metcalf in 1971, consisted of a 37 L (10 gallon) glass aquarium containing a raised area of

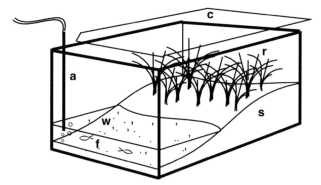

Figure 16.4. Diagram of a Metcalf model ecosystem (see Lee et al., 1976), containing sand (**s**), rice plants with insect larvae (**r**), water (**w**), fish and other aquatic biota (**f**), and an air inlet (**a**) and cover (**c**).

quartz sand at one end and an excavated flooded area at the other (Fig. 16.4). The raised area was planted with sorghum or another crop on which saltmarsh caterpillars *(Estigmene acrea)* were established, the radiolabeled chemical painted onto the leaves, and the environmental compartments sampled at intervals and analyzed. With experience, such a microcosm could be sustained for weeks without any other input except light until the foodchain eventually was broken (Lee et al., 1976).

An even more elaborate microcosm, designated by Nash (1983) as a "micro-agroecosystem," was a glass or plastic box about 1.5 m × 0.5 m at base and 1 m high containing 15 cm of soil and specific crop plants. The test chemical, usually a pesticide, was sprayed on the leaves, air drawn through the chamber, and air, soil, and plants sampled at intervals to determine distribution and fate.

In an interesting example of the relative environmental degradability and persistence of chemicals, Lu and Metcalf (1975) determined the uptake, distribution, and biodegradation of four simple aromatic compounds in the algae, crustaceans, insect larvae, snails, and fish of an aquatic model ecosystem. After 4 days, the persistence of each compound showed good correlation with the Hammett σ constant, as did the total amount of hydroxylated and conjugated metabolites:

$$\% \text{ Parent compound remaining} = 37.26\sigma + 54.14, \quad n\ 4,\ r^2\ 0.83 \quad (16.29)$$

$$\% \text{ In hydroxylated/conjugated form} = 46.23\sigma + 44.35, \quad n\ 4,\ r^2\ 0.99 \quad (16.30)$$

The results can be explained by assuming that electrophilic oxidation of the test compounds was the principal degradation route. Anisole (methoxybenzene), with the most electron-rich ring, was the most readily hydroxylated, while the relatively electropositive nitrobenzene was degraded the most slowly. Microbial degradation in the soil and water would relate to σ in the same way.

How well do microcosms meet our expectations? The smaller versions, at least, usually lack sunlight UV and other important environmental factors. They also involve only a limited population or community, and they seldom reach equilibrium. Most important, there is a strong tendency for people to think that microcosm results truly reflect what happens in the real world. If we recognize this prejudice, microcosms and the other physical models provide a lot of insight into chemical fate and ecological relations, and they always offer a better understanding of our dynamic chemical environment.

<hr>

16.7 | REFERENCES

Brown, D. S., and E. W. Flagg. 1981. Empirical prediction of organic pollutant sorption in natural sediments. *J. Environ. Qual.* **10:** 382–86.

Chiou, C. T., V. H. Freed, D. W. Schmedding, and R. L. Kihnert. 1977. Partition coefficients and bioaccumulation of selected organic chemicals. *Environ. Sci. Technol.* **11:** 475–78.

Chiou, C. T., P. E. Porter, and D. W. Schmedding. 1983. Partition equilibria of nonionic organic compounds between soil organic matter and water. *Environ. Sci. Technol.* **17:** 227–31.

Chou, J. T., and P. C. Jurs. 1979. Computer assisted computation of partition coefficients from molecular structures using fragment constants. *J. Chem. Inf. Comput. Sci.* **19:** 172–78.

Di Toro, D. M. 1985. A particle interaction model of reversible organic chemical sorption. *Chemosphere* **14:** 1503–38.

Exner, O. 1978. A critical compilation of substituent constants. In *Correlation Analysis in Chemistry* (N. B. Chapman and J. Shorter, eds.), Plenum Press, New York, NY pp. 439–540

Hansch, C., D. Kim, A. J. Leo, E. Novellino, C. Silipo, and A. Vittoria. 1989. Toward a quantitative comparative toxicology of organic compounds. *Crit. Rev. Toxicol.* **19:** 185–225.

Hansch, C., and A. J. Leo. 1979. *Substituent Constants for Correlation Analysis in Chemistry and Biology,* John Wiley & Sons, New York, NY.

Hansch, C., and A. J. Leo. 1995. *Exploring QSAR: Fundamentals and Applications in Chemistry and Biology,* American Chemical Society, Washington, DC.

Hansch, C., A. J. Leo, and D. Hoekman. 1995. *Exploring QSAR: Hydrophobic, Electronic, and Steric Constants,* American Chemical Society, Washington, DC.

Hansch, C., J. E. Quinlan, and G. L. Lawrence. 1968. The linear free energy relationships between partition coefficients and the aqueous solubility of organic liquids. *J. Org. Chem.* **33:** 347–50.

Irmann, F. 1965. A simple correlation between water solubility and structure of hydrocarbons and halohydrocarbons. *Chem. Ing. Tech.* **37:** 789–98.

Jaffe, H. H. 1953. A reexamination of the Hammett equation. *Chem. Rev.* **53:** 191:261.

Kamilet, M. J., R. M. Doherty, R. W. Taft, M. H. Abraham, G. D. Veith, and D. J. Abraham, 1987. Solubility properties in polymers and biological media. 8. An analysis of the factors that influence toxicities of organic nonelectrolytes to the golden orfe fish *(Leuciscus idus melanotus). Environ Sci. Technol.* **21:** 149–55.

Karickoff, S. W., D. S. Brown, and T. A. Scott. 1979. Sorption of hydrophobic pollutants on natural sediments. *Water Res.* **13:** 2541–48.

Kenaga, E. E. 1980. Correlation of bioconcentration factors of chemicals in aquatic and terrestrial organisms with their physical and chemical properties. *Environ. Sci. Tedchnol.* **14:** 553–56.

Lee, A.-H., P.-Y. Lu, R. L. Metcalf, and E.-L. Hsu. 1976. The environmental fate of three dichlorophenyl nitrophenyl ether herbicides in a rice paddy model ecosystem. *J. Environ. Qual.* **5:** 482–86.

Lu, P.-Y., R. L. Metcalf. 1975. Environmental fate and biodegradability of benzene derivatives as studied in a model aquatic ecosystem. *Environ. Health Perspect.* **10:** 269–84.

Lyman, W. J., W. F. Reehl, and D. H. Rosenblatt. 1990. *Handbook of Chemical Property Estimation Methods,* McGraw-Hill Book Co., New York, NY.

Mackay, D. 1984. Correlation of bioconcentration factors. *Environ. Sci. Technol.* **18:** 274–78.

March, J. 1992. *Advanced Organic Chemistry: Reactions, Mechanisms, and Structure,* 4th Ed., McGraw-Hill, New York, NY.

McCall, P. J., D. A. Laskowski, R. L. Swann, and H. J. Dishburger. 1983. Estimation of environmental partitioning of organic chemicals in model ecosystems. *Residue Reviews* **85:** 231–44.

Nash, R. G. 1983. Determining environmental fate of pesticides with microagroecosystems. *Residue Reviews* **85:** 199–218.

Schwarzenbach, R. P., P. M. Gschwend, and D. M. Imboden. 1993. *Environmental Organic Chemistry,* John Wiley and Sons, New York, NY.

Travis, C. C., and A. D. Arms. 1988. Bioconcentration of organics in beef, milk, and vegetation. *Environ. Sci. Technol.* **22:** 271–74.

Veith, G. D., K. J. Macek, S. R. Petrocelli, and J. Carrol. 1980. An evaluation of using partition coefficients and water solubility to estimate bioconcentraion factors for organic chemicals in fish, in *Aquatic Toxicology* (J. G. Eaton, P. R. Parrish, and A. C. Hendricks, Eds.), ASTM, Philadelphia, PA. STP 707, pp. 116–29.

Special Topic 16. QSAR and Toxicity

Quantitative structure–activity relations (QSAR) have been used to predict the toxic effects of chemicals in whole organisms, based only on log K_{ow}. For example,

Onset of narcosis in tadpoles: log $1/C$ = 0.90 log K_{ow} + 0.91,
n 57, r^2 0.93

LD_{100} in grain weevils by ketones: log $1/C$ = 0.60 log K_{ow} + 2.90,
n 4, r^2 0.98

LD_{100} in carp by alcohols: log $1/C$ = 0.81 log K_{ow} + 1.04, n 5, r^2 0.98

LD_{100} in dogs by alcohols: log $1/C$ = 0.57 log K_{ow} + 1.37, n 6, r^2 0.97

LD_{50} in guppies by phenols: log $1/C$ = 0.59 log K_{ow} + 2.66, n 19, r^2 0.95

However, what one sees here are only regression equations, based on limited sets of data. Despite their high correlation coefficients, they do not begin to reflect the complexity inherent in actual organisms. As Hansch points out, "one cannot equate a mouse with a bag of octanol and water." To even approach the real thing, a **"multivariate analysis"** is required that, in its ultimate form, must include terms for hydrophobicity (K_{ow} or π), reactivity (σ), and molecular size and shape such as the E_s values derived from Taft constants or the Verloop sterimol dimensions of molecular length (L) or radius (B) (Hansch and Leo, 1995):

$$\log 1/C = a \log K_{ow} + b\,\sigma\rho + c(E_s, L, B) + d \qquad (16.30)$$

For example, Eq. 16.31 uses σ to predict the LD_{50} of cholinesterase-inhibiting phenyl phosphate esters such as paraoxon, and Eq. 16.32 adds the sterimol radius to estimate carbonic anhydrase inhibition by substituted benzenesulfonamides.

$$\log 1/LD_{50} = 0.26\ \pi + 2.44\ \sigma - 0.61, \qquad n\ 8,\ r^2\ 0.98 \tag{16.31}$$

$$\log 1/k_i = 0.54\ \pi + 0.95\ \sigma - 0.35\ B + 6.29, \qquad n\ 31,\ r^2\ 0.84 \tag{16.32}$$

However, linear regression often fails at either very high or very low values of K_{ow}. The most accurate regression is more parabolic (Fig. 16.2B), perhaps due to metabolic degradation of substrate, limited solubility, or steric inhibition in binding to a receptor. An intact organism can be viewed as a series of connected compartments (Fig. 10.6), a **multicompartment model,** where the concentration in the last (nth) compartment, the receptor, is described by Hansch and Leo (1995):

$$\log C_n = a\ \log K_{ow} - 2a\ \log (K_{ow} + 1) + b \tag{16.33}$$

The parabolic relationship is modified in practice to a bilinear form (Fig. 16.2C), which has the advantage that its left side conforms to the more familiar dose–response plot of Fig. 8.1C. With β as a constant characteristic of the system under consideration, Eq. 16.33 takes the following form:

$$\log C_n = a\ \log K_{ow} - b\ \log (\beta\ K_{ow} + 1) + c \tag{16.34}$$

The LC_{50} (moles/L) of a series of 30 assorted organic chemicals in intact *Leuciscus idus melanotus,* a fish known as the golden orfe, is predicted well by Equation 16.35 (Hansch and Leo, 1995) based on the work of Kamlet et al. (1987).

$$\log 1/LC_{50} = 1.08\ \log K_{ow} - 1.13\ \log (\beta \times 10^{\log Kow} + 1) + 0.9,$$
$$r^2\ 0.92 \tag{16.35}$$

The evolution of a QSAR over a period of years has been described by Hansch and Leo (1995) and is illustrated by Eqs. 16.36–16.38. In the 1963 model of the plant growth stimulation by m-substituted phenoxyacetic acids (Eq. 16.36), C is the molar concentration that produces a 10% increase over controls in elongation of 3 mm sections of oat coleoptile in 24 hours. The QSAR was extended and improved in 1981 (Eq. 16.37), and the 1995 bilinear model (Eq. 16.38) shows the great improvement in correlation coefficient provided by the more complex computation.

$$\log 1/C = 3.24\ \pi - 1.97\ \pi^2 + 1.86\ \sigma_p + 4.16, \quad n\ 23,\ r^2\ 0.77 \tag{16.36}$$
$$\log 1/C = 1.04\ \pi + 0.59\ \sigma_m + 4.78L - 0.67L^2 - 3.87,$$
$$n\ 19,\ r^2\ 0.87 \tag{16.37}$$
$$\log 1/C = 1.25\ \pi + 0.97\ \sigma_m + 0.95L - 5.54\ (\beta \times 10^L + 1) + 1.39,$$
$$n\ 19,\ r^2\ 0.95 \tag{16.38}$$

Although such equations at first seem empirical, the high correlation coefficients actually reflect the underlying physical and chemical basis of intoxication. Lipophilic absorption, transport, reaction with the target, and biotransformation are all mass-driven and take place according to rate laws dictated by chemical structure. Intoxication indeed is based on chemistry. However, there seems certain to be enough biotic individuality, ecological complexity, and environmental variability, as well as plenty of interesting surprises, to keep environmental toxicologists and chemists busy long into the future.

Glossary

Abiotic: Nonbiological.

Adjuvant: An additive that enhances the effectiveness of a pesticide formulation.

Allelochemical: A natural substance produced by one organism that modifies the life or behavior of another.

Analyte: A substance being sought or measured by chemical analysis.

Anemia: A condition in which blood is deficient in red cells or in hemoglobin.

Carbanion: A negatively charged carbon, for example, $-CH_2{}^-$

Carcinogen: A substance that causes cancer.

Cardiac: Pertaining to the heart.

Cholinergic: Associated with the neurotransmitter acetylcholine.

Cirrhosis: Hardening of the liver, often associated with fat accumulation.

Cofactor: Also called an apoenzyme or coenzyme, the nonprotein reactant associated with an enzyme.

Cometabolism: Biotransformation not associated with the processing of nutrients.

Congener: A member of a structurally related series of compounds.

Conjugation: In organic chemistry, a series of alternating double and/or triple bonds; in biotransformation, a synthesis that masks a reactive functional group.

Coordination: The donation of electrons by one atom to another, as in the binding of a nitrogenous compound to a metal ion.

Coulomb: A unit of electrical charge equal to an ampere per second.

Deposit: An easily removed surface coating of a pesticide or other chemical.

Depuration: The process of elimination of a chemical from the body.

Dermal: Pertaining to the skin.

Dielectric. Electrical nonconductance of a substance, or the substance itself.

Drift: Also called *spray drift,* the atmospheric transport of a substance (pesticide).

Electrophile: An electron-seeking atom or group, for example, $CH_3{}^+$.

Endocrine: Related to internal secretion of hormones or to the secretory glands.

Extinction coefficient: Older term for molar absorptivity, signified by ϵ.

Free radical: A molecule or atom possessing an unpaired electron, usually signified by $\bullet$, as in the hydroxyl radical HO•.

Fugacity: Tendency of a chemical to escape from its existing phase into another.

Ganglion: An integrating concentration of neurons (nerve cells).

Genotoxic: Toxic by acting on genetic material, especially DNA.

Hemocoel: The blood of invertebrates such as insects or molluscs.

Hepatic: Pertaining to the liver.

Inebriation: Drunkenness.

Initiator: In chemistry, a free radical that generates the first step in a chain reaction; in toxicology, a substance that starts carcinogenesis, a primary carcinogen.

Intermediary metabolism: The biochemical process that converts nutrients into energy and structure.

Lactone: A cyclic ester.

Ligand: The electron donor in a coordination compound, for example, an amine.

Lipophilic: Having an affinity for lipid or fat.

LOEL: Lowest observed effect level.

Micelle: A molecular cluster in which the hydrophilic surface is electrically charged while the inner portion often represents hydrophobic aliphatic chains, for example, soap micelles with surface carboxyl anions and interior C_{14}–C_{18} alkyl chains.

Mineralization: The process of converting an organic chemical to inorganic, for example, oxidation of methane to carbon dioxide and water.

Necrosis: Localized death of tissue.

Neoplasm: A rapidly dividing growth, a tumor.

Neurological: Pertaining to the nervous system.

Neuromuscular junction: The point of attachment of a neuron to muscle fiber.

NOEL: No observable effect level.

Parenteral: Administered by injection into the body, such as into a vein (intravenous, i.v.) or the body cavity (intraperitoneal, i.p.).

Partition coefficient: The ratio of the concentration of a substance in one phase to that in an adjacent one.

PBPK model: Physiologically based pharmacokinetic model, a mathematical model of the absorption, disposition, and elimination of a chemical in the body.

Photolysis: The breakdown of a substance by visible or ultraviolet radiation.

Physiological time: A time scale perceived by another species, and not necessarily chronlogical (human) time.

Plasticizer: An additive used to soften a rigid plastic such as PVC or polyethylene.

Polar: In chemistry, referring to electrically charged species, such as a salt, compared to an electrically neutral *(nonpolar)* species such as hexane.

Postsynaptic: On the signal-receiving side of the gap *(synapse)* between neurons, compared to the other, *presynaptic* side.

Progressor: A chemical (carcinogen) that converts an initiated or promoted cell to a potentially malignant one. Also called a ***progressor agent.***

Promoter: A chemical (carcinogen) that increases the number of initiated cells.

QSAR: Quantitative mathematical relation of chemical structure to biological effect (**q**uantitative **s**tructure–**a**ctivity **r**elation).

Recalcitrant: Difficult to manage, unreactive.

Renal: Pertaining to the kidneys.

Semiconductor: A solid whose electrical conductivity lies between that of a metal and an insulator and is often altered by physical forces (*e.g.,* light).

Synergism: The effect of a mixture is greater than the sum of the individual effects of the components.

Teratogen: A substance causing birth defects.

Tolerance: In legal regulation, an acceptable level of chemical residue; in toxicology, temporary resistance to toxic effects.

Venom: A natural poison that is injected by one organism into another, such as by a bee.

Xenobiotic: A toxic substance foreign to the organism, or sometimes a normal metabolite, such as a vitamin, that is presented in an unusually high dose. Most xenobiotics are considered pollutants.

Index

Citations to figures, tables, and appendices appear in *italics*.

Abiotic reactions. *See* Transformations, abiotic
Absorption
  ionization and, 125
  mechanism, 123–24
Absorption, of chemicals, 122–25, *123,* 146, 147, *147, 217*
Absorption coefficient, 70
Absorption spectra. *See* Ultraviolet spectra
ACD (allergic contact dermatitis), 133–34, 234
Acetaminophen
  oxidation, 289
  toxicity, 289
Acetolactate synthase. *See* ALS
Acetylation, of amines, 96, *96,* 97, *114,* 115, *116, 121*
Acetylcholine (ACh)
  antagonists, 227, 240
  hydrolysis, 166, *167*
  as neurotransmitter, 130
  receptor, 163, *163,* 166, 168, 215
Acetylcholinesterase. *See* AChE
AChE
  active site, 166, *167,* 179
  in analysis, 18, 189, 190
  function, 130, 132
  inhibition, 10, 166–68, *167,* 229
  mode of action, 166, *167,* 179
Acid rain, 75
Aconitase, inhibition, 171
Actinometers, 71
Activation energy, 68–69
Acute toxicity. *See* Toxicity, acute
Adaptation, to toxicants, 140–41
Addiction, 131
S-Adenosylmethionine, methylation by, 96, *96,* 115, 212, 224
Adsorption, and desorption, 36, 45
Adsorption, from water, 45, *46*
Adsorption, isotherms, 46, *46*
Adsorption coefficients
  calculation, 46–47
  definition, 46

  measurement, 46, 47
  prediction, 309
  of various chemicals, *46, 50*
Advection 52, 53, 54, 58, 59–60
Aflatoxins, 113, 139
  from *Aspergillus, 240–41*
  metabolism, 240, *241*
  occurrence, 240–41
  toxic action, 181, *184*
  toxicity, 240
Aging, of phosphorylated AChE, 291–92
AHH receptor,170
A*h* receptor, 109, 170–71, *170*
ALA dehydratase, 217
Alcohol. *See* Ethanol
Aldehyde dehydrogenase, 170
Aldicarb, oxidation, 279, 286, 295
Alkaloids, definition, 226–27
Alkanes, bio-oxidation, 98–100, 119
Alkylating agents, 281–82
Alkylation, in carcinogenesis,181
  reactions, 87, 181, 281–82
Alkyl halides, reactivity, 282
Allelochemicals, ecotoxicology, 243–44
Allelopathy, 7, 244
Allergic contact dermatitis (ACD), 133–34, 234
ALS, inhibition, *174,* 175
Aluminum phosphide, 209
Amanita phalloides (Death's cap), 136, 236
Amanitins, from toadstools, 236, *237*
Ames test,150
Amines
  generation, *96,* 97, 106, *114, 116, 121*
  toxicity, 289
Amino acid biosynthesis, inhibition, 174–75, *174*
Amino acids, toxicity, 234–36, *235*
gamma-Aminobutyric acid. *See* GABA
Ammonia, atmospheric, *19,* 20
Anagyrine, 137
Analysis
  accuracy of, 15–16